Fundamentals of Protein Biotechnology

Bioprocess Technology

Series Editor

W. Courtney McGregor

Xoma Corporation
Berkeley, California

Volume 1	Membrane Separations in Biotechnology, *edited by W. Courtney McGregor*
Volume 2	Commercial Production of Monoclonal Antibodies: A Guide for Scale-Up, *edited by Sally S. Seaver*
Volume 3	Handbook on Anaerobic Fermentations, *edited by Larry E. Erickson and Daniel Yee-Chak Fung*
Volume 4	Fermentation Process Development of Industrial Organisms, *edited by Justin O. Neway*
Volume 5	Yeast: Biotechnology and Biocatalysis, *edited by Hubert Verachtert and René De Mot*
Volume 6	Sensors in Bioprocess Control, *edited by John V. Twork and Alexander M. Yacynych*
Volume 7	Fundamentals of Protein Biotechnology, *edited by Stanley Stein*

Volume 8 Yeast Strain Selection, *edited by Chandra J. Panchal*

Volume 9 Separation Processes in Biotechnology, *edited by Juan A. Asenjo*

Additional Volumes in Preparation

Large-Scale Mammalian Cell Culture Technology, *edited by Anthony S. Lubiniecki*

Extractive Bioconversions, *edited by Bo Mattiasson and Olle Holst*

Fundamentals of Protein Biotechnology

Edited by

STANLEY STEIN

Center for Advanced Biotechnology and Medicine
Piscataway, New Jersey

CRC Press is an imprint of the
Taylor & Francis Group, an **informa** business

CRC Press
Taylor & Francis Group
6000 Broken Sound Parkway NW, Suite 300
Boca Raton, FL 33487-2742

First issued in paperback 2019

ISBN-13: 978-0-8247-8346-4 (hbk)
ISBN-13: 978-0-367-40313-3 (pbk)

Library of Congress Cataloging--in--Publication Data

Fundamentals of protein biotechnology / [edited by] Stanley Stein.
p. cm. -- -- (Bioprocess technology; v. 7)
Includes bibliographical references.
Includes index.
ISBN 0-8247-8346-8 (alk. paper)
1. Proteins-- --Biotechnology. I. Stein, Stanley
II. Series
[DNLM: 1. Biotechnology. 2. Proteins-- --biosynthesis.
3. Recombinant Proteins-- --biosynthesis. W1 BI88U v. 7 / QU 55 F771]
TP248.65.P76F68 1990
660'.63-- --dc20
DNLM/DLC
for Library of Congress 90-3606
CIP

Visit the Taylor & Francis Web site at
http://www.taylorandfrancis.com

and the CRC Press Web site at
http://www.crcpress.com

Series Introduction

The revolutionary developments in recombinant DNA and hybridoma technologies that began in the mid-1970s have helped to spawn several hundred new business enterprises. Not all these companies are aimed at producing gene products or cell products, as such. Many are supportive in nature: that is, they provide contract research, processing equipment, and various other services in support of companies that actually produce cell products. With time, some small companies will probably drop out or be absorbed by larger, more established firms. Others will mature and manufacture their own product lines. As this evolution takes place, an explosive synergism among the various industries and the universities will result in the conversion of laboratory science into industrial processing. Such a movement, necessarily profit driven, will result in many benefits to humanity.

New bioprocessing techniques will be developed and more conventional ones will be revised because of the influence of the new biotechnology. As bioprocess technology evolves, there will be a need to provide substantive documentation of the developments for those who follow the field. It is expected that the technologies will continue to develop rapidly, just as the life sciences have developed rapidly over the past 10–15 years. No single book could cover all of these developments adequately. Indeed, any single book will be in need of replacement or revision every few years. Therefore, our continuing series in this rapidly moving field will document the growth of bioprocess technology as it happens.

The numerous cell products already in the marketplace, and the others expected to arrive, in most cases come from three types of bioreactors: (a) classical fermentation; (b) cell culture technology; and (c) enzyme bioreactors. Common to the production of all cell products or cell product analogs will be bioprocess control, downstream processing (recovery and purification), and bioproduct finishing and formulation. These major branches of bioprocess technology will be represented by cornerstone books, even though they may not appear first. Other subbranches will appear, and over time, the bioprocess technology "tree" will take shape and continue growing by natural selection.

W. Courtney McGregor

Preface

Biotechnology has become the catchword in the life sciences over the past several years. Universities and state governments are setting up biotechnology centers; there are biotechnology companies, as well as biotechnology divisions of major pharmaceutical companies. The most tangible products of biotechnology are proteins. Several proteins produced by recombinant DNA technology, such as interferon, growth hormone, and tissue plasminogen activator (TPA), are already being marketed for human therapeutic applications. It is expected that proteins will eventually constitute a major class of pharmaceutical products.

The purpose of this volume is to fill an educational gap concerning a fundamental understanding of proteins and their place in the current era of biotechnology. Although there are a number of books on the subject of proteins, they tend to be highly technical and geared toward the practicing protein chemist. Biochemistry textbooks are too focused on basic principles. Each chapter in this volume was written by a scientist whose work is directly relevant to that topic. Their mission was to (1) present the fundamental principles, (2) provide an overview, and (3) explain the practical implications in the context of biotechnology.

Chapter 1 is written by Sidney Udenfriend, an eminent scientist who, over four decades, has made important contributions to analytical biochemistry. His historical perspective illustrates the remarkable progress made in protein research.

Russ Lehrman, in Chapter 2, presents the fundamentals of protein structure. He covers the properties of the amino acid subunits and how they are arranged in proteins, from primary to quaternary level structure. Various methods of depicting three-dimensional protein structures are provided, as representative of illustrations found in the scientific literature.

The fundamental principles of protein biosynthesis are presented in Chapter 3. The individual steps of this process are detailed, from the organization of the genetic information to posttranslational modifications of the assembled protein. A discussion of considerations relevant to the production of recombinant proteins is included.

Considering the importance of separation techniques to virtually all aspects of protein purification and analysis, Chapter 4 describes principles and practices of liquid chromatography and electrophoresis. This prepares readers for the subject matter in some of the later chapters. Jodi Fausnaugh incorporates illustrations typical of those found in protein research.

In Chapter 5, Milton Hearn describes proteins from a functional viewpoint. Besides their role as structural components of cells and tissues, proteins as effectors of biological activity can be divided into six major groups. Details are given on the molecular events that lead to the observed physiological phenomena.

Chapter 6 presents the challenge of isolating new active proteins from biological sources. This process is documented using two examples from the author's personal experience. The complexities and surprises of natural systems are exemplified with the interferon and enkephalin families of proteins.

Chapter 7 elaborates on the interferon story depicted in the preceding chapters. Fazal Khan describes the isolation process for a recombinant human interferon. Production of clinical supplies of this rare natural protein represents a hallmark achievement in biotechnology.

Another example of recombinant DNA methodology is given in Chapter 8 by Barry Jones, who presents a mixture of genetic engineering, enzymology and synthetic chemistry to overcome technological barriers to the production of the peptide hormone, calcitonin.

Chapter 9 is a presentation of the approaches and techniques used for determination of primary structure. These principles are applicable to both natural proteins isolated in trace amounts, as well as to recombinant proteins produced in copious amounts. The analysis of a protein produced by recombinant DNA technology illustrates some concerns of the biotechnology industry.

As opposed to protein biosynthesis, chemical synthesis is often the method of choice for peptides smaller than 50 residues. Chet Meyers reviews the history and methodology of solid phase peptide synthesis in Chapter 10. Also discussed is the use of small peptides for clinical as well as research purposes, especially with regard to analogs having unusual modifications.

In Chapter 11, Sidney Pestka provides an overview of the application of recombinant DNA technology. The essential techniques used in cloning genes for specific proteins are explained. The author uses his own research experience on interferon to illustrate this technology.

In the final chapter, Jim Bausch is concerned with monoclonal antibodies. This type of protein deserves special attention, considering its unique characteristics and wide range of medical and research applications.

The content, style, and technical level of this volume are aimed at a more general audience. It is not intended to be a cookbook of procedures or an in-depth study on any particular subject, but is designed to provide a solid foundation for understanding proteins in their various aspects. Individuals in the pharmaceutical industry are considered to constitute a major segment of the readership. This volume may also be a useful textbook for an advanced biochemistry course. The editor and authors hope that they have fulfilled their objectives in an informative and enjoyable format.

Stanley Stein

Contributors

James N. Bausch Schering Corporation, Kenilworth, New Jersey

Joseph Bertolini Department of Biochemistry, Monash University, Clayton, Victoria, Australia

Jodi L. Fausnaugh Syntex Corp., Palo Alto, California

Milton T. W. Hearn Department of Biochemistry, Monash University, Clayton, Victoria, Australia

Fazal R. Khan Bioprocess Development Department, Hoffmann-LaRoche, Inc., Nutley, New Jersey

Barry N. Jones Sterling Research Group, Sterling Drug, Inc., Malvern, Pennsylvania

S. Russell Lehrman Upjohn Company, Kalamazoo, Michigan

Chester A. Meyers Squibb Institute for Medical Research, Princeton, New Jersey

Sidney Pestka Department of Molecular Genetics and Microbiology, Robert Wood Johnson Medical School, Piscataway, New Jersey

Vishva R. Rai Bioprocess Development Department, Hoffmann-LaRoche, Inc., Nutley, New Jersey

Peter G. Stanton Department of Biochemistry, Monash University, Clayton, Victoria, Australia

Stanley Stein Center for Advanced Biotechnology and Medicine, Piscataway, New Jersey

Sidney Udenfriend Laboratory of Molecular Neurobiology, Roche Institute of Molecular Biology, Nutley, New Jersey

Richard A. Wolfe Monsanto Company, St. Louis, Missouri

Contents

1

An Overview and Historical Perspective of Protein Biotechnology

Sidney Udenfriend

Roche Institute of Molecular Biology
Nutley, New Jersey

THE RESURGENCE OF PROTEIN CHEMISTRY

The techniques of molecular biology were introduced in the early 1970s. It appeared to many then that this spelled the end of protein chemistry. Why isolate a protein or peptide if all one had to do was clone the corresponding cDNA and sequence it? The rapidity with which proteins were cloned appeared to corroborate this sentiment. However, most of the proteins that were cloned initially, such as insulin and growth hormone, had already been totally characterized by chemical methods prior to the advent of molecular biology.

Molecular biologists, at the time (mid-1970s), were somewhat justified in bypassing protein chemistry because it was not useful in investigations of proteins of importance but of low abundance. Why was this so? Because at that time biological activities were already measurable in the picomole range and below. While molecular biology also operated in this range, protein chemistry was still operating in the nanomole range. This disparity in sensitivities led molecular biologists to develop ingenious methods to bypass protein isolation and characterization. Today with expression cloning methods it is possible to clone and sequence the cDNA of a protein merely through its biological or immunological activity.

However, over time, there was a resurgence of interest in protein chemistry. Microanalytical methods were introduced at every step, including chemical assay, isolation, amino acid assay, and sequencing. Today protein chemistry also operates in the picomole range, making it an equal partner with molecular biology for elucidating structure and function. Furthermore, protein chemistry contributes information that is not obtainable by molecular biology alone. First of all, obtaining even limited sequence information on an isolated protein permits the preparation of synthetic deoxynucleotides that can be of considerable help as probes in cloning the corresponding cDNA. This is particularly important in attempts to clone rare species of proteins such as hormones, growth factors, and receptors. Knowledge derived from protein sequencing is generally also more precise in determining the amino terminus of a protein than is cDNA sequencing alone. Finally, the active protein or peptide frequently undergoes considerable posttranslational processing that cannot be predicted from the cDNA; namely cleavage by peptidases, phosphorylation, glycosylation, subunit aggregation, and so on. DNA has been referred to as the blueprint of life. If this is so, then each protein coded for by a specific cDNA may be considered to be an individual structure. In real life we utilize the structure not its blueprint.

The advent of molecular biology and genetics brought with it a revolution in technology. Because of the high profile of molecular biology most individuals are aware of the advances in methods involving DNA and RNA. However, there have been comparable advances in the methodology of protein and peptide chemistry, although most of these are still not routinely available to biologists. Unfortunately, while our graduate schools are training molecular biologists in great numbers today, few are being trained in modern protein and peptide chemistry, particularly at the level of purification and analysis. It is hoped that this volume will help fill this educational gap.

AN HISTORICAL VIEW

Before detailing the different procedures of modern protein and peptide chemistry, it would be interesting to look at their history and development. It is only within the last 35 years that we have come to realize that each protein is composed of stoichiometric amounts of 20 different amino acids, and that the latter are further arranged in a fixed sequence that is unique for that protein. Until the late 1940s and early 1950s there were no analytical methods that were suitable for dealing with even large amounts of proteins or peptides. While resourceful biochemists at that time were able to purify some proteins that were major tissue constituents (i.e., hemoglobin, casein, insulin, etc.) they were not able to determine even their amino acid compositions.

The earliest methods for amino acid assay utilized "specific" precipitants for each of the amino acids in a protein hydrolysate. Such assays required gram quantities of protein and were not readily reproducible from laboratory to laboratory. Furthermore, they were not sufficiently specific, and reagents were not available for all the amino acids. In 1950, microbiological assays for amino acids were introduced. Certain bacteria could be grown in media with one of the amino acids missing and all the others in excess, so that, on addition of a protein hydrolysate, the organism grew in proportion to the concentration of that amino acid. For example, to measure leucine, a culture medium was used containing all the amino acids except leucine. On addition of a protein hydrolysate, growth was proportional to the leucine content. Assay of each of the amino acids for one protein hydrolysate required incubation of the organism with a specific medium along with controls, blanks, etc. Microbiological assay was more sensitive (100–200 μg) than precipitation methods, but it was time consuming and did not have the precision to yield convincing stoichiometric data. At about the same time the isotope derivative method for amino acid assay was introduced (1). Aliquots of a labeled hydrolysate were used to assay each amino acid individually by isotope dilution methods. This method was sufficiently precise and specific to show for the first time that all proteins then available, except collagen, contained less than 0.05 residue per mole of hydroxyproline. The method was sensitive (μg quantities of protein), but again laborious. It was not until Stein and Moore (2) introduced the ninhydrin amino acid analyzer and helped introduce commercial automated analysis that amino acid assay attained the sensitivity, precision, and relative simplicity required for chemical analysis. Of course, it was the introduction of column chromatography that made the amino acid analyzer possible. The introduction of commercial amino acid analyzers by Beckman Instrument Company and others finally gave protein chemists the precision necessary to prove convincingly that each protein possesses a unique amino acid composition.

The ninhydrin amino acid analyzer, little modified from the original Stein and Moore instrument, represented the dominant technology until about 10 years ago. However, the limits of sensitivity of most commercial amino acid analyzers until fairly recently was about 1 nanomole of each amino acid. The introduction of high-performance liquid chromatography (HPLC), newer fluorescent and colorimetric reagents, and modern computer technology has now pushed the limits of sensitivity to a few picomoles of each amino acid in a hydrolysate. There are now several procedures and instruments that routinely operate in the picomole range.

For analysis to be meaningful, it is necessary to purify a protein. In the 1960s Burgess et al. (3) and Schally et al. (4) purified and characterized thyroid-releasing factor (TRF) the first of many hypothalamic releasing factors. As it turned out, TRF represented less than 0.0004% of beef hypothalamic protein. To isolate sufficient amounts for analysis by the procedures available at the time (μmol quantities) these investigators had to start with several hundred thousand hypothalami (25–50 kg). Similar quantities of tissue were used in several trial runs prior to the final isolation. The large amounts of tissue were required because the overall yields were low by the procedures that were used at the time. In 1983, when Guillemin and colleagues (5) isolated growth hormone-releasing factor (GRF) they used only 7.2 g of tissue and isolated 1–5 nmol of peptide which provided more than enough material for chemical and biological characterization. More recently Esch et al. (6) in Guillemin's laboratory isolated and sequenced the follicle-stimulating hormone- (FSH) releasing peptide from a few liters of porcine follicular fluid. About 10 pmol was used to provide sufficient amino terminal sequencing to prepare a probe for cloning.

The practical advantages of this phenomenal decrease in scale of operation (micromoles to picomoles or 1,000,000-fold) should be pointed out. The use of less tissue and less reagent is important for financial reasons. Hundreds of thousands of beef hypothalami, the amount used for TRF isolation and the reagents, equipment, and manpower necessary to process such large amounts of tissue approached one million dollars. Modern micromethods are therefore highly cost effective. Another advantage is speed. A single run on an HPLC column takes an hour or two compared with days for older column methods. Furthermore, the efficiency of HPLC columns provides a higher degree of purification per run. Current micromethods for amino acid assay and sequencing are not only more sensitive, but also far more rapid compared with methods of only 10 years ago. In fact, the rate-limiting factor in most isolations is no longer the chemical methodology but the procedure required to monitor the biological properties of a newly discovered protein or peptide. Referring to advances in protein and peptide isolations over the years Roger Guillemin said, "Things have changed (and improved) over the years. The only thing that remains constant and an absolute requisite for all these purification procedures is a truly specific and reliable bioassay and people to do it right."

Isolation of proteins, until the 1950s, required precipitation procedures. Salts, solvents, and pH were used to produce differential precipitation of the desired protein in the presence of other proteins extracted from a tissue. Such procedures were obviously limited to major proteins and also relatively stable ones. Differential adsorption and elution was followed by column chromatography on various types of gels in the late 1950s and early 1960s. With detection by absorption at 280 nm, sensitivity was pushed down to the nanomole level. However, the limited resolving powers of gels required large columns and long running times (days). Volumes were accordingly large and elution was limited to aqueous solutions. HPLC radically changed the nature of protein and peptide purification.

The development of rapid methods for isolation of proteins and peptides was made possible by the smaller and smaller amounts required for sequencing. Before Sanger elucidated the primary structure of insulin in the early 1950s (7), no one had demonstrated that each protein had a unique sequence.

Although most scientists know that Sanger used overlapping peptides to attain the overall sequence of insulin, few realize that he never sequenced a peptide. He converted each of the two insulin chains to tri- and tetrapeptides by random partial hydrolysis in acid, separated peptides by paper chromatography, and determined the amino terminus and amino acid composition of each small peptide. From the amino acid compositions of these small randomly produced peptides Sanger was able to deduce the sequences of the A and B chains of insulin. This represented quite a feat! The first true sequencing from the amino terminus, utilizing the reagent phenylisothiocyanate was introduced by Edman and Begg (8). Largely as a result of advances in instrumentation, the Edman procedure has now achieved sensitivity in the low picomole range. Newer instrumentation has also made it possible to obtain long sequences in relatively short periods of time. With some commercial instruments, and with some proteins, it is now possible to obtain 40 to 50 sequences on as little as 100 pmol in about two days. Because of these advances in peptide chemistry, the Food and Drug Administration now requests partial sequence information on every batch of a recombinant protein product that is used clinically.

APPLICATION TO BIOTECHNOLOGY

Sequence data, whether obtained by protein or cDNA sequencing provide information for peptide synthesis. Recent developments in methods and instrumentation for peptide synthesis have also been formidable. The solid-state method, which is the heart of today's methodology in peptide synthesis, was slow to gain acceptance. When Merrifield first introduced this procedure in 1963 (9), synthetic peptide chemists of the day were not receptive. They acknowledged it as an ingenious concept but felt that its use would be limited because the partial reaction products that accumulated at each step, even though 99% complete, could not be separated from the desired peptide by the methods then available. It was generally believed that synthesis in solution would not be supplanted because it permitted isolation and purification of intermediates at each step. Merrifield's answer to this criticism by most of the peptide chemists of the day clearly shows his vision. ''Recent improvements in purification techniques suggest that future developments will allow the ready separation of product mixtures that presently appear intractable. In addition, new and refined analytical methods should permit much greater precision in assessing the homogeneity, composition, and conformation of synthetic product'' (10). HPLC was introduced not long after his remarks. Purifying even the largest peptides produced by solid-phase methods then became almost a trivial matter. Based on these synthesis and purification procedures, the production of large peptides (30–40 residues) for use as drugs is now economically feasible. Chemical synthesis of even larger peptide residues can now be carried out, but as yet, only recombinant DNA methodology can produce them in sufficient quantity and at a cost suitable for commercial applications.

Large peptides and proteins produced by recombinant DNA (rDNA) technology require purification from the normal products of the cells in which they are produced. Here, again, recent advances in purification procedures make this a relatively simple matter. The high specificity of monoclonal antibodies has been utilized to develop affinity chromatography procedures that, in one step, can isolate a cloned protein in 80% to 90% purity from a cell culture. For example, in the case of α interferon, one step of antibody affinity chromatography and a second of HPLC can yield a product that is over 99% pure and free of all detectable cell products (11). Such procedures have already been scaled up to produce and purify peptides and proteins for commercial use in 25–100g batches. Application of newer bioengineering principles to biological materials should permit scale-up to kilogram quantities.

Purification, characterization, and synthesis of a peptide or protein is quite an achievement. Besides having the polypeptide available for research and clinical study, it permits chemists and molecular biologists to investigate structure–function relationships. The purpose of structural modifications might be to increase the inherent biological activity, stabilize the molecule to the actions of tissue enzymes, or map the active site. A plethora of modifications to chemically synthesized peptides, including substitution with D-amino acids, is found in the scientific literature. Site-directed mutagenesis and other gene-splicing techniques can be used to generate a myriad of polypeptides with defined structural differences.

Sequencing information gives the chemist only a two-dimensional picture of a protein or peptide when the latter actually exist and function as three-dimensional entities. Anfinsen (12) showed that the information required to fold a protein into its unique, biologically active, three-dimensional conformation already exists in its primary structure. Attempts are being made to determine three-dimensional structure from known bond angles and bond energies. The introduction of supercomputers has made it practical to carry out the large number of calculations required for relatively small peptides. Two-dimensional nuclear magnetic resonance (NMR) is proving to be a powerful technique for this purpose. However, crystallization and x-ray analysis have already helped elucidate the three-dimensional structure of several larger proteins. Many more proteins are now under investigation. Aside from its scientific interest, three-dimensional structures of proteins will lead to the next major advances in biotechnology.

Some proteins produced by cloning procedures are active when administered to patients, namely, interferon, growth hormone, and insulin. However, other proteins now being produced serve as receptors on the cell surface for some biological messenger or for the entry of a virus. Still others may interact with the chromosomes or play a role in a body compartment that cannot be reached when administered. The three-dimensional structure of a protein will provide the modern pharmaceutical biologist with information that will permit him or her to determine the conformation of the few residues at the "active site" that endow the macromolecule with its unique activity. Small nonpeptide analogues based on the structure of the "active site" residues can then be synthesized by conventional chemical procedures. This new logic for pharmaceutical chemistry gained from probing the three-dimensional structure of proteins will provide the next generation of drugs and commercial biologicals. The antihypertensive drug, captopril, is an example of a synthetic nonpeptide analog of an active site that proved to be biologically active and reasonably stable to body enzymes. The primary, secondary, and tertiary structures of proteins have now become highly relevant to research and development programs. The value of such information is apparent from the numerous patents applied for and granted in this area.

CONCLUDING REMARKS

Peptide and protein chemistry has now come of age. It is already providing basic information for research into such areas as neuroscience, immunology, and infectious diseases, to name a few. However, it is ready to be exploited by industry to develop new drugs, diagnostic agents, and other types of biologicals. The different subspecialties of protein and peptide chemistry reviewed in this volume are of obvious interest to the specialist. However, those who are responsible for directing and coordinating research and development programs in industry should also be conversant with this important field and its many areas of expertise. Furthermore, it is time for the academic community to address the issue of the short supply of scientists trained in protein isolation and analysis, peptide synthesis, x-ray crystallography, and related areas. Such scientists are in great demand by industry and only a strong educational commitment can fulfill the present and future requirements.

BIBLIOGRAPHY

1. Keston, A. S., Udenfriend, S., and Cannon, R. K., *J. Am. Chem. Soc.* **68**, 1390 (1946).
2. Spackman, D. H., Stein, W. H., and Moore, S., *Anal. Chem.* **30**, 1190 (1958).
3. Burgus, R., Dunn. T. F., Desiderio, D., Ward, D. N., Vale, W., and Guillemin, R., *Nature (London)* **226**, 321 (1970).
4. Schally, A. V., Coy, D. H., Meyers, C. A., and Kastin, A. J. in *Hormonal Proteins and Peptides*, Vol. VII. Edited by C. H. Li. Academic Press, New York, 1979.

5. Guillemin, R., Brazeau, P., Böhlen, P., Esch, F., Ling, N., and Wehrenberg, W. B., *Science* **218**, 585 (1982).
6. Esch, F. S., Mason, A. J., Cooksey, K., Mercado, M., and Shimasaki, S., *Proc. Natl. Acad. Sci.* (USA) **83**, 6618–6621 (1986).
7. Sanger, F. and Tuppy, H., *Biochem. J.* **49**, 481 (1951).
8. Edman, P. and Begg, G., *Eur. J. Biochem.* **1**, 80 (1967).
9. Merrifield, R. B., *J. Am. Chem. Soc.* **85**, 2149 (1963).
10. Erickson, B. W. and Merrifield, R. B., *The Proteins,* Vol. III. Edited by H. Neurath and R. L. Hill. Academic Press, New York, 1976; p. 492.
11. Staehlin, T., Hobbs, D. S., Kung, H., Lai, C. Y., and Pestka, S., *J.Biol.Chem.* **256**, 9750–9754 (1981).
12. Anfinsen, C. B., *Science* **181**, 223 (1973).

2

Protein Structure

S. Russell Lehrman

Upjohn Company
Kalamazoo, Michigan

INTRODUCTION

Proteins are larger and more complex than the traditional pharmaceutical products, such as those used as antibiotic, antianxiety, and chemotherapeutic agents. Figure 1 illustrates the magnitude of these differences. For example, the benzodiazapines, a class of compounds which are useful in the treatment of anxiety, have molecular weights of about 300. Pharmaceuticals of similar size are typically characterized by chemical methods, including elemental analysis, nuclear magnetic resonance (NMR), mass spectrometry, (MS), and infrared (IR) spectrometry. If the compound has been crystallized, the relative positions of its atoms can be determined to a resolution of 0.1–0.2 angstroms (10^{-10} m). The macromolecular proteins, on the other hand, are extremely difficult to characterize in great detail. Molecular weights of monomeric proteins range from 5,000 to 200,000, and multisubunit proteins may weigh up to 1,000,000. Efforts to characterize proteins rely on different approaches, such as amino acid compositional analysis and sequence analysis by Edman degradation.

Because of their size, proteins are more flexible than classical pharmaceutical entities. This flexibility allows different proteins to fold into unique three-dimensional structures as they are biosynthesized within the cell. The study of protein conformation and folding requires considerable time and material, and interpretation of the experimental data requires careful analysis. The spectro-

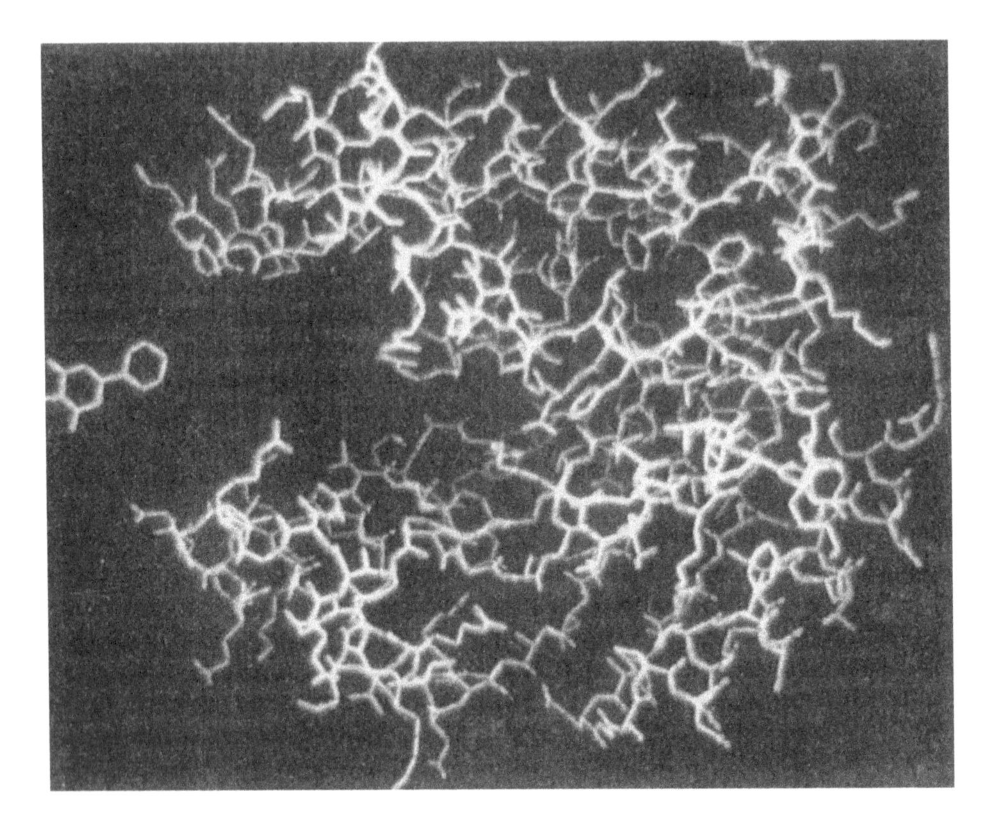

Figure 1 The structure of adenylate kinase, an enzyme which plays a role in cellular metabolism, and minoxidil, a pharmaceutical used in the treatment of hypertension.

scopic methods used in these efforts, although very useful, do not usually reveal the precise locations of atoms within a protein. Spectroscopic studies of proteins are important in pharmaceutical development, since a limited subset of the available protein conformations are biologically active.

The structures of about 350 proteins have been obtained using x-ray crystallography. For proteins, the best crystals diffract at a resolution of 1.5 A. Two-dimensional NMR is a new technique that is being used to obtain similar information for small proteins in solution.

The diversity of protein structures is reflected by a diversity of biological functions. Proteins may serve as enzymes, hormones, or structural components of hair and skin. They have been shown to regulate gene expression, transport ions across cell membranes, promote cellular attachment to solid surfaces, and bind a wide range of organic and inorganic compounds. Within each functional class of proteins, there is also a high degree of structural diversity. For example, both platelet-derived and epidermal growth factors are protein hormones. Whereas the former has a molecular weight of 32,000, 18 disulfide bonds, and a net positive charge, the latter has a molecular weight of 6054, three disulfide bonds, and a net negative charge.

Despite these complexities, all proteins contain unifying features which make their structures more comprehensible. This chapter introduces these concepts and illustrates them with proteins having realized or potential commercial utility.

BASIC ELEMENTS OF PROTEINS: THE AMINO ACIDS

Introduction

The basic building blocks of all proteins are the amino acids. As shown below, amino acids are compounds in which amine and carboxylic acid functional groups are covalently bound to a central alpha carbon atom (Fig. 2). In addition, the alpha carbon atom of each amino acid is covalently bonded to a hydrogen atom and a unique substituent (the "R" group). This latter substituent distinguishes the 20 naturally occurring amino acids (Fig. 3). Several of these properties are listed in Table 1.

In broad terms, amino acids are typically classified as (1) charged and polar, (2) uncharged and polar, or (3) nonpolar. These classifications are described in greater detail below. In other contexts, different groupings, such as aromatic or sulfur-containing, may be used.

Charged Polar Amino Acids

This group consists of amino acids which readily accept or release H^+ in physiological solutions. As indicated by their names, aspartic and glutamic acids are acidic (i.e., release H^+), while arginine, lysine, and, to a lesser extent, histidine are basic (i.e., bind H^+). These amino acids help stabilize protein structure through the formation of ion-pairs, which typically provide about 5 kcal/mol of stabilization energy. Although these amino acids are strongly hydrophilic due to the charged functional groups on their side chains,

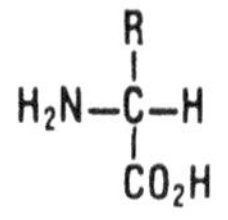

R
|
H_2N—C—H
|
CO_2H

Figure 2 The covalent structure of the naturally occurring amino acids. Each amino acid is distinguished by the composition of R (see Fig. 3).

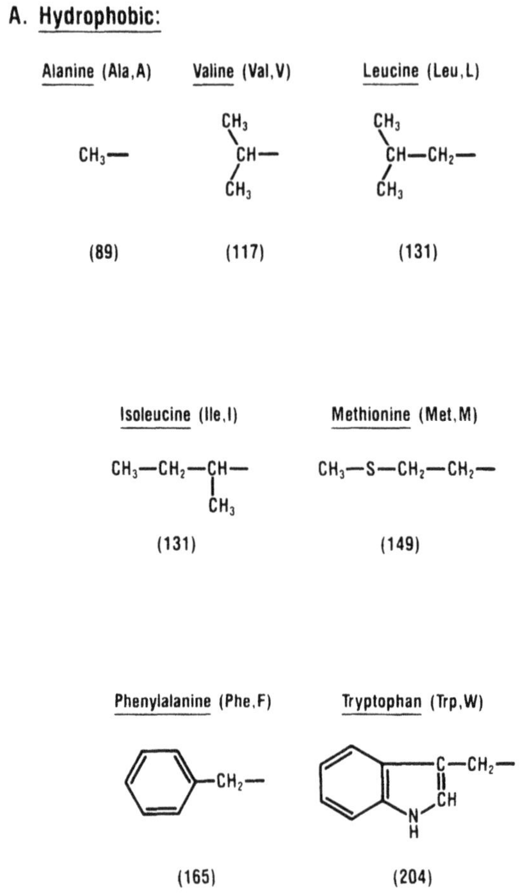

A. Hydrophobic:

Alanine (Ala,A)

$CH_3—$

(89)

Valine (Val,V)

CH_3 \ $CH—$ / CH_3

(117)

Leucine (Leu,L)

CH_3 \ $CH—CH_2—$ / CH_3

(131)

Isoleucine (Ile,I)

$CH_3—CH_2—CH—$ | CH_3

(131)

Methionine (Met,M)

$CH_3—S—CH_2—CH_2—$

(149)

Phenylalanine (Phe,F)

$—CH_2—$

(165)

Tryptophan (Trp,W)

$C—CH_2—$ ‖ CH N H

(204)

Figure 3 The side chains of the 20 naturally occurring amino acids. These have been classified as (A) hydrophobic, (B) hydrophilic and uncharged, and (C) hydrophilic and charged. The one and three letter codes for each amino acid are shown in parentheses. The molecular weight of each amino acid is indicated in parentheses below its corresponding structure.

B. Hydrophilic and uncharged:

Serine (Ser,S)

$HO-CH_2-$

(105)

Threonine (Thr,T)

$CH_3-C(OH)(H)-$

(119)

Tyrosine (Tyr,Y)

$HO-C_6H_4-CH_2-$

(181)

Glycine (Gly,G)

$H-$

(75)

Proline (Pro,P)

H_2C, $C H_2$, H_2C, $N H$ (ring)

(115)

Asparagine (Asn,N)

$NH_2-C(=O)-CH_2-$

(132)

Glutamine (Gln,Q)

$NH_2-C(=O)-CH_2-CH_2-$

(146)

Cysteine (Cys,C)

$HS-CH_2-$

(121)

C. Hydrophilic and charged:

Aspartic Acid (Asp,D)

$^{-}O-C(=O)-CH_2-$

(133)

Glutamic Acid (Glu,E)

$^{-}O-C(=O)-CH_2-CH_2-$

(147)

Histidine (His,H)

$HC=C-CH_2-$, HN^{+}, NH, $C H$ (ring)

(155)

Arginine (Arg,R)

$H_2N-C(=NH_2^{+})-NH-CH_2-CH_2-CH_2-$

(174)

Lysine (Lys,K)

$H_3\overset{+}{N}-CH_2-CH_2-CH_2-CH_2-$

(146)

the aliphatic portions of their side chains are sometimes observed to participate in hydrophobic interactions.

Uncharged Polar Amino Acids

The amino acids within this group are serine, threonine, tyrosine, asparagine, glutamine, glycine, proline, and cysteine. Serine, threonine, and tyrosine contain alcoholic (i.e., hydroxylic) functional groups, while glutamine and asparagine contain carboxyamides as part of their side chains. Therefore, these

Table 1 Selected Properties of Amino Acids

Amino acid	Molecular weight (Da)	Hydrophobicity[a]	Accessible surface area Å[b]	Occurrence in proteins[c] (%)	Miscellaneous
Isoleucine	113.17	0.73	175	4.6	
Phenylalanine	147.18	0.61	210	3.5	257 nm (197)
Valine	99.14	0.54	155	6.9	
Leucine	113.17	0.53	170	7.5	
Tryptophan	186.21	0.37	255	1.1	280 nm (5600)
Methionine	131.21	0.26	185	1.7	
Alanine	71.08	0.25	115	9.0	
Glycine	57.06	0.16	75	7.5	
Cysteine	103.14	0.04	135	2.8	pk_a 9.1
Tyrosine	163.18	0.02	230	3.5	275 nm (1420)
Proline	97.12	−0.07	145	4.6	
Threonine	101.11	−0.18	140	6.0	
Serine	87.08	−0.26	115	7.1	
Histidine	137.15	−0.40	195	2.1	pk_a 6.2
Glutamic acid	129.12	−0.62	190	6.2	pk_a 4.6
Asparagine	114.11	−0.64	160	4.4	
Glutamine	128.14	−0.69	180	3.9	
Aspartic acid	115.09	−0.72	150	5.5	pk_a 4.5
Lysine	128.18	−1.10	200	7.0	pk_a 10.4
Arginine	156.20	−1.76	225	4.7	pk_a 12

Amino acids have been ordered according to their hydrophobicities. The last column shows the pk_a values for those amino acids that have ionizable side chains or the maximum absorbance wavelengths (and molar absorptivities) for amino acids having aromatic sidechains molecular weights are per residue in proteins after subtractions of water (18Da).

[a]From Eisenberg et al., 1982.

[b]From Chothia, 1975.

[c]From Klapper, 1977.

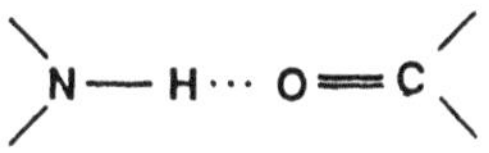

N—H··· O=C

Figure 4 The hydrogen bond is indicated by the dotted line. This type of bond forms when an electronegative atom, such as a carbonyl oxygen, interacts with another electronegative atom, such as an amine nitrogen, through a hydrogen atom bound to the latter heteroatom.

amino acid side chains are able to form hydrogen bonds, another type of the chemical interaction that is important in stabilizing protein structure. Hydrogen bonds form when two electronegative atoms share a hydrogen atom, creating a chemical bridge that brings them closer than would be predicted on the basis of their van der Waals radii. The atom which is covalently attached to the hydrogen atom is known as the hydrogen bond donor, and the other atom is known as the hydrogen bond acceptor. In addition to the hydrogen bonds involving amino acid side chains, amide protons and carbonyl oxygens of the protein backbone also participate in this type of chemical interaction (Fig. 4). Each hydrogen bond provides about 3 kcal/mol of stabilization energy.

Glycine is distinctive in that it possesses two hydrogen atoms linked to the alpha carbon, which makes it a highly flexible amino acid often found at bends in the protein chain. This flexibility is in contrast to proline, which is constrained by a carbocyclic ring joining the amine nitrogen to the delta carbon of its own hydrophobic side chain. Proline is more properly called an imino acid and constrains the peptide backbone and limiting flexibility of the protein chain. Although the proline side chain is aliphatic, as are the nonpolar amino acids discussed below, it is more polar because of its compactness. Proline and glycine often cause breaks in alpha helices, a type of secondary structure which will be described below.

Cysteine, another of the naturally occurring amino acids, contains a sulfhydryl group in its side chain. As shown below, sulfhydryls readily oxidize to form disulfide bonds (Fig. 5). The formation of disulfide bonds often plays

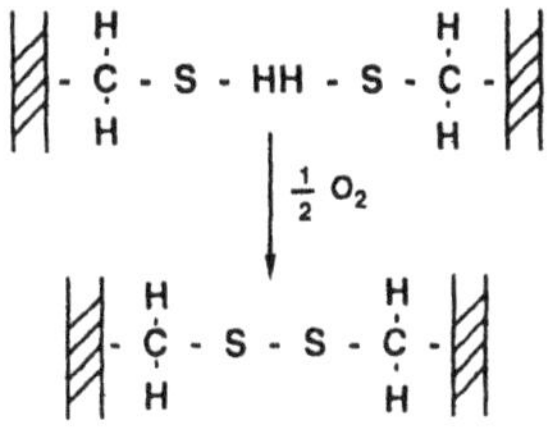

- CH₂ - S - HH - S - CH₂ -

$\frac{1}{2} O_2$

- CH₂ - S - S - CH₂ -

Figure 5 Oxidation of two cysteinyl residues results in the removal of two hydrogen atoms and the formation of a disulfide bond. The protein backbone is represented by the bars on either side of the sulfur atoms.

an important role in maintaining protein structure and function. In fact, one of the challenges in producing proteins through the use of genetic engineering is the formation of the proper (i.e., native) disulfide bonds.

Nonpolar Amino Acids

The third group of amino acids consists of alanine, valine, leucine, isoleucine, methionine, phenylalanine, and tryptophan. The side chains of alanine, valine, leucine, and isoleucine are straight-chain and branched aliphatic functional groups. The side chain of methionine contains sulfur, whereas phenylalanine and tryptophan possess aromatic functional groups. The principal mode of chemical interaction for this group of amino acids is through hydrophobic forces.

Hydrophobic bonds can be pictured as the segregation of water molecules and nonpolar amino acid side chains. This occurs because water molecules must be highly organized around nonpolar side chains in order to establish relatively low energy electronic interactions. These interactions do not compensate for the energy which is expended in establishing these interactions. Therefore, hydrophobic bonding occurs because of entropic rather than enthalpic forces. Hydrophobic bonds, which increase with the size of the nonpolar side chain, average about 3 to 5 kcal/mol. The major effect of hydrophobic bonding is to move nonpolar amino acids to the protein's interior.

Amino Acids Are Optically Active Compounds

The alpha carbon of each amino acid found in proteins is covalently bonded to four unique substituents to form what may be depicted as a tetrahedral pyramid. As noted above, the only exception is glycine, which is bonded to two hydrogen atoms. The four substituents can be distributed about the alpha carbon to form two distinct isomers (Fig. 6). Although these isomers have identical chemical properties, they rotate plane-polarized light in opposite directions. That is, if plane-polarized light is passed through solution which contains only one of these isomers, it will emerge in a different plane. A solution which contains the inverted isomer in equal concentration will rotate light by the same amount in the opposite direction. Because of this property, amino acids are called chiral compounds, that is, they possess optical activity.

The naturally occurring amino acids found in mammalian proteins almost always are in the "L" configuration (Fig. 6b). Each of the L-amino acids does not rotate plane-polarized light to the same extent, or even in the same direction. For example, at the same temperature and concentration, aqueous solutions of L-phenylanine and L-arginine rotate plane-polarized light −34.5 and +12.5°, respectively.

Amino acids retain their chirality following their incorporation into proteins. Maintaining optical purity of the amino acids within proteins is often

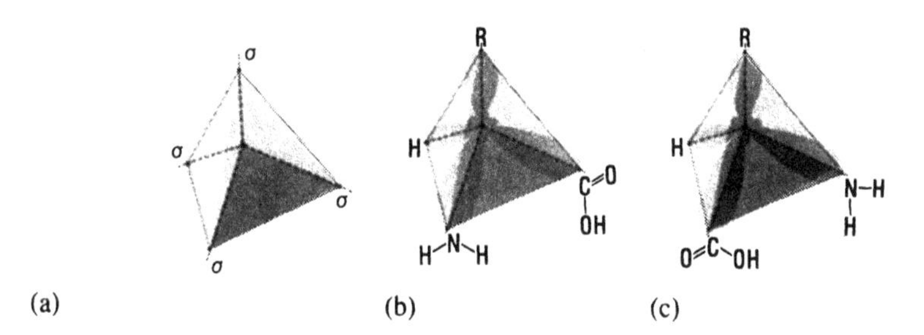

Figure 6 Amino acid chirality. (a) The four sp^3 hybridized orbitals of an amino acid's alpha carbon atom form a tetrahedral pyramid. These orbitals bond to four distinct groups to form (b) an L-amino acid, or, (c) a D-amino acid. L-amino acids are normally found in proteins.

important in maintaining biological function. For example, replacement with D- for L-aspartic acid in the dipeptide analog aspartame, produces a compound which possesses virtually no sweetness. Sometimes, however, selective changes in the chirality of amino acids enhances biological activity. This is because proteases (which are also optically active proteins) cleave D-amino acid-containing analogs more slowly. Metkephamid, an enkephalin analog which contains a D-amino acid, has enhanced activity due, in part, to its increased half-life:

H-Tyr-D-Ala-Gly-Phe-(N-methyl)-Met-NH_2
Metkephamid

Preclinical trials with this peptide have shown that it has significant analgesic activity, without many of the undesirable side effects of other peptide analgesics.

Links of the Protein Chain: The Peptide Bond

Proteins consist of amino acids linked together to form a linear chain. The linkage between amino acids, the peptide bond, results from the condensation of two amino acids as shown in Figure 7. Repeated condensations of this type produces a polymer of the following unit structure (Fig. 8). By convention, the first amino acid corresponds to the side chain R1 at the amino terminus (or N terminus) of the polypeptide chain. The last amino acid is referred to as the carboxy terminal (or C terminal) residue. The structural diversity of proteins, mentioned earlier, results from the length of this chain, as well as the identity and order of the amino acid residues (Ri) in the protein chain. For example,

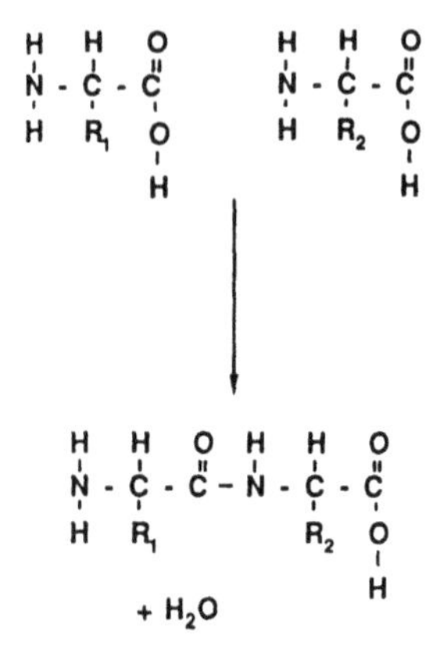

Figure 7 The condensation of two amino acids, with the elimination of water, to form a peptide bond.

the formation of random chains using all 20 amino acids in a 100 residue protein maximally results in 20^{100} unique compounds.

The four atoms comprising the peptide bond are planar or nearly planar; this configuration permits the delocalization of electron density (Fig. 9). Two planar configurations of these four atoms are possible, depending on the relative orientation of the carbonyl oxygen and the amide proton (Fig. 9). When the carbonyl oxygen and the amide proton are on opposite sides of the C—N bond, the configuration of the amide bond is said to be trans. Trans peptide bonds are sterically less hindered and are favored over cis peptide bonds by a ratio of 1000:1. Greater rotation about each bond to the alpha-carbons is possible. Accordingly, the protein backbone may be conceptualized as consisting of planar amide groupings arranged in different rotational orientations to one another via connections through the alpha-carbons.

For proline residues, however, steric hindrance has less impact on the orientation of the bond. Therefore, *trans*-proline is favored over *cis*-proline by

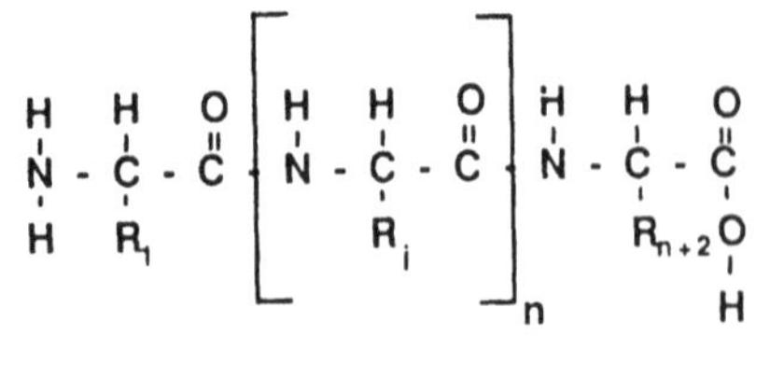

$$\mathrm{H_2N{-}CH(R_1){-}C({=}O)}\left[\mathrm{{-}NH{-}CH(R_i){-}C({=}O){-}}\right]_n\mathrm{NH{-}CH(R_{n+2}){-}C({=}O){-}OH}$$

Figure 8 The linkage of amino acids through peptide bonds to form a protein backbone.

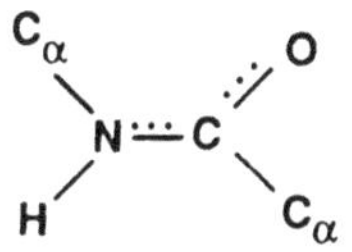

C_α O
N⋯C
H C_α

Figure 9 Delocalization of the nitrogen p orbital, and the carbonyl pi orbital electrons. This delocalization can only occur when the peptide bond assumes a planar configuration. With the exception of proline, the amide proton and the carbonyl oxygen are almost always found to be trans to each other.

only 4:1 (Fig. 10). Cis peptide bonds are most often found preceding proline residues within beta turns. The loss of rotational freedom around the alpha-carbon of proline plays a significant role in the three-dimensional structure of a protein (see below).

THE FOUR LEVELS OF PROTEIN STRUCTURE

Introduction

Protein structure consists of four levels. The primary structure of a protein describes the order of amino acids along its linear chain. The genetic information in the DNA (deoxyribonucleic acid) chain determines the identity and placement of amino acids within the protein chain. (The biosynthetic process is presented in the following chapter.) After the amino acids are linked together, alterations of amino acid side chains along the protein chain may take place. These are known as postribosomal (or posttranslational) modifications.

Once assembled, portions of the peptide backbone fold into regular conformations. Regular conformations contain well-defined, repeating torsion angles

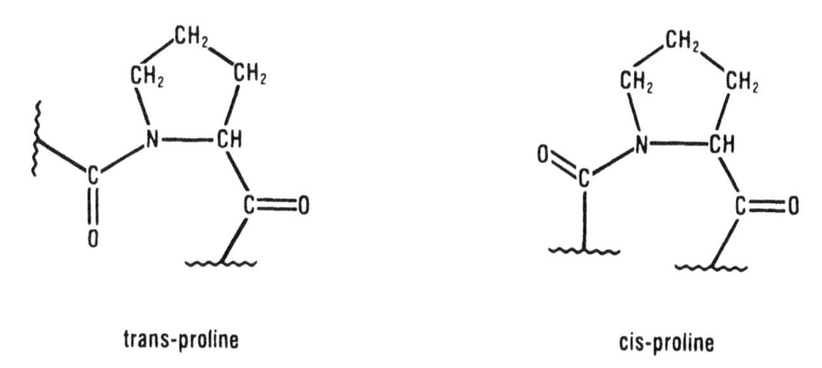

Figure 10 Peptide bonds involving proline are sometimes found in the cis conformation. The ratio of trans:cis isomers in globular proteins is about 4:1.

and are stabilized by hydrogen bonds. Such segments form what is called the secondary structure of the protein. The principal conformations of secondary structure are alpha helix, beta sheet, and beta or reverse turn. Other elements of secondary structure, such as the 3_{10} helix, are less commonly observed. These segments of secondary structure often fold together to form supersecondary structures.

Until recently, nonregular secondary structure has been simply called random coil. Examination of protein crystallographic structures has shown that portions of this "random coil" can be defined as turns or omega loops. The latter element of secondary structure can be identified using stringent criteria.

The tertiary structure of a protein is its overall shape as defined by the relative three-dimensional location of its amino acids. Protein tertiary structure consists of the regular and irregular secondary and supersecondary structures mentioned above. Proteins comprising several peptide chains held together by specific, noncovalent contacts have quaternary structure.

Primary Structure

As noted above, the primary structure of a protein is predetermined by specific genes which are translated into the amino acid sequence according to the genetic code. Once formed, a protein's primary structure leads to the other levels of structure mentioned above. Studies of ribonuclease, conducted by Anfinsen and colleagues initially demonstrated this relationship by showing that ribonuclease, which had been treated to destroy all secondary and tertiary structure, could refold to the native enzyme. However, this example cannot be generalized. The folding into the proper three-dimensional structure of a pro-

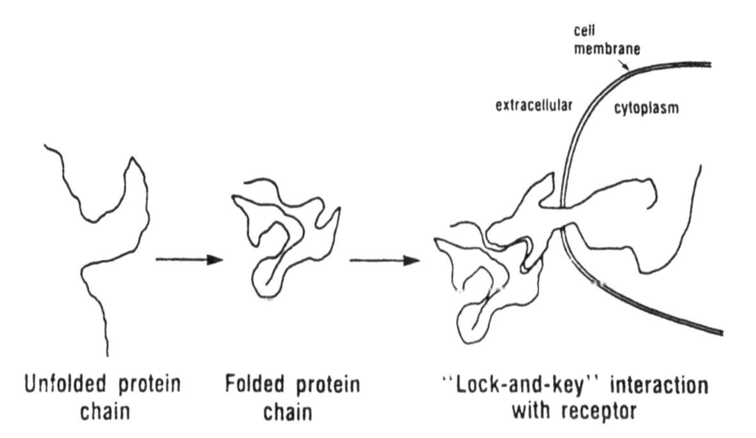

Figure 11 Protein hormones must, in many cases, fold into their native structure in order to productively bind to their receptors.

tein also depends on the environment prevailing during the biosynthetic process. Once a protein is denatured, it is often not a simple matter to regain the native structure, as is the case with ribonuclease (Fig. 11).

Is the biological activity of a protein observed in the absence of secondary and tertiary structure? This question has been addressed for a large number of proteins by determining various biological activities of proteins which have been denatured or fragmented. Some peptide hormones, such as beta-endorphin, calcitonin, and vasoactive intestinal peptide, contain short fragments of five to seven residues, which by themselves retain significant biological activity. Because of their length, these peptides are not likely to contain significant regular secondary structure. Fragments of many intermediate-sized protein hormones and enzymes have not been observed to have the biological activities attributed to the intact protein. In these cases, it appears that induced or native three-dimensional structure is required to form a region which will interact with hormone receptor or enzyme substrate (Fig. 11).

Protein primary structure sometimes determines its ultimate destination. For example, when initially synthesized in vivo, proteins that are to be secreted from a cell contain an amino-terminal signal sequence that has groupings of basic (1 to 3 residues), hydrophobic (the next 14 to 20 residues), and hydrophilic (variable length) regions. After a protein is transported across the cellular membrane, these signal sequences are enzymatically removed, generating the mature protein. Recent studies have indicated that other signal sequences assist the transport of proteins across nuclear, mitochondrial, and lysosomal membranes.

Postribosomal Modifications

The primary sequence of a protein sometimes helps determine the specificity of alterations which occur after the protein sequence is assembled. For example, the addition of carbohydrate to eukaryotic proteins through asparagine residues (i.e., N-linked) occurs only when the residue to be glycosylated is separated from serine or threonine residues by one amino acid residue as shown here:

–Asn–X–Ser (or Thr)–,

where X may be virtually any amino acid residue. The signal sequences for attachment of carbohydrate to the hydroxyl group of Thr or Ser (i.e., O-linked) have not been established.

Erythropoietin, a protein hormone which promotes the maturation of red blood cells, contains about 50% carbohydrate, linked to asparagine, serine, or threonine. Although removal of carbohydrate does not affect the bioactivity of all proteins, it is essential for the in vivo activity of erythropoietin. Another

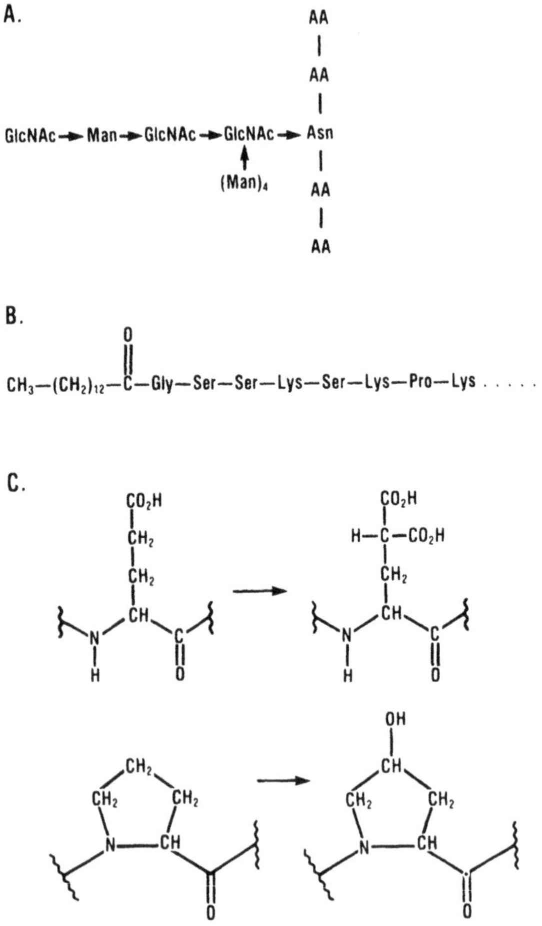

A.

GlcNAc → Man → GlcNAc → GlcNAc → Asn (with $(Man)_4$ → second GlcNAc; Asn within chain AA—AA—Asn—AA—AA)

B.

$CH_3-(CH_2)_{12}-C(=O)-$Gly—Ser—Ser—Lys—Ser—Lys—Pro—Lys . . . . .

C.

Figure 12 Examples of postribosomal modification. (A) Addition of carbohydrate to asparagine residues. Random amino acids of the protein sequence are denoted by AA. (B) Fatty acid acylation of the N terminus. (C) Modifications of amino acid side chains, such as the modification of glutamic acid to form carboxy-glutamic acid, and 4-hydroxy-proline from proline.

example of postribosomal modification, which is principally directed by primary structure, is observed with collagen. In this protein, the second proline residue in the repeating tripeptide, -(Pro-Pro-Gly)$_n$- is typically converted to 3- or 4-hydroxyproline.

Many other postribosomal modifications have been observed, and a partial list is shown in Figure 12. For example, some proteins must be specifically cleaved in order to generate biological activity. This is true for proinsulin, a protein containing 82 residues. The translation product is enzymatically cut at two positions within the peptide chain to yield the mature 51-residue heterodimer, in which the two chains are held together by disulfide bonds (see Fig. 12). Similarly, trypsin, an enzyme which cleaves proteins at the C-terminal side of the peptide bond of arginine or lysine residues, is activated by the removal of two dipeptides from its nascent protein chain. Two proteins, p21 and pp60*src*, which have been shown to correlate with malignant cellular transformation in vivo and in vitro, require the addition of fatty acids at specific positions in their N-terminal regions. Many of the protein factors which participate in the blood coagulation cascade, require the synthesis of γ-carboxy glutamic acid from glutamic acid within their N-terminal regions.

The formation of disulfide bonds is one of the most common postribosomal modifications. Selective disulfide bond formation appears to occur when the three-dimensional structure of the protein brings cysteines into close proximity, possibly with the assistance of nearby amino acid residues. The disulfide bonds observed in mature proteins are not necessarily those which initially form during protein folding. For example, bovine pancreatic trypsin inhibitor, which contains three disulfide bonds, has been shown to form transient disulfide bonds on route to its final structure.

Secondary Structure

The principal examples of regular secondary structure are alpha (α) helix, beta (β) sheet, and the reverse turn (Fig. 13). Irregular conformations of protein structure have been categorized as "random coil," turns, and omega loops. As pointed out by Jane Richardson, random coils are neither random nor coiled. This section further describes the types of secondary structure.

Alpha Helix

Alpha helices are a common element of protein secondary structure. Nearly a third of all amino acid residues within proteins participate in these structures. The formation of alpha helices was initially proposed by Linus Pauling and associates, and their existence was confirmed shortly thereafter in the laboratory of Max Perutz. Alpha helices are segments of amino acids which are organized into a cylindrical shape with amino acid side chains oriented away from the surface of the cylinder. These structures are stabilized by hydrogen bonds which form between the carbonyl moiety of amino acid residue "i" and the amide proton of residue "i + 3." These amide bonds are roughly parallel to the cylindrical axis (Fig. 13a). Each turn of the helix contains 3.6 amino acid residues and covers a distance of 5.41 angstroms. Within proteins, an average stretch of alpha helix contains 10 to 15 amino acid residues, equivalent to 3 to 4 helical turns.

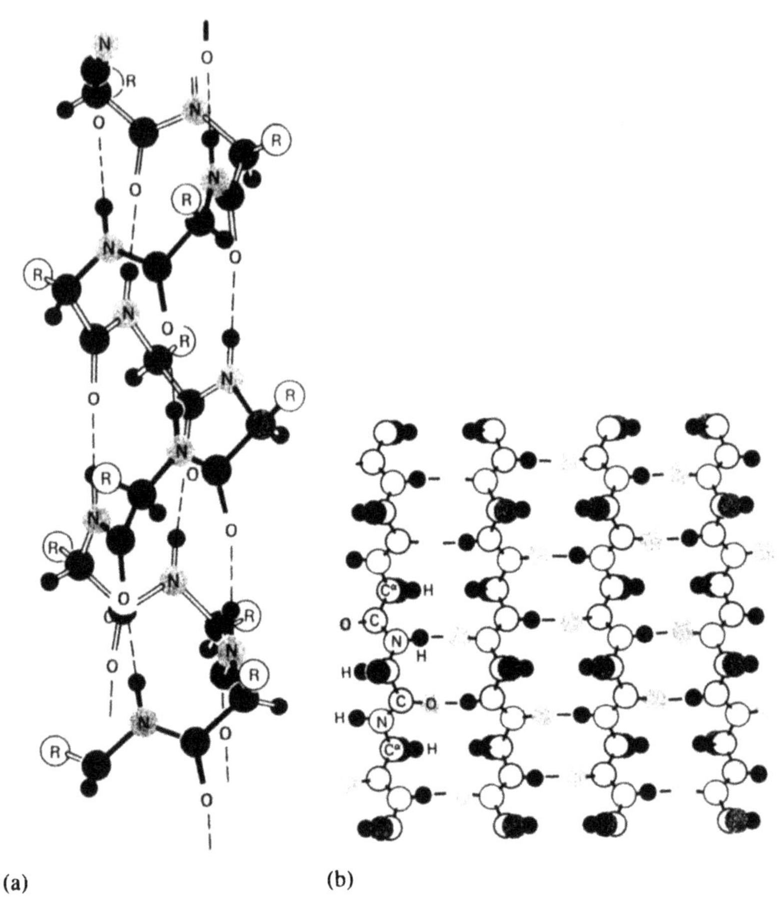

(a) (b)

Figure 13 The three main types of secondary structure. (a) The alpha helix. In this figure, the peptide backbone is illustrated as a ball and stick model with large and small circles used to represent carbon and hydrogen atoms, respectively. The dashed lines illustrate the hydrogen bonds, which form along the peptide backbone. (b) An antiparallel beta sheet. In this figure, the large circles represent the amino acid side chains. Unlike alpha helices, beta sheets are stabilized by hydrogen bonds which form between separated strands. (c) The three types of reverse turn. The reverse turns are distinguished by the position of the hydrogen bond which stabilizes the conformation, and by the relative orientation of the amino acid side chains. (All structures reprinted with permission from Creighton, 1984).

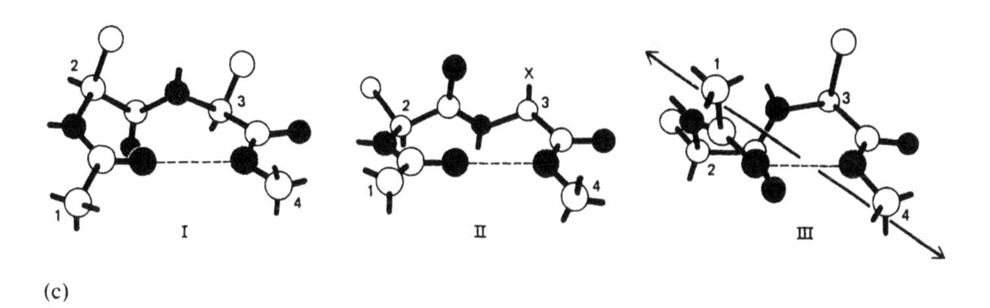

(c)

Alpha helices have been examined to determine if some amino acid residues prefer to form this type of secondary structure. These studies show that while any amino acid residue can be part of an alpha helix, those with extended side chains (such as leucine or methionine) are more frequently found within alpha helices, while compact or constrained residues (such as glycine or proline) are found much less often.

As shown in Figure 13a, all hydrogen bonds within an alpha helix are aligned with hydrogen bond donors and acceptors oriented in the same manner. Each hydrogen bond is a dipole in which the amide nitrogen of the helix is electropositive and the carbonyl is electronegative. Collectively, these partial electrical charges form a helix dipole which helps destabilize this form of secondary structure. The presence of oppositely charged residues near the ends of alpha helices help to stabilize this type of secondary structure by offsetting the helix dipole. Recent efforts in the laboratory of Robert Baldwin have shown that negatively charged, acidic residues near the N terminus of synthetic ribonuclease S peptide analog, help stabilize the alpha helix. Similarly, the placement of basic residues near the C terminus of these compounds helps stabilize these structures.

Many alpha helices segregate their hydrophilic and hydrophobic amino acids on opposite faces of the helical barrel (Fig. 14). This property of alpha helices is called amphiphilicity (or amphipathicity) and has attracted much interest in recent years. Amphiphilic helices of hormones such as calcitonin and bovine growth hormone appear to play direct roles in receptor interaction or in stabilizing bioactive conformations.

Alpha helices play important structural and functional roles in proteins. For example, x-ray crystallographic studies of porcine growth hormone show that this protein contains four alpha helices (Fig. 15). The antiparallel, up-and-down orientation of the helices permits partial cancellation of the helix dipoles. Data from several laboratories suggest that the segment comprising residues 96 to 133, which includes the third helix of the bundle, forms part of the receptor binding site. In contrast, interleukin-2 (IL-2) contains a four-helix bundle which stabilizes protein tertiary structure, but is not directly involved in recep-

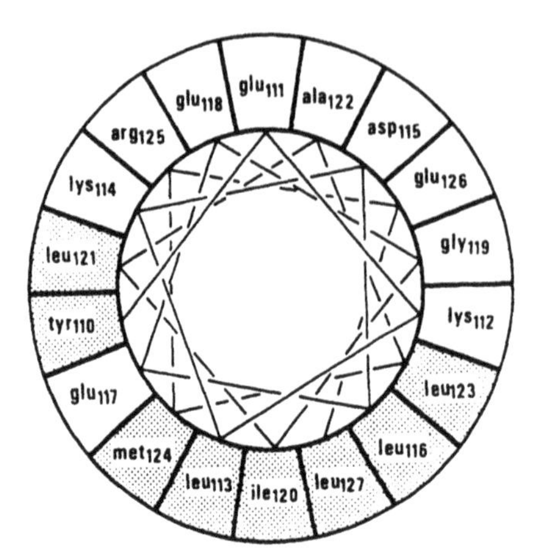

Figure 14 Helical wheel analysis of the amphiphilic alpha helix of bovine growth hormone found between residues 110 and 127. In this representation, the alpha helix is viewed down its axis, with the internal lines indicate the linkage of amino acids along the peptide backbone. The amino acid sidechains point away from the axis of the barrel. Hydrophobic regions have been shaded. (From Brems et al., 1987).

tor binding (Fig. 16). Instead, two additional helices which extend away from the helical core of the protein appear to be more actively involved in the biological activities of this protein. As proof, deletion of 10 amino acid residues from the N terminus of interleukin-2 reduces its ability to induce the proliferation of thymocytes by 30–50%. Deletion of the next ten amino acids, which contains a portion of the A helix, renders the protein biologically inactive.

The 3_{10} Helix

Another helix, known as the 3_{10} helix, is observed much less frequently than the alpha helix. This conformation is stabilized by a hydrogen bond which encloses a ring of ten atoms and contains 3 amino acid residues. The 3_{10} helix is observed infrequently because the hydrogen bond which forms the ring is nonlinear, the orientation providing the maximum stabilization energy, and since destabilizing steric interactions often occur between the amino acid side chains which form the helix. Nevertheless, 3_{10} helices are sometimes observed at the N and C termini of alpha helices.

Beta Sheet

Approximately 28% of all amino acid residues within proteins form beta sheets. Beta sheets contain a minimum of two strands, stabilized by hydrogen bonds to form a sheetlike structure (Fig. 13b). Within globular proteins, beta sheets often contain two to six segments, each of three to ten residues. Although the strands can be oriented in parallel or antiparallel, the latter orientation is preferred. As shown in Figure 13b, each residue forms two hydrogen bonds, with the amide proton and carbonyl oxygen bonded to hydrogen bond donors and acceptors on the adjacent strands. These interactions typically produce a pleat with a righthanded twist. Neighboring amino acid side chains point in opposite directions from these pleats.

Amino acid residues which are branched at the beta carbon (for example, isoleucine, threonine, and valine) are included within beta sheets more often than are other residues. Aspartic acid, lysine, and asparagine are residues less frequently found within beta sheets.

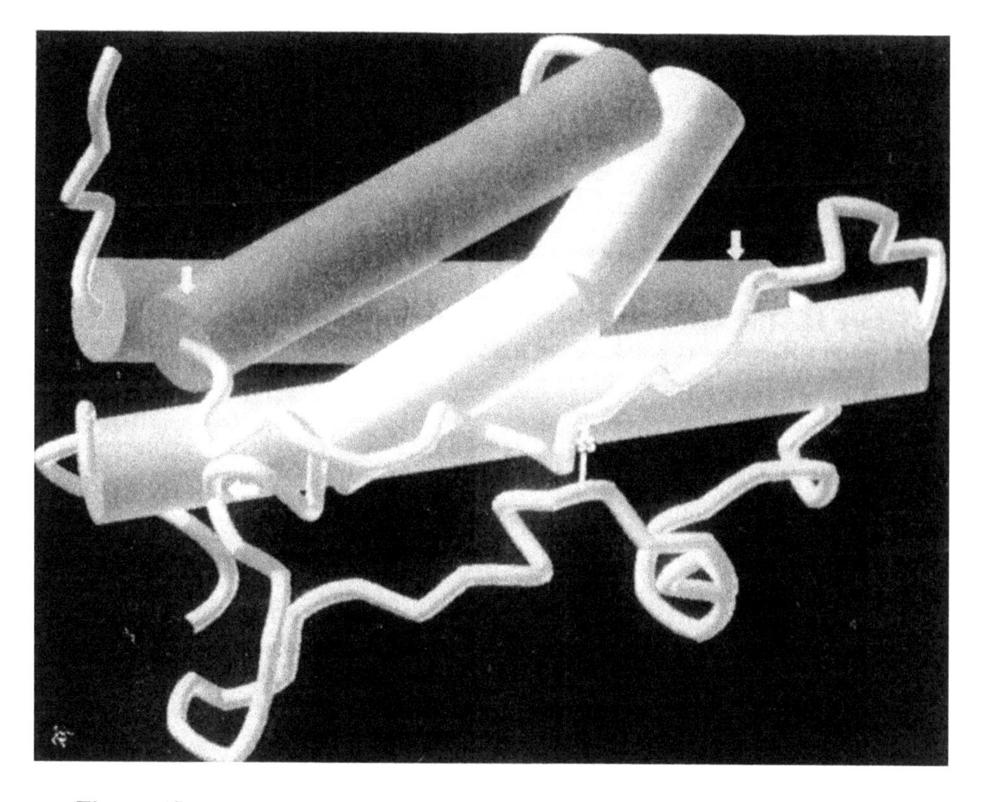

Figure 15 The structure of porcine growth hormone. Alpha helices are drawn as cylinders and the ribbon follows irregular structure within the protein. Although the neighboring alpha helices are oriented antiparallel to each other, the elongated connecting sequences permit the A-B and C-D helices to assume a parallel alignment. One of the two disulfides of porcine growth hormone is visible. (From Abdel-Meguid et al., 1987, reprinted with permission.)

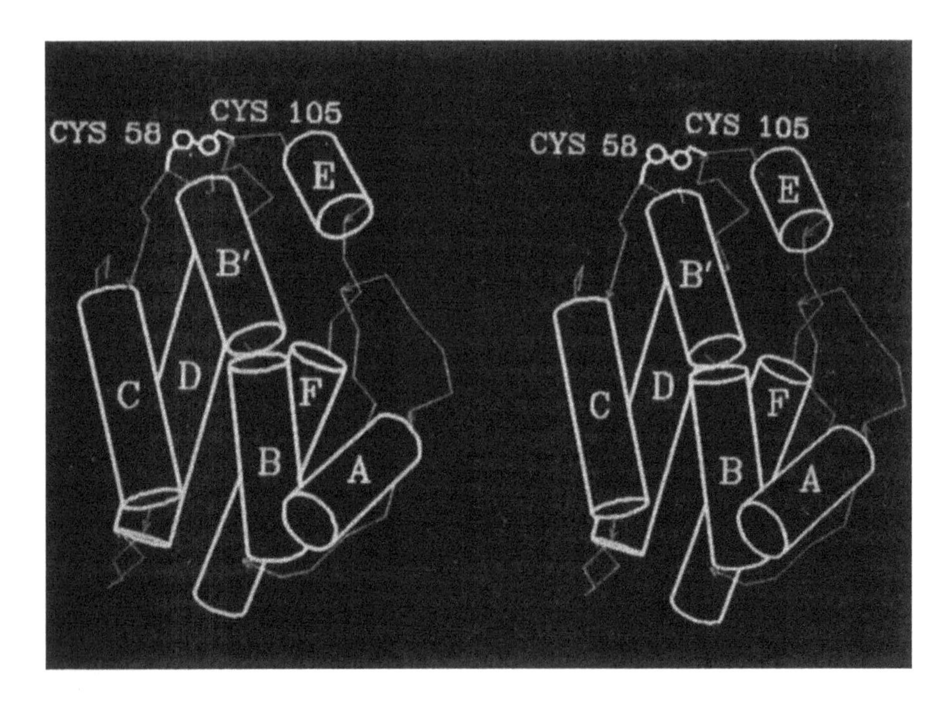

Figure 16 The structure of interleukin-2, based on x-ray diffraction studies determined to a resolution of 3.0 Å. (From Brandhuber et al., 1987.)

The classic example of a protein of high beta sheet content is silk fibroin, a structural protein of about 400,000 Da. A principal constituent of this protein is made up of the following peptide subunit:

-(Gly-Ala)$_2$-Gly-Ser-Gly-Ala-Ala-Gly-(Ser-Gly-Ala-Gly-Ala-Gly)$_8$-Tyr-.

This sequence is iterated up to 50 times in different sections of the protein to form extended beta-sheet structure. The glycine residues are mainly found on one surface of this protein, while the alanine and serine residues are located on the opposite face. The beta sheet surface which contains the glycine residues, as might be expected, permits closer contact with other beta sheets than the surface which contains the alanine and serines.

The Beta, or Reverse Turn

This structural unit produces reversals in the peptide chain, often at the protein surface. In contrast with alpha helices and beta sheets, which are variable in length, all reverse turns contain four amino acid residues. The three types of well-defined reverse turns, types I, II, and III, collectively account for about 25% of all amino acid residues within proteins. They are all stabilized by hydrogen bonds, which form between the amide proton of the first amino acid residue, and the carbonyl of the fourth amino acid residue of the turn (Fig. 13c).

The various types of reverse turns are primarily distinguished by the orientation of the i + 1 and i + 2 amino acid side chains. In type I and II beta turns, the amide bonds which form between these amino acid residues are reversed. Therefore, in the type I reverse turn, the amide proton of the i + 2 residue is situated between the side chains of the i + 1 and i + 2 residues, and the carbonyl of the i + 1 residue extends in the opposite direction. This geometry is favorable for all naturally occurring amino acids except proline. In the type II reverse turn, the amide bond is reversed, placing the i + 1 carbonyl between the two amino acid side chains. For this reason, only glycine can be sterically accommodated in the i + 2 position of type II reverse turns. Because of the relatively low steric repulsion between backbone and side chain moieties, the type I reverse turn is observed more than twice as often as types II or III reverse turns. The type III reverse turn is actually one turn of a 3_{10} helix, discussed above. In this type of reverse turn, the i and i + 1 residues are rotated such that the i + 1 side chain is parallel to the plane of the turn and the side chain of the i residue extends above the plane.

One Type of Nonregular Structure: The Omega Loop

Omega loops are compact structures located on the surface of proteins, without obvious interactions with other protein segments. They are so named since idealized loops resemble the Greek letter omega (Ω). Omega loops contain between 6 and 16 amino acid residues, do not contain alpha helix or beta sheet conformations, and are set off from neighboring protein segments by two residues which are separated by distances of from 3.7 to 10 angstroms. Using these criteria, a large number of proteins including immunoglobulins, superoxide dismutase, and thermolysin have been shown to contain omega loops. Because they are located on protein surfaces, omega loops have been proposed to play roles in molecular recognition processes which precede glycosylation, phosphorylation, supramolecular assembly, and protein transport.

Supersecondary Structure

The elements of regular conformation described above are often not isolated within proteins, since they are stabilized by neighboring secondary structure. The association of discontiguous protein segments which contain regular conformations to form larger units is called supersecondary structure. For example, two protein segments which contain regular conformations could pair to form alpha-alpha, beta-beta, or alpha-beta structures. Similarly, three segments may be grouped together as alpha-alpha-alpha, alpha-beta-alpha, beta-alpha-beta, and beta-beta-beta. All of these supersecondary structures are observed in proteins except for those which include only one strand of beta sheet.

One example of supersecondary structure, the beta-alpha-beta configuration is found in proteins such as triose phosphate isomerase and tryptophan synthetase. This structural unit has a strong tendency to be righthanded, possibly because the beta strands prefer to form a righthanded twist. The connection of the two beta strands by the alpha helix permits the former structural units to align in parallel. The coupling of two beta-alpha-beta folding units is also commonly observed and has been named the Rossman fold. This structural subunit is commonly found in dehydrogenases where it binds to nicotinamide adenosine dinucleotide (NAD). One of the beta-alpha-beta subunits binds to the nicotine moiety while the other binds to the adenosine moiety of NAD. Some of the dehydrogenases which contain the Rossman fold arose as the result of convergent evolution. Therefore, it appears that the Rossman fold is particularly well suited for binding to this dinucleotide, which is an important cofactor for many enzymatic processes.

Two types of supersecondary structure contain only beta strands. One of these is the beta meander, otherwise called the straight up-and-down beta sheet. Beta meanders have been observed within enzymes such as the serine proteases, lactate dehydrogenase, and staphylococcus nuclease. This structural subtype resembles the beta-alpha-beta subtype, but with the central beta strand substituting for the alpha helix in the latter folding subunit (Fig. 17a). In the beta meander, about two-thirds of the possible backbone hydrogen bonds form. Since this degree of hydrogen bonding correlates with the degree of hydrogen bonding within a three-stranded alpha helix, these structures have similar stability. A second type of supersecondary structure which contains only beta sheet is the Greek key (Fig. 17b), so named because the strand pattern resembles that found on ancient Greek pottery. There are at least 20 known examples of widely differing structures which contain the Greek key folding unit, including the variable, or antigen-binding region of immunoglobulin proteins. These latter beta strands are removed from their adjoining beta strands and are oriented in an antiparallel manner. The Greek key tends to be righthanded with respect to its crossover strands, a property which is not well understood.

Folding Domains and Tertiary Structure

The tertiary structure of a protein defines the average three-dimensional position of all amino acids with respect to each other, and represents the sum of regular and irregular conformations along the protein chain. A protein's ter-

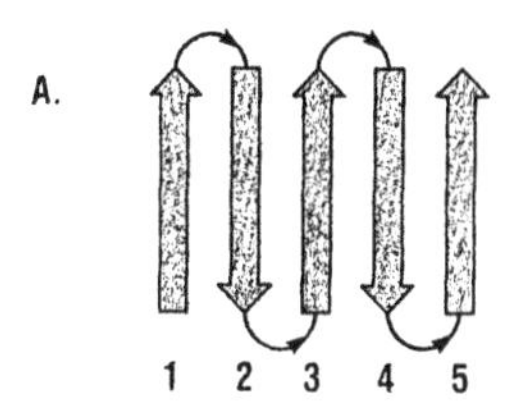

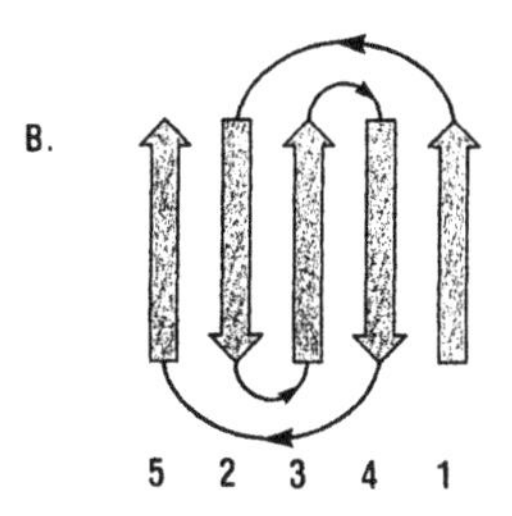

Figure 17 Schematic representation of the main types of beta sheet. The arrows point from the amino to the carboxyl end of each beta strand. The (A) beta meander and (B) Greek key.

tiary structure results from the combined effect of the chemical forces between its amino acid residues, such as hydrogen bonding, charge-charge, van der Waals, and hydrophobic interactions. When properly folded, proteins are typically 10–15 kcal/mol more stable than when they are fully denatured. This is not a large difference, considering that the energy gain on the formation of a single hydrogen bond is about 3 kcal/mol. Therefore, much of the stabilization energy gained from interactions mentioned above is balanced by the loss in entropy that results from confining a protein in a set of low energy conformations. The limited stability of protein tertiary structure is reflected in the ease of denaturation, which can be accomplished by alterations in temperature, pH, or by treatment with chaotrophs, such as urea or guanidine. Despite the ease of denaturation, protein tertiary structure is usually essential for maintaining biological activity.

Quaternary Structure

One well-known example of a multisubunit protein is hemoglobin. It consists of two identical alpha and two identical beta subunits. Each chain further possesses a noncovalently attached prosthetic group, heme. The three-dimensional configuration of this tetrameric protein is depicted in Figure 18.

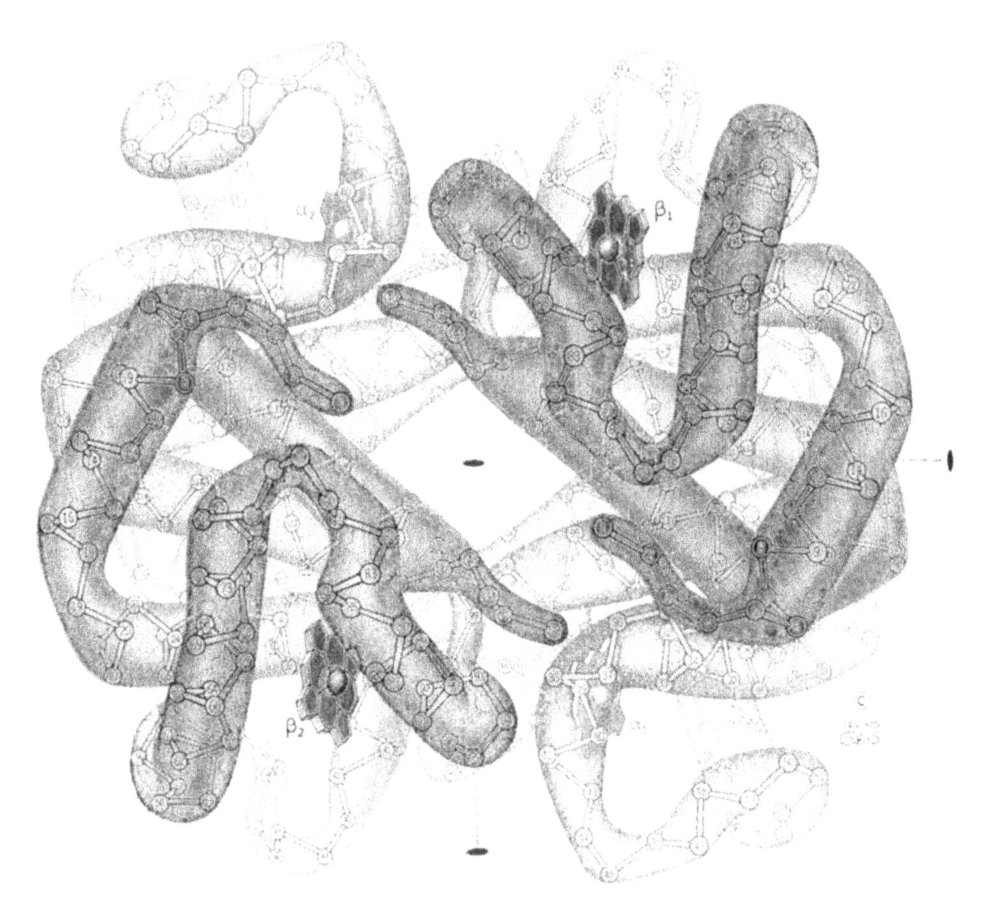

Figure 18 The quaternary structure of hemoglobin. In this representation, the alpha carbon atoms of each amino acid have been numbered according to their position in the sequence. The other backbone atoms have been omitted for clarity. Each chain has been drawn as an elongated coil, clearly indicating where stabilizing interactions occur. For example, contacts between all chains, except the alpha-1 and alpha-2 chains, help stabilize this multimeric protein. (Copyright by Irving Geis. Reprinted with permission.)

STRUCTURAL CATEGORIES OF PROTEINS

Introduction

Proteins can be categorized on the basis of their secondary and supersecondary structure. These are: (1) antiparallel alpha, (2) parallel alpha/beta (3) antiparallel beta, and (4) small irregular. Richardson, who has contributed significantly to the understanding of protein tertiary structure, has described these concepts in *The Protein Folding Problem*. Below is a brief description of each domain type and illustrative examples.

Antiparallel Alpha Proteins

This category includes proteins which contain alpha helices to the exclusion of other types of secondary structure. The helices of these proteins are typically aligned in an antiparallel manner (Fig. 19). Two proteins that exemplify this configuration are myohemerythrin and cytochrome b_{562}.

To illustrate this structural category, we return to our discussion of porcine growth hormone. As mentioned previously, porcine growth hormone consists of four alpha helices connected by one short and two extended fragments (Fig. 15). The extended peptide segments which connect helices A with B, and C with D, permit these helical pairs to be oriented in parallel in opposite corners of the four helix bundle. The short segment between helices B and C forces these helices into an antiparallel orientation. The result is a barrel-like protein in which all neighboring helices are antiparallel.

Parallel Alpha/Beta Proteins

The beta-alpha-beta folding motif discussed above is the main structural element of these proteins. In triose phosphate isomerase, the substructures are

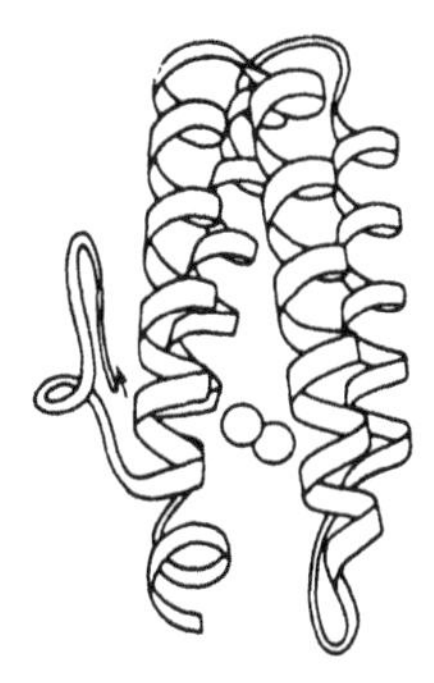

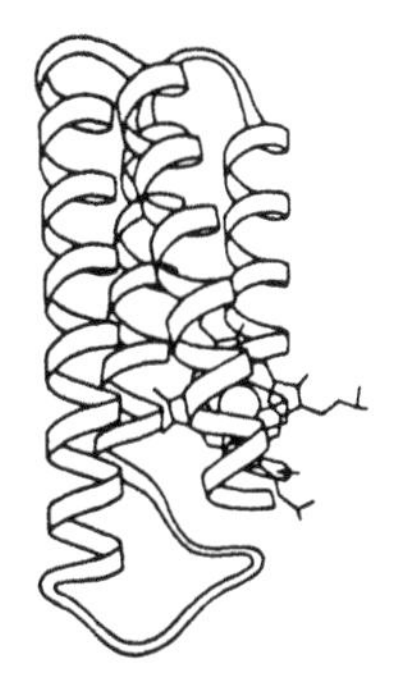

Figure 19 The tertiary structures of myohemerythrin and cytochrome b_{562}. Alpha helices are represented as helical ribbons, or spirals. Unlike porcine growth hormone (see Fig. 15), the helices which are contiguous in the primary sequence are antiparallel. Within the myohemerythrin structure, the two circles represent two iron atoms complexed by the helices. The helices of cytochrome b_{562} complex an iron-containing heme. (From Richardson, J. S., 1984, reprinted with permission).

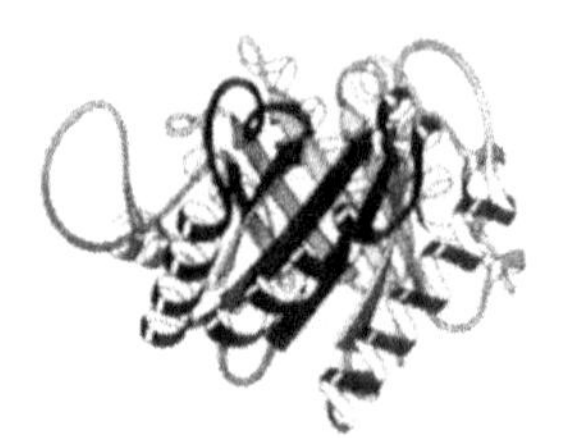

Triose Phosphate Isomerase Triose Phosphate Isomerase

Figure 20 Top and side views of the triose phosphate isomerase structure. Top view shows that the strands of beta sheet of this enzyme form a symmetrical core, which is surrounded by alpha helices. These clearly defined regions of secondary structure interact along an internal surface within the molecule. The side view more clearly shows the relative orientation of the alpha helices and beta sheets within this structure. (From Richardson, J. S. 1984, reprinted with permission.)

linked together such that the beta strands form a flat sheet surrounded by a ring of alpha helices on the outer surface of the protein (Fig. 20). When the beta-alpha-beta subunits are connected end-to-end in this manner, they are referred to as a singly wound parallel beta sheet. When an extended peptide segment permits the crossing of two beta-alpha-beta folding subunits, the resultant beta sheet is said to be doubly wound. The Rossman fold, described earlier, is an example of a double-wound parallel beta sheet. Other examples of doubly wound proteins include hexokinase and flavodoxin.

Antiparallel Beta Proteins

This type of protein is analogous to the antiparallel alpha structure described above. The most common form of antiparallel beta proteins contain beta strands which are organized into Greek key structures as discussed above. One prominent example of an antiparallel beta domain is the variable region of immunoglobulin (i.e. V_L). This region is directly involved in antigen-binding specificity.

Small Irregular Proteins

These proteins often contain a high level of cysteine or prosthetic groups. They have a lesser degree of secondary structure than the other classes described above. Any regular secondary structure that is found in these proteins is typically distorted, possibly because their small size precludes a large number of intrachain stabilizing interactions.

An example of a small irregular protein is insulin, a hormone that is important in the regulation of glucose metabolism. This protein contains A and B chains of 21 and 30 residues, respectively (Fig. 21a). These chains are connected by disulfide bonds which form between residues A7 and B7, and between residues A20 and B19. In addition, an intrachain disulfide bond within the A chain joins residues 6 and 11. The disulfide linkages had been established in the precursor protein, proinsulin prior to proteolytic maturation.

The A chain contains two segments of antiparallel alpha helix, spanning residues 2 to 8, and 13 to 20. The major component of secondary structure within the B chain is an alpha helix which extends from residues 9 to 19. A sharp turn between residues B20 to 23 facilitates the formation of close contacts between residues B15 and B24 and residues B11 and B26.

The intrachain disulfides bury a portion of this protein to form a small hydrophobic core. This core is important for maintaining the protein's structural integrity, as reflected in the strong homology among insulins from different species. These invariant residues include the A6 and 11 cysteines, the A2 isoleucine, and leucine residues at positions A16, B11, and B15.

Numerous structure/function studies have been carried out on insulin. These studies have revealed, for example, that the sequential removal of five residues from the C terminus of the B chain (i.e., residues B26 through 30) result in small incremental losses of insulin bioactivity, but that removal of residues B24 and 25 results in a significant loss of bioactivity. Insulin analogs missing B24 and 25 retain only 0.5% of that protein's glucose uptake activity in vivo. Similarly, removal of a limited number of residues from the N terminus of the B chain does not have a major impact on insulin bioactivity. In contrast, a marked decrease in bioactivity results when the A1 glycine residue is replaced by other amino acid residues. These results suggest that insulin reacts with its receptor through amino acid residues which have been brought together according to the constraints of the tertiary structure.

THE DE NOVO DESIGN OF PEPTIDES AND PROTEINS

Our increased understanding of protein structure has encouraged scientists to design and synthesize model sequences having predefined secondary and tertiary structure. For example, Kaiser and associates have designed peptides that include the amphiphilic alpha helix found within calcitonin, but which are minimally homologous. Nevertheless, these peptides, which are slightly more helical than calcitonin, bind to the calcitonin receptor and induce hypocalcemic activity in vivo.

Richardson and associates designed a series of peptides containing a four-strand, up-and-down antiparallel beta barrel. One compound, betabellin-2, is a 62 residue peptide containing two identical 31 residue sequences linked end-

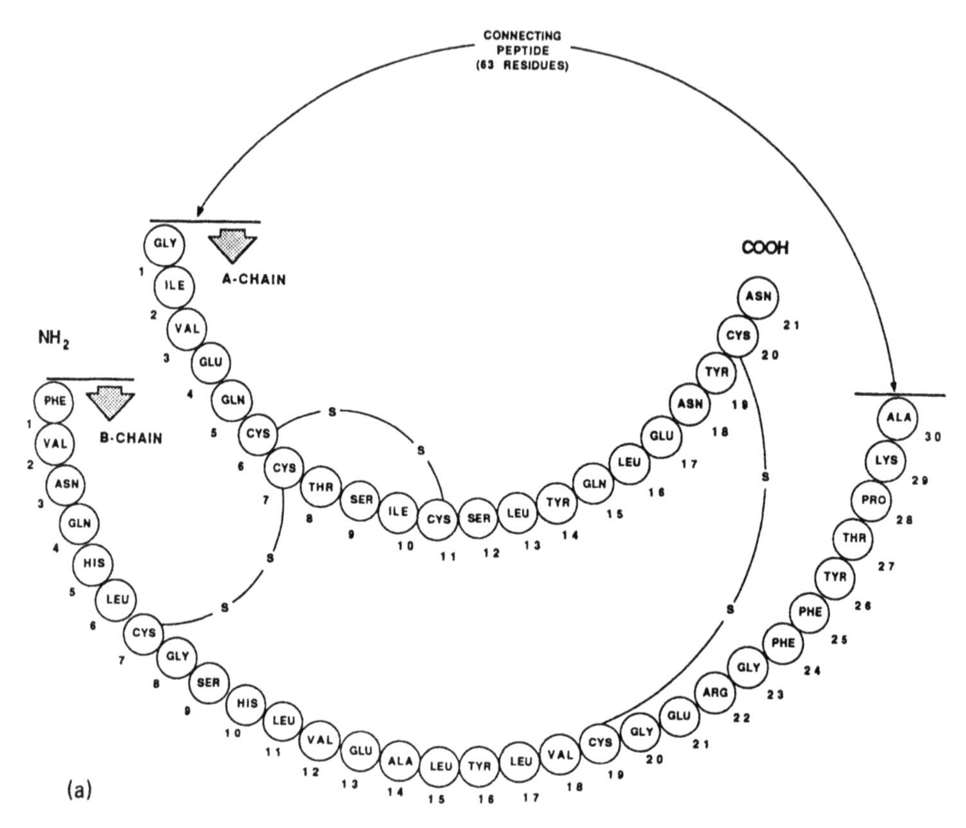

Figure 21 (a) The primary structure of insulin. The proinsulin peptide which connects the B-chain Ala-30 and the A-chain Gly-1 residues is illustrated here. (b) The tertiary structure of insulin. This picture was obtained on an Evans and Sutherland PS 300 graphics terminal.

to-end by an organic crosslinker. Betabellin-2 was designed to have a high propensity for antiparallel beta sheet formation, interspersed with –Pro–Asn– sequences at defined positions. The latter segments were included to encourage the formation type I beta turns. These beta turns are crucial in forming antiparallel beta strands within the peptide. Other modifications were made in order to optimize sidechain packing, to avoid destabilizing steric interactions, and to help increase hydrophilic interactions between water and amino acids which would likely reside on the surface of the peptide. Spectroscopic studies of this peptide indicate that it contains 40% beta sheet and 60% disordered coil. No alpha helix was detectable. Studies such as these may provide further insight into the relationship between primary, secondary, and tertiary structures.

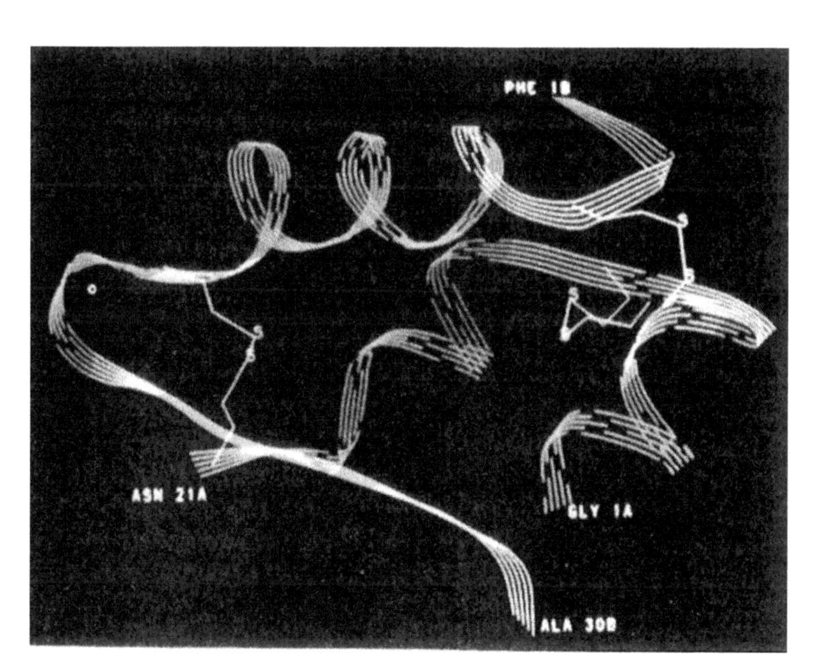

(b)

These examples have been presented to illustrate some of the concepts being studied using artificial proteins. It is worth noting how methods of protein design differ among scientists. There are many other ongoing studies utilizing synthetic and theoretical protein design which are providing useful information about the underlying principles of protein structure.

The concepts and illustrative examples presented in this chapter indicate that the study of protein structure is a complicated field, but one in which many basic concepts have already been defined. Considerable effort remains expended in order to refine our understanding of protein structure which will ultimately permit accurate predictions of tertiary structure based on primary sequence. This understanding will undoubtedly aid in furthering the utility of proteins for industrial and medical applications. The new field of rational drug design employs a combination of theoretical considerations and empirical three-dimensional structural information, with the goal of producing peptide mimetics as therapeutic products.

ACKNOWLEDGMENTS

I would like to thank Mr. Irving Geiss, Dr. A. McKay and Dr. S. S. Abdel-Meguid for providing artwork and Drs. J. A. Stodola and G. M. McClune for their helpful comments.

NOTE ADDED IN PROOF

It was recently noted by A. S. Edison (Trends in Biochemical Science (1990) 15, 216) that the β-sheet shown in Figure 13b contains D-, rather than L-amino acid residues. An illustration of β-sheet sheet structure, drawn with the correct handedness, can be found in the TIBS reference, along with a brief and interesting account of the error.

REFERENCES

1. Abdel-Meguid, S. S., Shieh, H. S., Smith, W. W., Payringer, H. E., Violand, B. N., and Bentle, L. A., Proc. Natl. Acad. Sci. USA 84, 6434–6437 (1987).
2. Brandhuber, B. J., Boone, T., Kenney, W. C., and McKay, D. B. *Science* 238, 1707–1709 (1987).
3. Brims, D. N., Plaisted, S. M., Kauffman, E. W., Lund, M., and Lehrman, S. R. *Biochemistry* 26, 7774–7778 (1987).
4. Chothia, C. *J. Mol. Biol.* 105, 1–14 (1975).
5. Creighton, T. E. *Proteins: Structures and Molecular Properties*, W. H. Freeman and Co., New York, 1984.
6. Dickerson, R. E. and Geis, I *The Structure and Action of Proteins*, W. A. Benjamin, Inc., Menlo Park, CA, 1969.
7. Eisenberg, D., Weiss, R. M., Terwillinger, T. C., and Wilcox, W. *Faraday Symp. Chem. Soc.* 17, 109–120 (1982).
8. Havel, H. A., Chao, R. S., Haskell, R. J., and Thamann, T. J. Investigations of protein structure with optical spectroscopy: Bovine growth hormone. *Anal. Chem.* 61, 642–650 (1989).
9. King, J. Deciphering the rules of protein folding. *Chem. Eng. News*, 32–54, April 10, 1989.
10. Klapper, M. H. *Biochem Biophys. Res. Commun.* 78, 1018–1024 (1977).
11. Oxender, D. L. and Fox, C. F. (eds.). *Protein Engineering*, Alan R. Liss, Inc., New York, 1987.
12. Richardson, J. S. What do the folds in proteins look like? *The Protein Folding Problem*, D. B. Wetlaufer (Ed.). Westview Press, Boulder, CO, 1984, pp. 1–28.
13. Schulz and Shirmer *Principles of Protein Structure*, Springer-Verlag New York Inc., New York, 1979.

3

Protein Biosynthesis

Richard A. Wolfe

Monsanto Company
St. Louis, Missouri

Stanley Stein

Center for Advanced Biotechnology and Medicine
Piscataway, New Jersey

INTRODUCTION

Proteins are biochemical substances that mediate virtually all the processes of a living organism. At a first approximation (for simple proteins), the linear sequence of amino acids (the primary structure) determines the final three-dimensional structure, and, hence, all the functional properties of the fully processed molecule (1). However, many complex proteins are first synthesized as pre-pro-proteins, whose three-dimensional structures are also solely determined by the amino acid sequence, and these molecules are subsequently "processed" to their final biologically active forms. The degree and type of processing varies with both the identity of the protein and the cell performing the biosynthesis. One outstanding feature of biotechnology is the ability to produce in simple microorganisms (such as the bacterium *Escherichia coli*) large quantities of molecules that have the same primary structure as those that are naturally present in higher animals in only miniscule amounts. The basic principles for the biosynthesis of the primary structure of proteins, which allow for this versatile feature of biotechnology, are presented in this chapter.

The processing of the pre-pro-protein into its mature form differs between simple organisms and the cells of higher organisms such as mammals. Many simple proteins, such as the much touted interferons and interleukins, are, however, processed similarly in human cells and bacteria. Thus, some recombinant proteins produced in simple organisms are, for all practical purposes, identical to the naturally occurring molecules, and they have been demonstrated to be equally efficacious in pharmacological studies or as biocatalytic agents (enzymes). Other more complex proteins are modified in a manner that cannot, as yet, be reproduced by genetic engineering of lower organisms. Since the manufacture of proteins via biosynthesis in microorganisms is, at today's state of technology, at least two orders of magnitude less expensive than the same process in mammalian cell bioreactors, these simple organisms are utilized whenever possible.

As more is discerned about the mechanisms of pre-pro-protein processing, it seems likely that the ability to perform at least some of these processing steps can be engineered into the faster growing, less demanding, simple organisms. Clearly, this is a long-term goal that will not be realized in the near future. Therefore, expression systems developed from the cells of higher organisms (including humans) will be the only practical method for producing complex posttranslationally modified proteins throughout the next decade. Other approaches may prove to be scientifically and economically sound. Analogs of complex natural proteins having similar functional properties have been produced in *E. coli* by recombining "fragments" of the natural species through molecular cloning techniques [e.g., synthetic single chain antibodies (2)]. This approach obviates the need for processing of a complex pre-pro-protein.

THE GENETIC INFORMATION

Proteins are synthesized in cells, one amino acid residue at at time, proceeding from the amino terminal. The particular sequence of amino acids in a polypeptide chain is dictated by the sequence of nucleotide bases in the corresponding mRNA (messenger ribonucleic acid) chain. mRNA, as the word messenger indicates, represents a copy of the original genetic information in the DNA of the cell. Copying this genetic information from DNA to RNA by the enzyme RNA polymerase is referred to as transcription and is illustrated for prokaryotes (simple organisms without a nucleus) in Figure 1. In this process, the enzyme scans the DNA until it recognizes a promoter region on either one of the DNA strands. All promoters contain some common, but not perfectly identical, nucleotide sequences, termed consensus sequences that represent the recognition signals for the RNA polymerase.

The cell has a control mechanism for determining which genes are to be transcribed. This is based on sequences of nucleotides that lie either within or near the promoter (for prokaryotes these regions of the gene have been termed "operators"). Regulatory proteins can either repress or enhance transcription, according to environmental factors, by interacting with these regions of the gene. A given gene may contain multiple regulatory sequences that interact

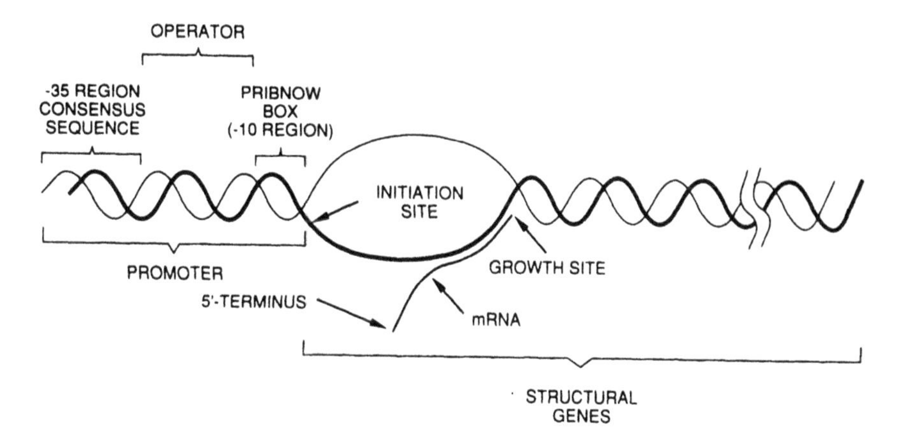

Figure 1 The process of transcription in prokaryotes. The enzyme, RNA polymerase (not shown in this figure for the sake of clarity) scans the double-stranded DNA until it recognizes a promoter region. The recognition features are the consensus sequences, centered at about −35 bases and −10 bases (also called the Pribnow box) from the initiation site of RNA synthesis. One of the strands of the double-stranded DNA is drawn bolder to indicate that the genetic information is being read on only that strand. The ability to transcribe the structural gene(s) is regulated by the operator region, which functions by interacting with a specific regulatory protein (not shown). The various regulatory proteins for the different operators in the genome of the organism transduce environmental factors by interacting with appropriate metabolites.

with either the same, or different, regulatory proteins. In most cases, transcriptional regulation is a finely tuned balance between repression and enhancement that is altered by environmentally controlled changes in the availability of the repressor and enhancer proteins.

After binding to the promoter, the RNA polymerase unwinds a section of the double-stranded DNA and begins synthesis of the RNA chain. The RNA sequence is dictated by the DNA sequence, according to the Watson-Crick base-pairing rules (A, T, G, and C in the DNA pair with U, A, C, and G, respectively, in the growing RNA chain). The polymerase moves along the DNA chain unwinding new sections as the growing molecule peels off and the previously transcribed region rewinds into a double helix.

The genetic information coding for the biosynthesis of an individual protein is called a structural gene. In prokaryotes it is common that a series of structural genes are in tandem, under the control of a single promoter, and this is called an operon. An operon encodes a group of proteins that are responsible for a particular metabolic function and, therefore, are to be synthesized in a coordinated manner.

The transcription process in eukaryotes (organisms with the DNA segregated into a nucleus) is far more complicated, corresponding to their greater metabolic complexity. In addition, many structural genes are split into multiple units that are separated from each other by extra (intervening) nucleotides (introns) that do not encode amino acid sequence information. The other regions of the structural gene that do contain the sequence information are termed exons. The introns can play a role in the regulation of transcription, and regulated variations in the way that the cell deals with, or processes, the information contained within these complex genes can lead to multiple forms of a protein.

In eukaryotes, the entire structural gene is transcribed from the DNA to produce nuclear precursor messenger RNA (pre-mRNA) and then "processed" into mRNA. A major portion of this processing occurs on a large ribonucleoprotein complex, the spliceosome (3). The large portions of the RNA that are encoded by the introns are "spliced out" of the molecule, and the remaining portion of the RNA that encodes the sequence data for the protein is further modified and transported to the cytoplasm. The pattern of splicing the segments of pre-mRNA encoded by exons into mRNA is regulated, and can change in response to changes in a cell's environment. In fact, not all of the exon-encoded sequences are necessarily utilized to form mRNA; in some extreme cases the information contained in entire exons are left out of the "finished mRNA." Thus, through "tissue-specific alternative splicing" the actual form of a protein can alter in response to changes in a cell's surroundings (4).

Numerous RNA copies may be transcribed from a single gene and numerous copies of a protein may be translated from each mRNA molecule. The process of reading the mRNA seqeunce is called translation. The RNA message is read in the 5′ to 3′ direction, corresponding to the amino-terminal to carboxy-terminal direction of the protein. This genetic message is read as a triplet code. Since there are four different choices for a nucleotide base at each of the three positions in a triplet code, there are 64 possible codons.

The genetic code, which describes the correspondence between the codons and their translated amino acids, is given in Table 1. This code is universal, from the simplest bacterium to complex multicellular organisms, and this principle allows the expression of human genes for proteins (the structural genes) in bacteria, yeast, insect cells, etc. In the genetic code (Table 1), most amino acids are represented by two or more codons. There are three stop codons, which do not specify any amino acid and, hence, terminate biosynthesis of the polypeptide chain.

All proteins are initiated with the amino acid methionine (Met), which is represented by only one codon, AUG. Since there may be several methionines present in a protein, there must be a mechanism by which the proper AUG

Table 1 The Genetic Code

First position (5′ end)	Second position				Third position (3′ end)
	U	C	A	G	
U	Phe	Ser	Tyr	Cys	U
	Phe	Ser	Tyr	Cys	C
	Leu	Ser	Stop	Stop	A
	Leu	Ser	Stop	Trp	G
C	Leu	Pro	His	Arg	U
	Leu	Pro	His	Arg	C
	Leu	Pro	Gln	Arg	A
	Leu	Pro	Gln	Arg	G
A	Ile	Thr	Asn	Ser	U
	Ile	Thr	Asn	Ser	C
	Ile	Thr	Lys	Arg	A
	Met	Thr	Lys	Arg	G
G	Val	Ala	Asp	Gly	U
	Val	Ala	Asp	Gly	C
	Val	Ala	Glu	Gly	A
	Val	Ala	Glu	Gly	G

Ribonucleotides are encoded in mRNA as triplet words. The initiating AUG determines the reading frame of the code.

codon is used for initiation. The genetic component for identification of the initiating position was found to be a short sequence of bases, referred to as the Shine-Dalgarno sequence. It is found in the mRNA about 10 bases upstream (on the 5′ side) of the AUG codon which is to be used for initiation.

To clarify these concepts, the following mRNA sequence:

5′AUGUUACGGCGAUGA-3′

would be translated into the polypeptide sequence:

NH_2-Met-Leu-Arg-Arg-COOH.

Of course, such small polypeptides cannot be biosynthesized on ribosomes (described below). The minimal length of a polypeptide chain is on the order of 80 amino acids.

MECHANISM OF BIOSYNTHESIS

The Interpreter, tRNA

How is the information contained as a triplet code of nucleotide bases deciphered and converted into amino acid sequence information? The set of molecules performing this function is tRNA (transfer RNA), schematically illustrated in Figure 2. These molecules are about 90 nucleotides in length and have base-pairing patterns that, from a two-dimensional perspective, have a cloverleaf pattern. Each tRNA has an anticodon triplet, which recognizes the corresponding codon for an amino acid. The anticodon triplet for $tRNA^{Met}$ is 3′-UAC-5′, which is complementary to 5′-AUG-3′ (A pairs with U and G pairs with C, in opposite polarity). (The reader is referred to biochemistry textbooks, listed at the end of this chapter, for background on DNA structure and function.) On the opposite side of the tRNA (see Fig. 2) is a site for attachment of an amino acid. Every tRNA has the sequence 5′-CCA-3′ at its 3′-terminus and the amino acid becomes covalently attached to the 3′-hydroxyl group (in equilibrium with the 2′-hydroxyl group) of the 3′-terminal A. However, $tRNA^{Met}$, which has the 3′-UAC-5′ anticodon, will only couple to the amino acid methionine at its 3′-terminal A. This specificity is determined by an enzyme, aminoacyl-tRNA synthetase. There is a different aminoacyl-tRNA synthetase corresponding to each of the 20 amino acids taking part in protein biosynthesis.

To review these concepts, three functional features of tRNA have so far been described. There is an amino acid attachment site which is the same for all tRNA molecules; there is an anticodon site, which is different for each tRNA; there is a recognition site on each tRNA for its corresponding aminoacyl-tRNA synthetase. Aminoacyl-tRNA $synthetase^{Met}$ will recognize only $tRNA^{Met}$ and will charge it only with the correct amino acid, methionine. An aminoacyl-tRNA synthetase for a particular amino acid must be able to recognize several different tRNA molecules, because there are frequently several different tRNAs for any one amino acid. This multiplicity is partially due to the different anticodons necessary for the reading of the multiple mRNA codons for a given amino acid. To further complicate matters, there may be more than one specific aminoacyl-tRNA synthetase for each amino acid. Despite this complexity, incorrect charging of a tRNA is rare. The tRNAs, charged with their appropriate amino acids, are now ready to participate in the biosynthetic process.

The Ribosome

The subcellular organelle for protein biosynthesis is called the ribosome. This structure is composed of two substructures, the large and small ribosomal subunits, each containing numerous different proteins, as well as ribosomal RNA.

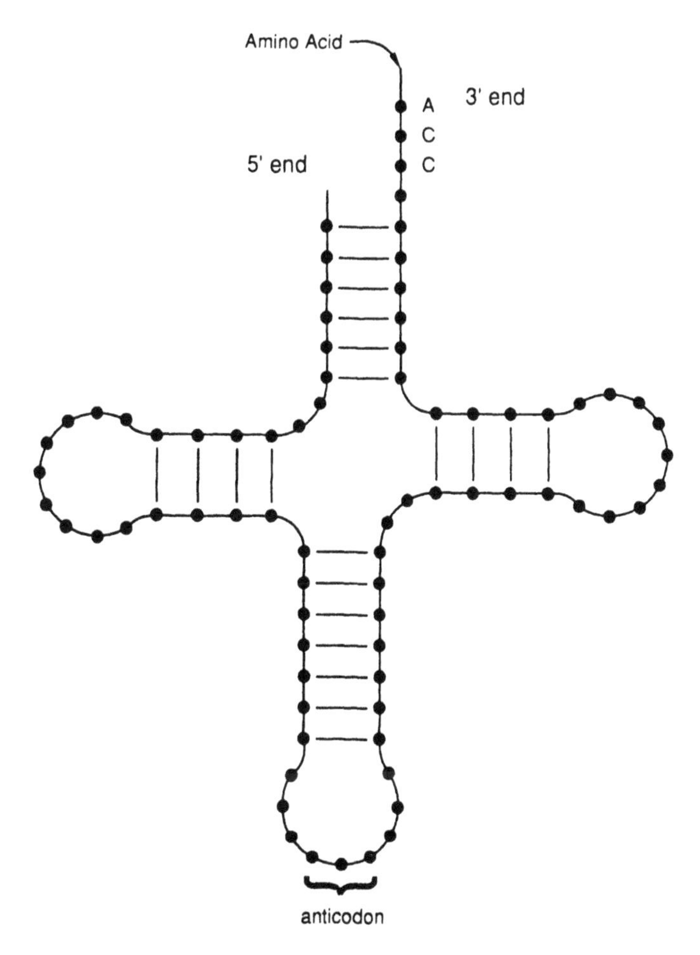

Figure 2 Primary structure of tRNA. Each dot represents a nucleoside in the chain. The general appearance is that of a cloverleaf. In three-dimensional space, the double-stranded portions, indicated by the cross lines, assume a double helical configuration and the molecule is folded like the letter C, with the amino acid acylation site and the anticodon site pointing toward each other. A unique feature of tRNA is the presence of unusual bases, especially in the two side loops.

An mRNA strand and a charged initiating $tRNA^{Met}$ attach to a small subunit, which in turn attaches to a large subunit. Essentially, the ribosome aligns the charged tRNAs alongside the mRNA, according to codon–anticodon recognition, and causes peptide bonds to form between the amino acids attached to each tRNA. This is a sequential process involving only two tRNAs at a time.

The process of eukaryotic polypeptide polymerization is illustrated in Figure 3. The mRNA to be translated binds to a groove in the ribosome, possibly

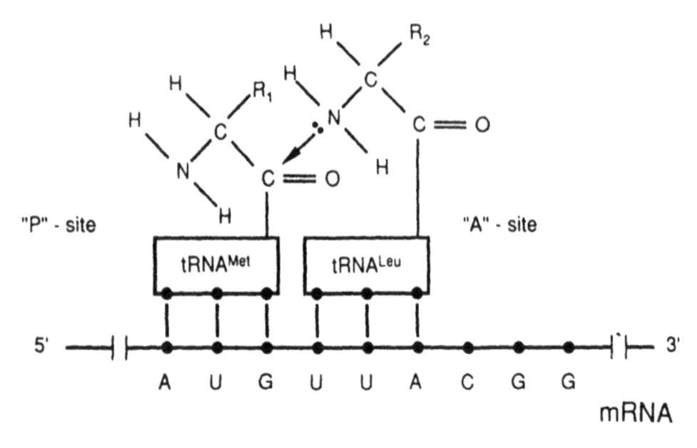
"P" - site
"A" - site
tRNA^Met
tRNA^Leu
5'
3'
A U G U U A C G G
mRNA

(a)

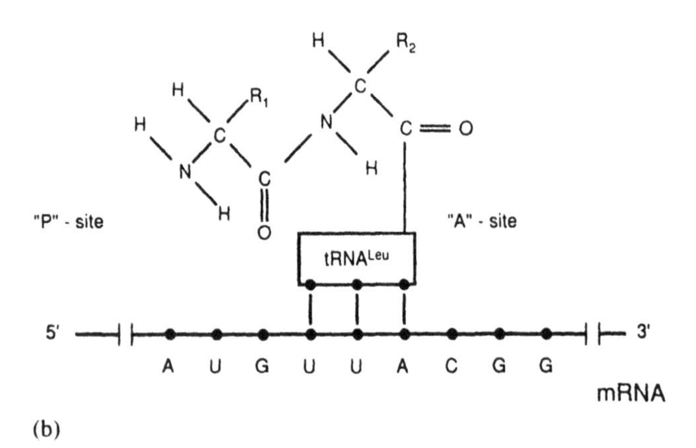
"P" - site
"A" - site
tRNA^Leu
5'
3'
A U G U U A C G G
mRNA

(b)

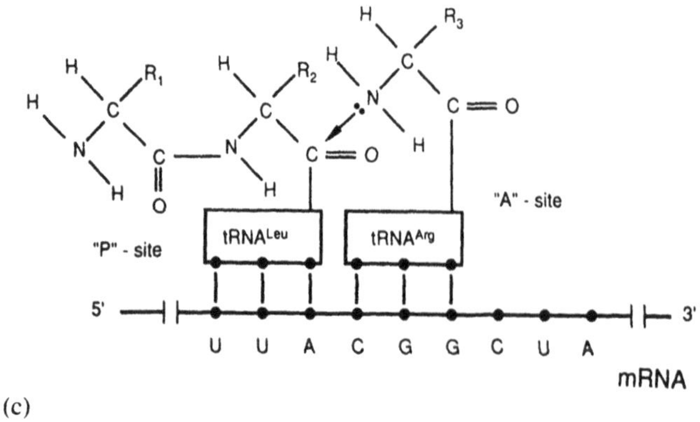

"A" - site
"P" - site
tRNA^Leu
tRNA^Arg
5'
3'
U U A C G G C U A
mRNA

(c)

at the interface between the two ribosomal subunits. Attachment occurs at the region of the initiating AUG, most likely guided by interaction of the Shine–Dalgarno sequence (discussed earlier) with a complementary sequence in the ribosomal RNA (rRNA). There are two tRNA docking sites, termed P for peptidyl and A for amino acid, on the ribosome.

The initiating charged tRNA, Met-$tRNA^{Met}$, binds at the P site (Fig. 3a). The charged tRNA which bears the anticodon to the next codon in the mRNA, UUA in this illustration, will bind at the A site. Upon nucleophilic attack of the amino group of the second amino acid, leucine, the Met-$tRNA^{Met}$ acyl bond is broken and the methionine forms a peptide bond with leucine (Fig. 3b). The $tRNA^{Met}$, which is no longer charged with its amino acid, is released from the P site. The dipeptide methionyl-leucine, still attached to $tRNA^{Leu}$, is translocated to the P site, leaving the A site ready for occupation by the tRNA bearing the anticodon for the third amino acid, in this illustration arginine (Fig. 3c). At the same time, the mRNA is translocated so that the codon designating the third amino acid will be available for binding the anticodon of the tRNA charged with the third amino acid. This process of elongation is repeated until a stop codon is reached. The initiation, elongation, the termination steps require additional protein factors that are not part of the ribosomal structure.

Prokaryotic versus Eukaryotic Initiation

In prokaryotes, biosynthesis is initiated with $tRNA^{fMet}$, which is acylated with *N*-formylmethionine, rather than with methionine. The formyl group almost always is removed from the amino terminus of prokaryotic proteins. Only about half the proteins in prokaryotes have methionine as the amino terminal amino acid. Thus, posttranslational removal of the formyl group and the methionine residue are both common occurrences. Formyl methionine and its cor-

Figure 3 Process of polypeptide formation. (a) The complex consisting of an mRNA attached to a ribosome, a charged initiating tRNA in the "P" site and a charged second tRNA in the "A" site. The second codon on the mRNA specifies the amino acid leucine. The amino group of leucine participates in a nucleophilic attack at the carbonyl-carbon of the methionine in the formation of the peptide bond. (b) The uncharged $tRNA^{Met}$ is released from the ribosome. The dipeptide, attached to the tRNA of the second amino acid, translocates from the "A" site to the "P" site and the mRNA indexes a distance of three nucleosides along the ribosome. (c) The next codon on the mRNA is then aligned with the "A" site, which now accepts the next charged tRNA. After peptide bond formaiton, the tripeptide–$tRNA^{Arg}$ complex translocates to the "P" site. This process is repeated until a stop codon is encountered and polypeptide chain elongation is terminated.

responding tRNA are not involved in eukaryotic protein biosynthesis (with the exception of the subcellular organelles, mitochondria, and chloroplasts).

Eukaryotic proteins are synthesized with an N-terminal extension, typically about 25 mostly hydrophobic amino acids, referred to as a leader sequence. The immature protein (termed "preprotein") containing the leader sequence is posttranslationally processed by enzymatic removal of the entire leader sequence, which includes the initiating methionine. Many eukaryotic proteins contain sorting signal sequences that direct the newly synthesized preprotein into specific subcellular organelles such as the nucleus mitochondria or Golgi complex.

The nature of the amino terminus is a major concern in the production of recombinant mammalian proteins in simple organisms. When a mammalian gene is recombined into a bacterial host, the protein product often has an additional residue of methionine at its amino terminal. This situation is unpredictable. The efficiency of the enzymatic removal of the initiating methionine appears to depend mainly on the adjacent sequence of the protein, although fermentation conditions may be a factor. In any event, recombinant proteins must be analyzed for the presence of this additional methionine residue (5). Its presence on even a small fraction of the protein product is a concern, since the recombinant product is no longer identical to the mammalian protein.

POSTTRANSLATIONAL MODIFICATIONS

Only a limited number of mammalian proteins have been correctly manufactured in simple organisms through genetic engineering. Among these are insulin, the interferons, and the interleukins. Other proteins that have complex pre-pro-structures that are posttranslationally modified by distinct, sequential, multistep processes can only be manufactured via recombinant mammalian cell cultures. Our understanding of the underlying biochemical mechanisms of these processes is limited, but rapidly expanding.

Eukaryotes have an internal membrane network called the endoplasmic reticulum (ER), as depicted schematically in Figure 4. Proteins that are destined to be secreted and/or posttranslationally modified by an ER-dependent process have signal peptides. The first 25 to 30 amino acids at the amino terminal of these proteins comprise a signal which determines whether the protein will be channeled into the lumen (internal space) of the ER. This takes place as the protein is being synthesized on the ribosome. After the signal peptide has been synthesized the ribosome/mRNA/polypeptide complex "pauses" until the complex is translocated to the ER.

This temporary cessation of biosynthesis is mediated by a cellular signal recognition particle (SRP) that binds to the signal sequence as soon as the growing polypeptide chain becomes long enough to extend beyond the ribo-

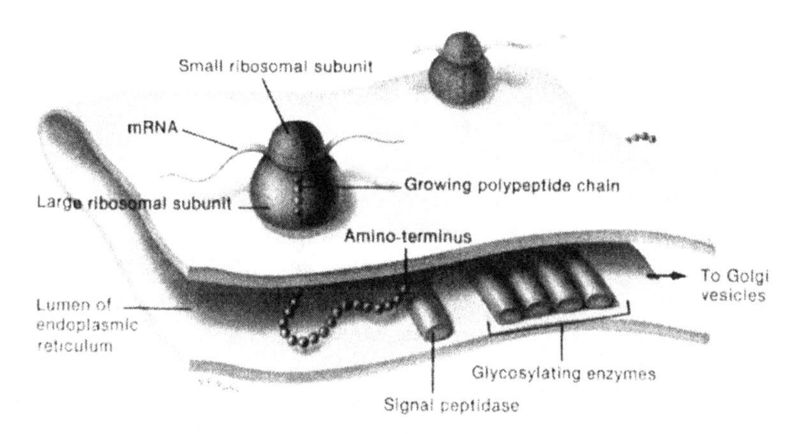

Figure 4 Posttranslational processing in the endoplasmic reticulum. Depending on the particular mRNA undergoing translation, the ribosome may become attached to the outer surface of a ductlike network in eukaryotic cells called the endoplasmic reticulum (ER). The growing protein chain is extruded through the membrane into the lumen of the ER. Various enzymes are encountered which can modify the protein cotranslationally. A common occurrence is the removal of the leader sequence (i.e., the first 25 amino acid residues, also referred to as a signal sequence). A group of enzymes may also glycosylate the nascent protein through a multistep process. Another modification is the covalent attachment of myristic acid to the nascent protein. Posttranslational events such as these are dictated by both the primary structure of the protein itself and the capabilities of the cell synthesizing the protein. After being processed through the ER, proteins are transported in vesicles to a membranous organelle called the Golgi apparatus, wherein other modifications, as well as packaging into secretory vesicles may occur. Prokaryotic cells do not have an ER. Consequently, a human protein produced in bacteria will lack any modifications normally present in that protein.

some. This complex then binds to a membrane receptor of ER (SRP receptor, docking protein) in a manner that causes the displacement of the signal recognition particle from the polypeptide. Protein synthesis resumes, and the signal peptide extends through the membrane into the lumen of the ER. As the polypeptide chain extends, the protein is "cotranslationally translocated" across the membrane.

Protein synthesis continues, and the newly synthesized amino acid chain accumulates within the ER. Several ribosomes can translate a given mRNA simultaneously, and when this occurs multiple ribosomes are bound to the ER membrane. In fact, actively synthesizing cells have so many ribosomes bound to the ER surface that the ultrastructural appearance of the organelle actually changes. The organelle is then termed "rough ER." On the other side of the membrane (lumen), the growing polypeptide chains begin to fold according to the information contained in their primary structure (amino acid sequence, see above). In most cases, the signal peptide is removed by a signal peptidase while the protein is still being synthesized.

Maturation of the protein, or posttranslational processing, proceeds at protein-specific rates. Thus different proteins may be processed differently, and the rates of processing vary. The processing step that is actually rate limiting does, in fact, depend on the protein.

A large portion of the proteins synthesized in the cells of higher animals are made as proproteins. The proprotein is larger than the protein, as it contains either an N-terminal extension, a C-terminal extension, or an additional internal polypeptide that is removed by proteolysis to produce the finished protein. Insulin, for example, folds to its three-dimensional structure which is then stabilized by the formation of disulfide bonds (see below). An internal polypeptide is "clipped" out of the molecule, and this leaves a two-chain crosslinked protein that is released from the pancreas as active insulin (see Chap. 2).

Insulin is "packaged" into secretory vesicles that release their contents in response to circulating glucose concentrations. Therefore, a "preproinsulin" is synthesized containing an N-terminal extension (signal peptide) which directs the molecule to the appropriate subcellular organelles for processing and storage. The "presequence" is removed as a part of the maturation of the protein. Proteins destined for other organelles such as the nucleus or mitochondria contain other presequences that mediate their subcellular localization.

Modification such as the attachment of carbohydrates to the proteins (glycosylation) take place via sequential passage through cellular organelles that evolved to mediate the procedures. Preproproteins are inserted into the endoplasmic reticulum (ER) during synthesis, enzymatically modified, transported in vesicles from the ER to the first level of the Golgi, modified by other protein-specific processes, repackaged, and transferred to subsequent "levels of Golgi" for further processing. Thus, the sequential addition/removal of specific oligosaccharide residues can be controlled, at least in part, by the physical separation of the processing enzymes and the transport of the partially processed protein between these cellular compartments. This compartmentalization of the posttranslational processing apparatus may prove to be the major obstacle in the path toward the production of complex mammalian proteins in simple organisms.

Disulfide Bonds

Disulfide bond formation constitutes a common and essential posttranslational event. The amino acid cysteine, which has a thiol group on its side chain will bond to another cysteine to form a disulfide linkage. The proper formation of these linkages is a critical factor in the final three-dimensional structure of the protein. Human interleukin-2 has a disulfide linkage between residues 58 and 105; the third cysteine residue at position 125 has a free sulfhydryl group (5). The reduced protein apparently takes on a three-dimensional structure, which brings the appropriate cysteine residues into near proximity. The formation of disulfide bonds then locks this structure together.

A more complex structure, such as that of bovine pancreatic trypsin inhibitor (6 cysteines), is believed to fold and form the disulfide crosslinks found in the native form through several intermediate stages. The apparent free energies, transition states, and relative stabilities of the eight possible intermediates have been calculated from analysis of folding rates and intermediates detected in different redox environments (6). Surprisingly, only the native form of the molecule is made in vivo.

Prokaryotes have a high internal reducing environment which may not allow disulfide bond formation. Therefore, the disulfide bonds are allowed to form in many recombinant proteins after extraction from the bacterial host, and the correct formation of these crosslinks is another matter of concern. On the other hand, the maturation of proteins synthesized into the eukaryotic ER appears to proceed only after proper folding (and disulfide bond formation) has occurred.

Glycosylation

Cytoplasmic proteins are never glycosylated; only membrane or secreted proteins are processed in this manner (except in rare pathological states). Protein-linked sugars are localized to the extracytoplasmic face of cellular membranes. The sugars are added to the nascent protein in the terminal ER and Golgi apparatus. The physiological role of the protein-linked oligosaccharide is not as yet clear, but in many cases the total size of the oligosaccharide groups rivals that of the protein itself. It is known, however, that the immunogenicity of proteins is altered by the protein-linked sugars, and that the stability/half-life of glycoproteins in situ are dependent on the identity of the oligosaccharides.

These secretory and membrane glycoproteins are transported from the rough endoplasmic reticulum through the Golgi, and during this process are significantly altered. Recent evidence suggests that in many cases this can be a distinctive and rate-limiting step in protein maturation. Different proteins move through this pathway at various rates, and extensive folding and covalent modifications including glycosylation occur. The studies indicate that a protein will not move from the ER to the Golgi until it has achieved a proper conformation, and when this transport occurs it is via a transport vesicle. The different glycosyl transferases reside in specific regions to which the maturing protein is sequentially exposed, and in this manner an ordered carbohydrate moiety can be constructed.

Two distinct types of sugar–protein linkages occur. O-linked sugars attach through the hydroxyl group (of serine, threonine, and in collagen hydroxylysine), and typically an N-linked oligosaccharide is attached to an asparagine side chain. The enzymatic recognition signals for the latter type linkage are the amino acid sequences –Asn–X–Thr– or –Asn–X–Ser–, where X may be any of a number of amino acids. A 14-sugar oligosaccharide (3 glucose, 9 mannose, and 2 *N*-acetylglucosamine molecules) is linked to the asparagine, and is then modified into various structures by a variety of enzymes. In contrast, the O-linked sugars are added individually via repetitive transfer from nucleotide sugars (7).

An example of a glycoprotein of current biomedical interest is human immune (gamma) interferon. The molecular weight of the translated protein is about 17,000. The two major forms isolated from induced peripheral blood lymphocytes have apparent molecular weights (by sodium dodecyl sulfate polyacrylamide gel electrophoresis) of about 21,000 and 25,000 (8). There are two potential glycosylation sites in the protein. It was determined that the 21,000 dalton species is monoglycosylated, whereas the 25,000 dalton species is diglycosylated. It is also likely that there exists some microheterogeneity due to differences in the extent of glycosylation at either site.

In the case of immune interferon, glycosylation does not appear to influence the in vitro antiviral or antiproliferative activities, as demonstrated with the recombinant protein produced in *Escherichia coli* (8). The in vivo effects of glycosylation are unknown, although the carbohydrate moieties may prevent denaturation of the biologically active three-dimensional conformation, reduce protease degradation, or alter the clearance rate from the blood. Recombinant, human immune interferon is currently being pursued in the clinical setting. Time will tell whether there are any functional differences in the recombinant protein produced in *E. coli* or mammalian cell culture. Recombinant proteins may also be produced in other types of cells, such as yeast or insect cells, but there are major differences in the type and extent of carbohydrates added at the glycosylation sites.

Other glycoproteins such as erythropoietin, EPO, have significantly reduced efficacy if they are not glycosylated. In this case the recombinant protein is equally active as its natural counterpart in vitro, but the pharmacokinetics are significantly altered (9). The half-life of the nonglycosylated protein in situ is significantly less than the "properly" glycosylated EPO. Therefore, companies currently developing manufacturing processes for clinical quantities of EPO are utilizing mammalian cell constructs that appropriately posttranslationally modify the protein.

An additional complication with respect to protein glycosylation is the strong evidence that the glycoproteins with identical primary structure are glycosylated differently if they are made by cells in different parts of the body.

Evidence for this tissue-specific glycosylation exists for only a few proteins, and may not be extrapolable to glycoproteins in general. However, it is well documented that the oligosaccharide patterns of proteins made by cancer cells (transformed) differ from those of proteins synthesized by "normal" cells (10).

Other Posttranslational Events

Numerous modifications can occur posttranslationally in the biosynthesis of proteins. Proteolytic processing to the biologically active species may occur in the Golgi or in secretory granules. Often, the biologically active molecule is a smaller species derived from a large proprotein. Neuropeptide Y is cleaved from a proprotein by a multistep process that results in an amidated C-terminal tyrosine (see Chap. 8). This modification is mediated by an enzyme complex that recognizes the sequence –Tyr–Gly–Lys–Arg–, and the nitrogen from the glycine ends up amidating the tyrosine (11). Glutamic acid residues can be γ-carboxylated by a vitamin K-dependent enzyme complex, and β-hydroxylation of aspartic acid residues also occurs. The functional assembly of the major type of structural protein, collagen, requires enzymatic conversion of proline and lysine to their hydroxylated forms. Other modifications such as sulfation of tyrosine and carbohydrate residues are well documented (12,13).

Another common event is phosphorylation of threonine, serine, and tyrosine residues. Phosphorylation may be considered to be a regulatory process of biological function rather than related to biosynthesis. The point to be made from this discussion is that the mature, biosynthetic, protein product may have structural modifications. Although the same basic mechanism of protein biosynthesis applies in all living organisms, these posttranslational alterations must be carefully considered in the production of recombinant proteins.

Recombinant proteins are sometimes engineered with an unnatural leader sequence in order to get them secreted from the microorganism. The gene for human granulocyte/macrophage colony stimulating factor (GM-CSF) has a leader sequence which allows for secretion of the recombinant protein in mammalian cells (14). The same protein has been produced in yeast. In order to get secretion into the medium, the alfa mating factor sequence, as well as a processing signal to remove the leader sequence, was engineered into the gene (15). Similarly, the same structural gene has been attached to DNA that encodes for an *E. coli* leader sequence which directs the protein toward secretion to the outer membrane of the cells (16). Thus, knowledge of the biosynthetic mechanisms and signals have been utilized advantageously to produce quantities of proteins that could not have been isolated from natural sources, or simple recombinant cell constructs. Future advances in our understanding of these principles may permit us to engineer into the cells of "simple" organisms at least some aspects of the complex protein processing machinery present in the cells of higher animals.

REFERENCES

1. King, J. Deciphering the rules of protein folding. Chem. Eng. News., (April 10, 1989) 32–54.
2. Bird, R. E., Hardman, K. D., Jacobson, J. W., Johnson, S., Kaufman, B. M., Lee, S., Lee, T., Pope, S. H., Riordan, G. S. and Whitlow, M. Single-chain antigen-binding proteins. Science (1988) 242:423–426.
3. Ruby, S. W. and Abelson, J. An early hierarchic role of u1 small nuclear ribonucleoprotein in spliceosome assembly. Science (1988) 242:1028–1034.
4. Streuli, M. and Saito, H. Regulation of tissue-specific alternative splicing: exon-specific cis elements govern the splicing of leukocyte common antigen pre mRNA. EMBO J (1989) 8 (3):787–796.
5. Lahm, H. W. and Stein, S. Characterization of recombinant human interleukin-2 with micromethods. J. Chromatogr. (1985) 326:357–361.
6. Creighton, T. E. and Goldenberg, D. P. Kinetic/Role of a meta-stable native-like two-disulphide species in the folding transition of bovine pancreatic trypsin inhibitor. J. Mol. Biol, (1984) 179:497–526.
7. Darnell, J., Lodish, H., and Baltimore, D. Molecular Cell Biology. Scientific American Books, Inc. (1989), second edition.
8. Rinderknect, E., O'Connor, B. H., and Rodriguez, H. Natural human interferon-gamma. J. Biol. Chem. (1984) 259:6790–6797.
9. Fukuda, M. N., Sasaki, H., Lopex, L., Fukuda, M. Survival of recombinant erythropoietin in the circulation: the role of carbohydrates. Blood (1989) 73 (1):84–89.
10. Parekh, R. B., Tse, A. G., Dwek, R. A., Williams, A. F. and Rademacher, T. W. Tissue-specific N-glycosylation, site-specific oligosaccharide patterns and lentil lectin recognition of rat Thy-1. EMBO (1987) 6:1233–1244.
11. Andrews, P. C., Brayton, K. A., Dixon, J. E. Posttranslational proteolytic processing of precursors to regulatory peptides. Experientia Suppl (1989) 56:192–209.
12. Kornfeld, R. and Kornfeld, S. Ann. Rev. Biochem (1985) 54:631–664.
13. Farquhar, M. G. Ann. Rev. Cell. Biol (1986) 1:447–448.
14. Wong, G. G., Witek, J. S., Temple, P. A., Wilkens, K. M., Leary, A. C., Luxemberg, D. P., Jones, S. S., Brown, E. L., Kay, R. M., Orr, E. C., Shoemaker, C., Golde, D. W., Kaufman, R. J., Hewick, R. M., Wang, E. A., and Clark, S. C. Human GM-CSF: Molecular cloning of complementary DNA and purification of the natural and recombinant proteins. Science (1985) 228:811–814.
15. Cantrell, M. A., Anderson, D., Cerretti, D. P., Price, V. McKereghan, K., Tushinski, R. J., Mocjhizuki, D. Y. Larsen, A., Grabstein, K., Gillis, S., and Cosman, D. Cloning, sequence and expression of a human granulocyte/macrophage colony-stimulating factor. Proc. Natl. Acad. Sci. USA (1985) 82:6250–6254.
16. Libby, R. R., Braedt, G., Kronheim, S. R., March, C. J., Urodal, D. L., Chiaverotti, T. A., Tushinski, R. J., Mochizuki, D. Y., Hopp, T. P., and Cosman, D.

Expression and purification of native human granulocyte-macrophage colony stimulating factor from an *E. coli* secretion vector. DNA (1987) 6:221–229.

SUGGESTED READING

Darnell, J., Lodish, H., and Baltimore, D., *Molecular Cell Biology* Scientific American Books, Inc., W. H. Freeman, New York, 1989.

Richardson, C. C., Boyer, P. D., Dawid, I. B., and Meister, A., (Eds.). *Annual Review of Biochemistry*, Vol. 56. Annual Reviews, Inc., Palo Alto, CA, 1987.

Pollitt, S. and Inouye, M. Structure and function of the signal peptide. In *Bacterial Outer Membranes as Model Systems*. M. Inouye, Ed. John Wiley, New York, 1986, pp. 117–130.

Bendig, M. M. The production of foreign proteins in mammalian cells. *Gen. Eng.* 7, 91–127 (1988).

Marino, M. H. Expression systems for heterologous protein production. *BioPharm* 2, 18–29, (1989).

4

Protein Purification and Analysis by Liquid Chromatography and Electrophoresis

Jodi L. Fausnaugh

Syntex, Inc.
Palo Alto, California

INTRODUCTION

Of particular interest to biochemists is the separation and identification of biochemical compounds. Two common techniques employed in this endeavor are chromatography and electrophoresis. It is generally acknowledged that chromatography was developed in 1903 by the Russian botanist Mickhail Tswett, who used a chalk column to separate the pigments in green leaves. The word "chromatography" was used by Tswett to describe the colored zones that moved down the column. However, in 1855, Karl Runge, a German chemist described a method of separating inorganic materials by paper chromatography. It was not until the 1930s and 1940s that chromatography developed as a practical technique. The work of Martin and Synge in 1941 on partition chromatography was a major advancement in the field. Their development later won them the Nobel Prize. In the late 1960s, chromatography advanced further with the development of high-performance liquid chromatography (HPLC).

Electrophoresis can be, in many cases, a complementary technique to chromatography. Most biopolymers are electrically charged and will, therefore, migrate within an electric field. The movement of these molecules through a solvent in an electric field is defined as electrophoresis. As with chromatography, electrophoresis can be used to determine the molecular weight of a protein, to separate molecules on the basis of net charge or shape and to determine changes in the charge of individual amino acids.

There are two general categories of application in the separation of biopolymers: analytical and preparative, and, essentially, all the procedures described in this chapter can be used in either manner. The purpose of analytical separations is to obtain information, whereas a product is to be obtained in a preparative application. Sometimes the same procedure simultaneously provides analytical data as well as a product. A general overview of the principles of chromatography and electrophoresis, as used in protein biotechnology, are presented as a background to the specific examples described in several chapters of this volume. This chapter provides only a fundamental description of the relevant principles. A greater depth of understanding may be gained from the suggested readings listed at the end of this chapter.

LIQUID CHROMATOGRAPHY (LC)

The ability of a chromatographic media to separate the components of a sample mixture is controlled by the distribution of the components between a solid stationary phase and a liquid mobile phase. The stationary phase is generally made up of small particles of a relatively inert material, such as cellulose, agarose, polystyrene, or silica. These particles may be derivatized with various functional groups to give the stationary phase certain desirable properties. (Further details are presented in the section on Modes of Separation.) The stationary phase is packed into a column, through which the mobile phase flows. The individual components in the sample mixture are carried through the column by the mobile phase liquid. They elute (i.e., emerge) from the end of the column depending upon their selective retardation caused by diffusion or interactions with the stationary phase.

Originally, stationary phases were packed into large-diameter glass columns and operated under atmospheric pressure. These stationary phases were of fairly large particle sizes and the analysis times were relatively long. In order to increase the speed and resolution of the separations, high performance liquid chromatography (HPLC) was developed. The stationary phases in HPLC are made of more rigid and smaller particles than in "classical" LC. These smaller stationary phases require significant force to propel the mobile phase through the column.

Hardware

Before proceeding, it would be helpful to look at the components of an LC instrument (Fig. 1). The column is considered the "heart" of the system. It may be a stainless steel tube capable of withstanding several thousand psi (pounds per square inch) as in HPLC or may be made of plastic or glass for classical LC. The type and quality of the stationary phase in the column is

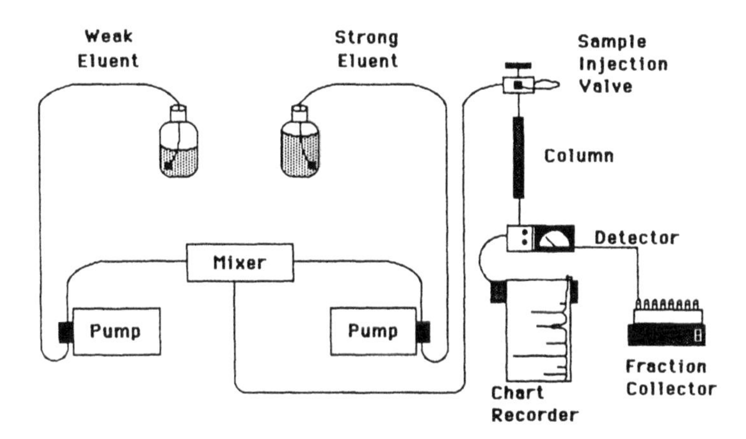

Figure 1 Schematic illustration of a liquid chromatography instrument. In this particular example, two pumps are used to mix the mobile phases in different proportions in order to generate a gradient during the chromatographic run. The pumps or the entire system would be under the control of a microprocessor.

critical to the separation. Often the column is jacketed so that it may be heated or cooled to an appropriate temperature.

In classical LC, gravity or a peristaltic pump is used to force the mobile phase through the column. In HPLC, a pump capable of delivering mobile phase at a pressure of up to 6000 psi is used. In isocratic elution, a single eluent is employed and delivered by a single pump. In gradient elution the eluting strength of the mobile phase is increased during the chromatographic run in order to shorten the elution time of strongly retained sample components. This may be done by means of a step or continuous gradient. In a step gradient, the mobile phase is abruptly changed from one solvent to another, whereas in a continuous gradient, the eluting strength of the mobile phase is gradually changed by mixing a second solvent with the first in varying proportions during the course of the chromatographic run. Each eluent may have an independent pump or, alternatively, proportioning valves may be used with a single pump.

The sample mixture is generally loaded onto the head of the column by means of a loop valve for HPLC and with a simple pipette for classical LC. In a loop valve, the sample is introduced into a stainless steel loop at atmospheric pressure by means of a syringe. At this time, the mobile phase is bypassing the loop as it flows through alternate ports of the valve and on through the column. The valve is then rotated to bring the sample loop in line with the stream of mobile phase and the column. The mobile phase sweeps the sample out of the loop and carries it to the column. The same principle applies to automatic injectors.

The column effluent is monitored for the various sample components using a detector that measures a physical property of the sample; most commonly this is absorbance of light at a specific wavelength. Other detectors measure, for example, fluorescence, refractive index, or conductivity. The type of detector used depends on the physical properties of the sample mixture and the level of sensitivity required. In certain analytical applications, a reagent is added to the column effluent, using an additional pump, prior to detection. The reagent reacts with the sample components producing a product which can be detected. The electronic signal from the detector is passed to a chart recorder or computer. The separated components of the sample mixture are then collected by a fraction collector, if desired.

Major advances have been made in the last decade with regard to the quality of the stationary phases in HPLC. The acronym HPLC can also be interpreted as "high-pressure" or "high-priced" liquid chromatography. In HPLC, the stationary phase matrix is, generally, composed of silica, polystyrene, or other polymers that are small (varying between 3–10 μm), uniformly sized, spherical, and totally porous (to achieve high surface area). At the present time, several new polystyrene and silica matrices are being evaluated that are nonporous. High surface area for these new stationary phases is achieved through a further reduction in the size of the particle. The components of the chromatographic apparatus (Fig. 1) must all be compatible with the high-pressure and high-resolution properties of the HPLC column. Another popular version of this technique, known as "FPLC," (fast protein liquid chromatography) uses a polymer matrix in place of silica and a system of totally biocompatible hardware (glass and inert tubing instead of stainless steel). The cost of HPLC or FPLC columns may preclude large-scale separations, although HPLC columns having loading capacities of several grams are used in certain production applications.

Modes of Separation

Adsorption or Normal Phase

In adsorption or normal-phase chromatography, binding to the stationary phase is due to polar interactions, especially hydrogen bonding. The stationary phase in this mode of chromatography is more polar than the mobile phase. This was one of the first modes of chromatography developed and as such was labeled normal phase to distinguish it from a later type of chromatography called reverse phase. Typical stationary phases are silica, alumina, hydroxylapatite, and silica coated with a polymer layer bearing hydroxyl, alkylnitrile, or alkylamine groups. Adsorption chromatography is often used with a nonaqueous mobile phase to separate small organic molecules. Typical mobile phase solvents for normal-phase chromatography include hexane, methylene chloride, tetrahydrofuran, and methanol. Aqueous eluents tend to cause deterioration of the matrix. However, adsorption chromatography has been found to be quite useful in particular protein separations.

Hydroxylapatite has found the most use among normal-phase stationary phases for protein chromatography. This is a crystalline form of calcium phosphate [$Ca_{10}(PO_4)_6(OH)_2$] and is prepared from $CaHPO_4 \cdot 2H_2O$. The stationary phase binds molecules which interact with calcium such as DNA, RNA, and phosphoproteins. However, it also binds most proteins even though they are not phosphorylated. Interaction between the stationary phase and the protein sample components occurs at low ionic strength and elution is achieved by raising the phosphate concentration or the ionic strength. The mechanism of interaction between the protein and stationary phase is not understood, but in many cases hydroxylapatite will resolve protein mixtures that other modes of chromatography cannot.

Reverse Phase

In this mode, the stationary phase has a hydrophobic layer on its surface. The functional group is most often a linear alkyl chain of 3–18 carbon atoms covalently attached either directly to a matrix or through a polymeric coating on the matrix (Fig. 2). Other stationary-phase ligands include phenyl and diphenyl. The most popular matrix for reverse-phase chromatography (RPC) is silica, although several commercial columns are available based on organic polymeric matrices (e.g., polystyrene and TSK PW). Because residual Si-O^- groups cause ionic interactions with sample components, silica stationary phases are exhaustively "end-capped" (further derivatized) with small silane groups (e.g., trimethylchlorosilane) if there is no polymer coating on the silica (Fig. 2). Binding of sample components to the stationary phase is due to hydrophobic interactions. Reverse- (or reversed-) phase chromatography derives its name from the fact that the elution properties are opposite to those used in adsorption (or normal phase). The stationary phase is less polar than the mobile phase and the strength of the eluent decreases with its polarity. RPC has become, in general, the most commonly used mode for all separations. With relatively small molecules, such as amino acid derivatives and peptides, elution is usually accomplished by increasing the concentration of an organic solvent (e.g., acetonitrile) in buffered water or dilute aqueous acid.

Consid... tremely ionic nature of many biomolecules, a related ...hy, known as ion-pair chromatography (IPC), has been ..., a lipophilic counter ion is added to the mobile phase. ... RPC column. It is thought that the counter ion either ... the sample components rendering them hydrophobic ...ly with the stationary phase making it ionic. The elu... ...PC are usually the same as those for RPC. Typical

...ol ...veral

A.

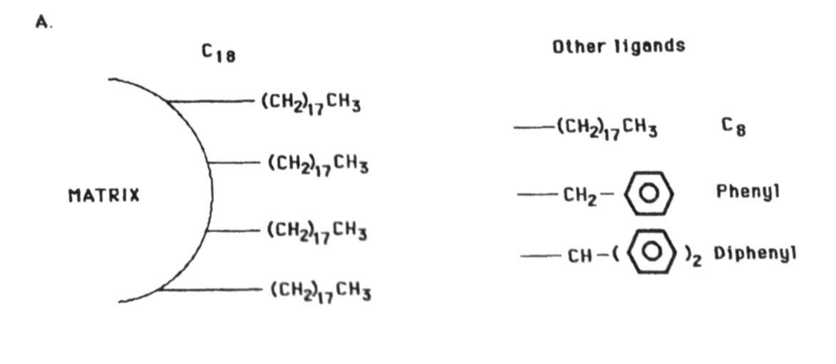

B.

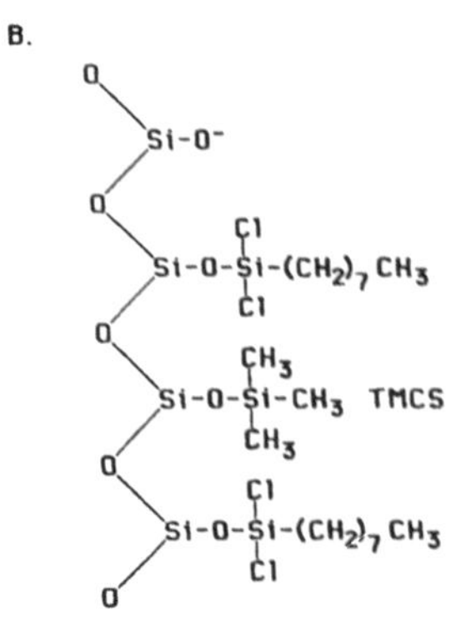

Figure 2 Schematic illustration of the surface of a reverse-phase stationary phase. (A) Different ligands found on reverse-phase supports, (B) chemical linkage of ligands to a silica matrix and the structure of end-capped silica.

pairing agents include quaternary and tertiary amines, perfluorinated carboxylic acids and perchloric acid. In many cases the absence of an ion-pairing agent will result in loss of retention or severe tailing of the ionic sample components.

The most common organic solvents used in RPC are shown below in order of decreasing elutrophic strength.

propanol > acetonitrile > methanol > water

In eluting proteins from an RPC column, a gradient of 0.1% trifluoroacetic acid (TFA) in water to 0.1% TFA in 60%–90% isopropanol is generally employed. This mobile phase also is used in eluting very hydrophobic polypeptides. For polypeptides of less hydrophobic character, acetonitrile or metha is used in place of the isopropanol. A typical RPC chromatogram of protein standards is shown in Figure 3.

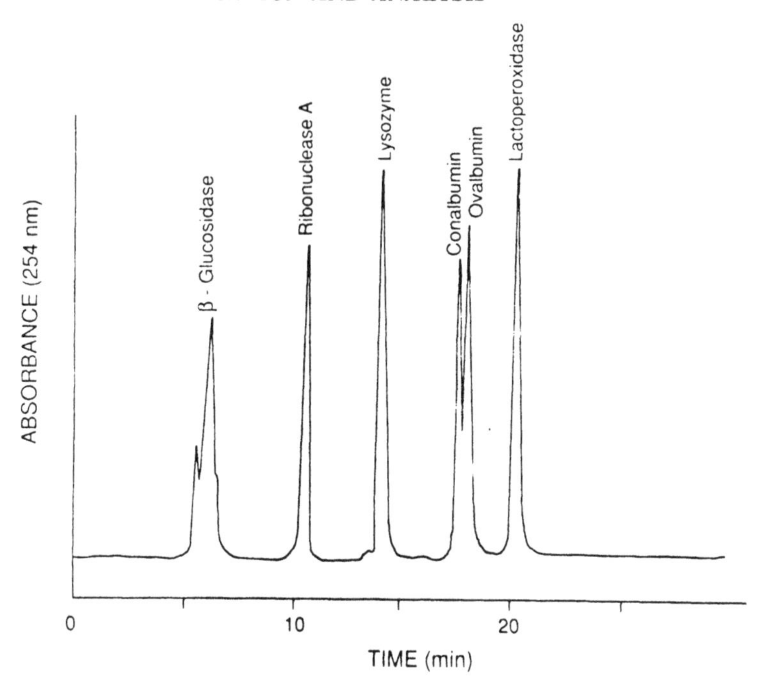

Figure 3 Chromatogram of a standard protein mixture on a reverse-phase C8 column. The proteins were injected onto a Synchropak C8 column (5 × 0.46 cm) and chromatographed in a 20-min linear gradient from 0.1% TFA to 0.1% TFA in isopropanol:water (60:40). The flow rate was 1.0 ml/min and the column effluent was monitored at 254 nm.

An important application of RPC is the purification of peptides resulting from proteolytic digestion. Figure 4 shows a chromatogram of the peptide fragments produced by the digestion of the protein lysozyme with the proteolytic enzyme, trypsin. The separation is on a C8-RPC (alkyl chain of eight carbon atoms) column. Since the RPC mobile phases are volatile, the peptides are easily prepared for further analysis.

Hydrophobic Interaction

Proteins above a molecular weight of about 20,000 either may not be soluble in the mobile phase required for elution in RPC or may become denatured (i.e., lose their biological activity) in the presence of such solvents. In such cases, where the retention of biological activity is important, the technique known as hydrophobic interaction chromatography (HIC) is employed. The HIC stationary phase is similar to that of RPC, except that the hydrophobic functional groups are more distantly spaced (i.e., of a lower surface density of alkyl chains) and tend to be of a shorter chain length. This results in decreased interactions between the protein and the stationary phase. In HIC, high-salt-concentration mobile phases are used to promote the hydrophobic interactions. Elution is accomplished by decreasing the salt concentration of the mobile phase instead of using an organic solvent as in RPC. The most popular salt used in HIC mobile phases is ammonium sulfate at concentrations up to 3 M in 100 mM buffer. Salts such as sodium chloride do not have the ionic strength at moderate concentrations to induce a hydrophobic interaction.

HIC has gained in popularity since the introduction of several high-performance HIC stationary phases. Retention on an HIC stationary phase is easily manipulated through both mobile-phase and stationary-phase variables. The stationary phase variables include both ligand chain length and ligand density. Protein retention increases with an increase in both of these values. In general, a chain length of greater than five will result in irreversible binding of the protein to the stationary phase. In this case, an organic solvent would be required to effect elution. Phenyl lignads are common as HIC stationary phases. The phenyl ligand has the approximate hydrophobicity of a chain length of four. Mobile-phase variables include salt concentration, pH and salt type. Retention increases with salt concentration and the use of a salt of higher

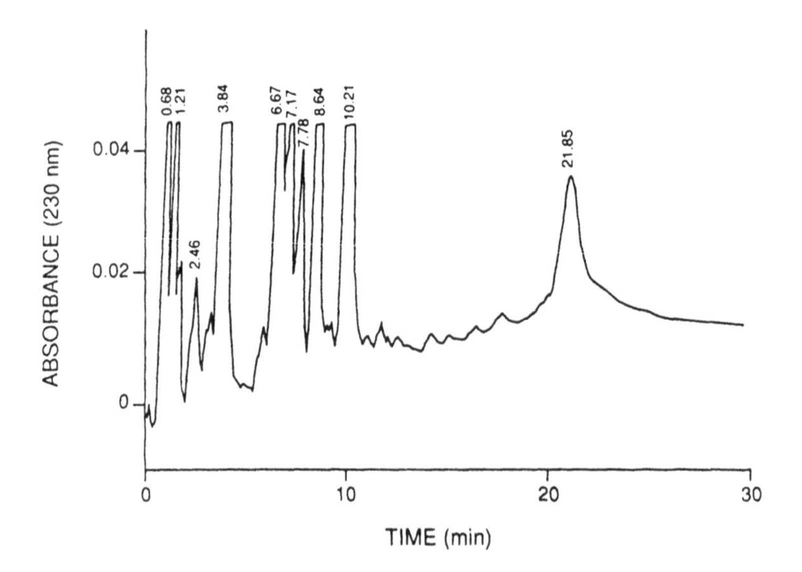

Figure 4 Chromatogram of a trypsin digest of lysozyme. The digest mixture was injected onto a Synchropak C8 column and chromatographed in a 20 minute gradient from 15 to 70% mobile phase B. Mobile phase A was 0.1% TFA and mobile phase B was 0.1% TFA in acetonitrile:water (70:30). The flow rate was 1.0 ml/min and the column effluent was monitored at 230 nm.

"salting-out" ability. Retention appears to follow the lyotropic or Hoffmeister series with respect to the type of salt used. The lyotropic series rates salts according to their effect upon the "orderedness" of water. A salt which causes a decrease in the entropy of water has a higher salting-out ability and a higher ranking in the lyotropic series. The effect of pH on protein retention appears to be dependent upon the proteins in the sample mixture. In general, protein retention is at a minimum around neutral pH and increases as the pH becomes more basic or acidic.

Several applications have shown HIC to have tremendous resolving power, even to the extent of separating proteins having only one amino acid difference. One example involves three isozymes of lysozyme isolated from Peking duck egg whites (designated A, B, and C). When the related lysozymes were analyzed on a phenyl-HIC column, the substitution of an arginine for proline at position 79 on duck C lysozyme decreased the retention of that protein over duck B lysozyme (Fig. 5). Further analysis of lysozymes from related bird species showed that only amino acid substitutions in a limited area of the lysozyme surface affected retention. This area has been designated as the contact surface area between the protein and the stationary phase and extends in the three-dimensional structure from residues 41 to 102 and from 75 to the α-helical region starting with residue 89. The three-dimensional structure of hen egg white lysozyme is given in Figure 6.

As stated above, recovery of enzymatic activity is an important consideration in the use of reverse-phase columns. When the recovery of enzymatic activity of α-chymotrypsin was measured after chromatography on an RPC and an HIC column, recovery exceeded 86% by HIC and only 54% by RPC. With many proteins this difference in preservation of biological activity is even more pronounced. Enzyme recovery also exceeded 90% for α-amylase, lactic dehydrogenase, and lysozyme and 95% for β-glucosidase after chromatography on an HIC column. These measurements were made for protein sample sizes ranging from 10 to 200 μg. It has also been shown that both the organic solvents used in RPC, as well as the actual interaction of the protein with the stationary phase, are detrimental to the structure of the protein. This interaction appears to lead to an unfolding of the protein upon the surface of the stationary phase. Except in rare cases where the protein structure is not stable in high-salt-concentration mobile phases, recovery of enzymatic activity after HIC usually exceeds that found after RPC. Protein mass recovery by HIC usually exceeds 85%, while the value for RPC is often lower.

Size Exclusion (Gel Permeation) Chromatography

This mode differs from all others in that chemical interactions between the sample components and the stationary phase are undesirable and generally do not occur. Size exclusion chromatography (SEC) or gel permeation chroma-

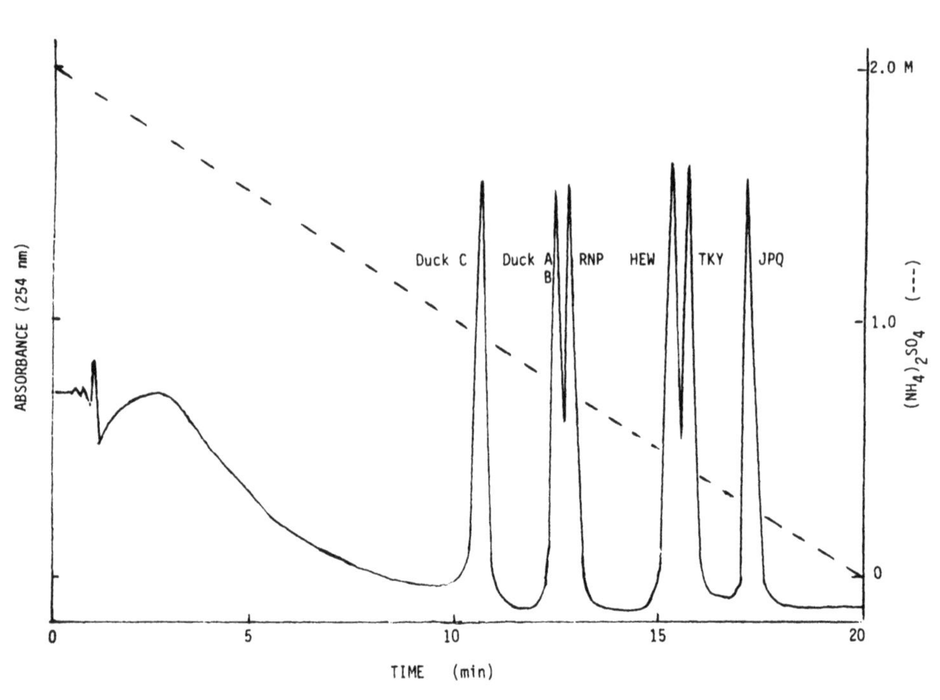

Figure 5 Chromatogram of a mixture of avian lysozymes on an HIC column. The mixture was chromatographed on a TSK-phenyl 5PW column in a 20 minute linear gradient from 100 mM Tris buffer, pH 8.0 containing 3M ammonium sulfate to 100 mM Tris buffer, pH 8.0. The flow rate was 1.0 ml/min and the column effluent was monitored at 254 nm. Duck A, Duck B, and Duck C, isozymes of lysozyme found in Peking duck; RNP, ring-necked pheasant lysozyme; HEW, hen egg white lysozyme; TKY, turkey lysozyme; JPQ, Japanese quail lysozyme.

tography (GPC) columns contain porous particles with a selected range of pore diameters. A sample component will diffuse into those pores having a diameter greater than that of the component. Large molecules are excluded from some of the pores and elute quicker than small molecules which permeate more of the pores. The mechanism of separation for SEC is not based strictly upon molecular weight, but rather upon the shape of the molecule. Spherical molecules have a much smaller effective diameter than do elongated molecules of the same molecular weight. Thus, the elongated molecule would elute earlier from the SEC column, since it would be excluded from more of the pores than the spherical molecule. Molecules with diameters sufficiently large enough to prevent entry into any of the stationary-phase pores are said to be totally excluded from the column and elute in the void volume (V_o). A molecule capable of diffusing into all of the pores of the stationary phase or pore volume (V_p) is defined as being totally included and elutes in a volume equivalent to

Figure 6 Three-dimensional structure of the peptide backbone of hen egg white lysozyme. External hydrophobic amino acids are marked with an asterisk.

the total permeation volume ($V_o + V_p$). Approximate exclusion molecular weights corresponding to porosities of 6 nm (60 Å) and 75 nm (750 Å) are 10^4 and 10^6, respectively.

Elution volume (V_e) can be related to the void and the pore volume, by the following equation:

$$V_e = V_o + K_d V_p$$

The distribution coefficient, K_d, is defined as the ratio of the pore volume accessible to the solute (V_a) to the total pore volume.

$$K_d = V_a/V_p$$

The value of K_d can range from 0 for very large molecules to 1 for very small molecules.

When a series of proteins of varying molecular weights are eluted from a size exclusion column, a calibration curve can be constructed by plotting the log of the molecular weight versus the elution volume of each species. A typical calibration curve is shown in Figure 7. If the pore distribution of the matrix is large, the calibration curve will have a very steep slope. This means that the molecular weight range that can be separated on the column is wide. However, there will be less resolution between species of closely related molecular weight. A narrower pore distribution will give greater resolution of close molecular weight species, but the molecular weight range will be small.

The above equations dealing with elution from an SEC column assume that the column is behaving ideally, that is that there is no interaction between the sample molecules and the stationary phase. In cases where there is an interaction, the elution volume will be larger than that predicted by the calibration curve. Usually, the interaction is adsorption such as an electrostatic interaction seen with residual Si-O^- groups of silica or glass or a hydrophobic interaction with the stationary-phase coating.

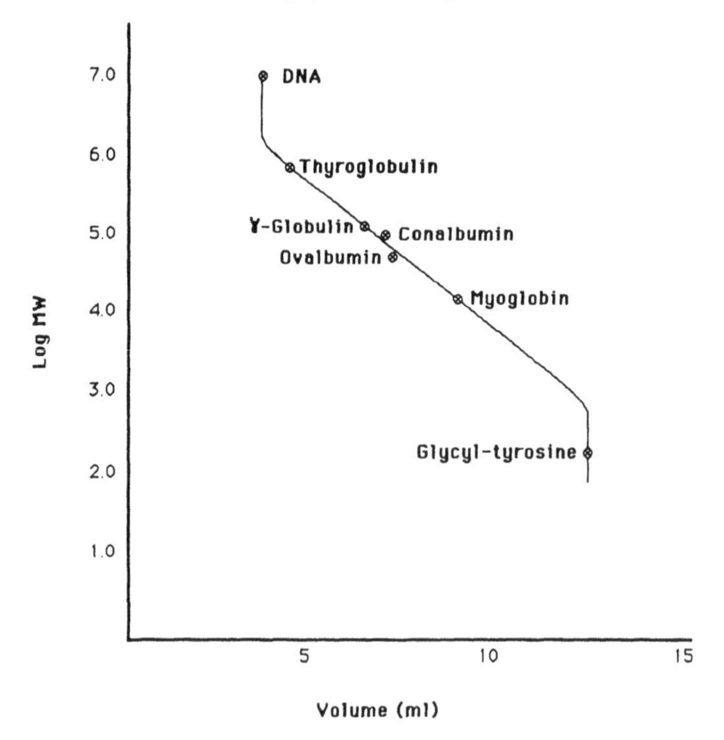

Figure 7 Calibration curve of a TSK 3000 SW size-exclusion column. The standard proteins were injected onto the column in 100 mM sodium phosphate buffer, pH 7.0 at a flow rate of 0.50 ml/min. The column effluent was monitored at 254 nm.

In general, an interaction with the stationary phase is only detrimental when the elution volume will be used to determine the molecular weight of the sample molecule. "Nonideal" size exclusion can be used to improve the resolution between certain molecular species of interest. Pfannkoch et al. have measured the K_d of a series of small molecules (oxalic acid, citric acid, glycyltyrosine, lysine, and phenylethanol) as a function of the ionic strength of the mobile phase. At low ionic strength, citric and oxalic acid eluted from a silica-based SEC column before the total permeation volume, thus appearing to have a higher molecular weight. Lysine, however, was retained past the total permeation volume and glycyltyrosine eluted in the permeation volume. It was concluded that the stationary phase was slightly anionic. The shorter elution times of oxalic and citric acid were due to ion exclusion, while the longer retention of lysine was due to an ionic interaction. Phenylethanol was retained past the permeation volume at all ionic strengths, but did show an increase in retention with ionic strength. This indicated a hydrophobic interaction with the stationary phase. Pfannkoch et al. have shown that all commercially available SEC columns exhibit some anionic and hydrophobic interactions.

Size exclusion stationary phases are synthesized from a variety of materials. Stationary phases for classical types of chromatography include dextrans, agarose, polystryene, and polyacrylamide. HPLC size exclusion stationary phases are usually based on silica, although several rigid organic polymers are available such as the Toyo Soda PW, hydroxylated polyether-based material. The silica stationary phases generally are coated with a hydrophilic inert polymer to block the natural negative charge and to protect the silica from degradation at basic values of pH.

SEC columns are operated under isocratic elution conditions. In general, mobile phases consist of a buffer at fairly low ionic strength (e.g., 100 mM). The pH of the mobile phase will depend on the type of stationary phase and the sample components to be resolved. Figure 8 shows an example of a standard mixture eluted from an SEC column. Silica-based stationary phases should be operated at a pH ranging from 2 to 8. Many additives can be used in SEC mobile phases such as those to stabilize enzymes or to prevent protein aggregation. In the case of proteins and polypeptides of limited solubility in aqueous mobile phases, such as membrane-associated proteins, sodium dodecyl sulfate, guanidine hydrochloride, or a mixture of dilute acid (e.g., 0.1% TFA) with acetonitrile or isopropanol (20–50%) can be used in the mobile phase.

Ion Exchange

The stationary phase in ion-exchange chromatography (IEC) bears a positive (anion exchange) or negative (cation exchange) charge due to derivatization of

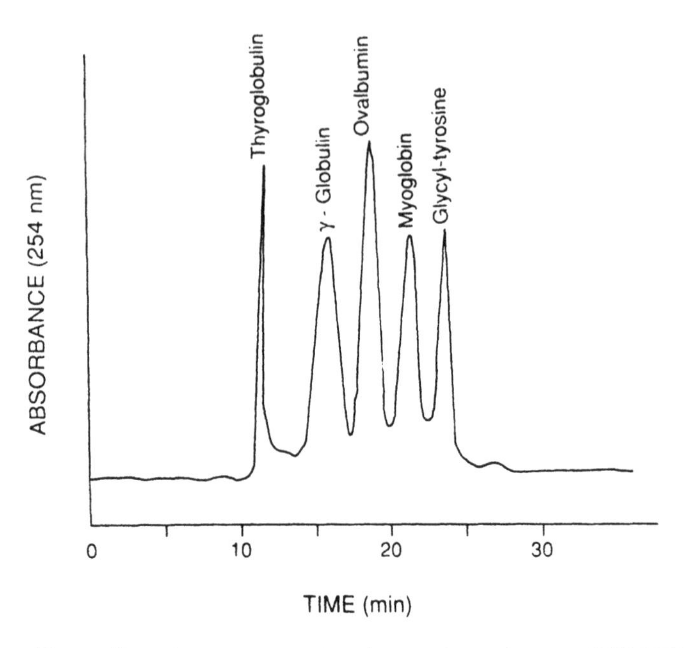

Figure 8 Chromatogram of a mixture of proteins on a TSK 3000 SW size exclusion column. The proteins were injected onto the column in 100 mM sodium phosphate buffer, pH 7.0 at a flow rate of 0.50 ml/min. The column effluent was monitored at 254 nm.

the stationary phase matrix with an appropriate functional group. Typical functional groups for cation exchange stationary phases are sulfonic or carboxylic acids, whereas an anion exchange stationary phase may have a primary, tertiary, or quaternary amine (Fig. 9). The ligand is always accompanied by a counter ion which may be one of several different ions. The counter ion of the stationary phase can strongly influence the chromatography due to the difference in the strength of the ionic interaction between the stationary phase and different counter ions. The sample component binds to the stationary phase through ionic interactions at low ionic strength by displacing the stationary phase counter ion. Elution may be effected by neutralizing the charge on either the stationary phase or the sample component or by the exchange of ions from the mobile phase. This is accomplished by altering the pH or increasing the ionic concentration of the mobile phase. Typical salts used in IEC include sodium chloride, magnesium chloride, potassium phosphate, and ammonium sulfate.

Proteins are composed of amino acids having anionic, cationic, and hydrophobic side chains. The ionization state of these amino acids depends upon the

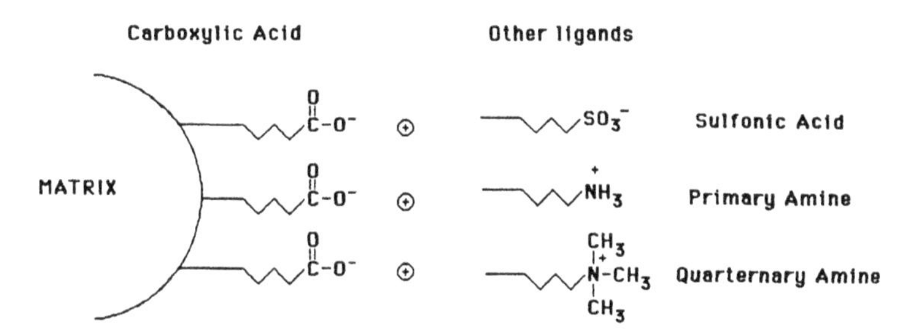

Figure 9 Schematic illustration of the ligands found on an ion exchange stationary-phase surface.

pH of the solution. At acidic pH, the amino groups of lysine, arginine, and histidine bear a full positive charge, while glutamic and aspartic acid are neutral due to protonation of the carboxylic acid group. At basic pH, the amino groups become deprotonated and neutral, while the carboxylic acid groups bear a negative charge. The pH at which the positive and negative charges on the protein are equal is known as the isoelectric point. This variable is unique to the individual protein. At a pH value below the isoelectric point, a protein should be retained by a cation exchange stationary phase. In the same manner, a protein will be retained by an anion exchange column at a pH above its isoelectric point. IEC has not proved to be as simple as the above explanation of retention would suggest. This "net charge model" predicts that a protein would not be retained on either an anion exchange or a cation exchange column at its isoelectric point. In fact, most proteins are retained by columns up to a whole pH unit past their isoelectric points. These discrepancies between experimental and predicted data can be resolved if one considers that the charges on a protein surface are rarely evenly distributed. This charge asymmetry can account for the presence of a highly charged region on the protein which can interact with the stationary phase even at the isoelectric point. Further discussion of the mechanism of ionic interaction can be found in the Principles of Chromatography section of this chapter.

The most common stationary phase materials for IEC are based on polystyrene, silica, and other hydrophilic polymers such as Monobeads. Polystryene-based ion-exchange resins are produced by the copolymerization of styrene and divinylbenzene. The higher the percentage of divinylbenzene crosslinker, the stronger is the mechanical strength of the resin. The percentage of divinylbenzene is usually in the range of 4–8%. However, as the percentage of crosslinker increases, the porosity of the resin decreases, providing less surface area for sample binding and smaller pores for the sample to penetrate. This leads to decreased column efficiency due to poor mass transfer. Recently, Polymer Laboratories has introduced a large-pore polystyrene HPLC anion exchange column for the separation of biopolymers. The matrix contains 1000 angstrom pores for efficient mass transfer of large biomolecules. The matrix is coated with a polymer containing quaternary amines to give the stationary phase its ionic character. This column may be used in the analysis of proteins or oligonucleotides. The popular FPLC ion exchange columns from Pharmacia are also polymer based. Pharmacia offers both cation exchange (Mono S) and anion exchange (Mono Q) stationary phases based on the monodisperse hydrophilic polymer particles called Monobeads. Other HPLC ion exchange columns are silica based. These stationary phases have high mechanical strength, large surface areas and are available in a range of pore sizes. The major limitation of most silica-based columns is that the operating pH range is only from 2–8. At extremes of pH the silica matrix is subject to dissolution by the mobile phase.

Affinity

Affinity chromatography (AC) takes advantage of the specific and selective binding between two biomolecules. Examples are the interaction between an enzyme and its cofactor and an antibody and its antigen. An affinity column is prepared by immobilizing one of the interacting molecules (known as the ligand) on a suitable matrix such as agarose or silica. A solution containing the molecule to be isolated is passed over the affinity column under conditions in which the interaction between the ligand and the sample occurs. The column is then washed to remove unbound or non-specifically bound molecules. The molecule of interest is then eluted from the column. Elution is generally achieved through a change in the pH of the mobile phase or the utilization of chaotropic agents. A good affinity stationary phase should have the following characteristics: (1) the underlying matrix itself should not bind molecules to a significant extent, (2) the ligand should tightly bind the molecule to be purified, (3) the molecule to be purified should not be destroyed during elution, and (4) the ligand should not be altered during the binding of the sample molecule.

The ligand can be coupled to the matrix by any of a number of methods. There are several preactivated matrices commercially available. Among these are Ultraffinity-EP from Beckman Instrument Company (Palo Alto, CA) and Affi-Gel 10 from Bio-Rad Laboratories (Richmond, CA). The Ultraffinity-EP stationary phase has a silica matrix with a hydrophilic "spacer arm" terminating with an active epoxide function. The Affi-gel stationary phase is a derivatized, crosslinked agarose bead containing an *N*-hydroxysuccinimide ester as the activated functional group. The solution containing the ligand is passed through the prepacked column or mixed with the gel batchwise to effect immobilization of the ligand.

An affinity column can be made to be extremely selective, such as the coupling of a monoclonal antibody to the matrix to make an immunoaffinity stationary phase. In addition to the methods outlined above, immunoaffinity columns can be produced by coupling protein A from *Staphylococcus aureus* cells to a matrix. Protein A binds the Fc portion of immunoglobulin G molecules. This leaves the antigen-binding site (Fab) free to bind antigens. An antibody solution is simply passed through the column and then the antibody is crosslinked to the stationary phase, usually with carbodiimide.

Another type of affinity column is group selective rather than molecule selective. An example of this is Concanavalin A-Sepharose from Pharmacia (Piscataway, NJ). Concanavalin A belongs to the group of proteins known as lectins which have binding sites for carbohydrate moieties on glycoproteins. Passage of the sample mixture through the column would lead to the selective retention of a specific group of glycoproteins (e.g., mannose-containing). Elution of the glycoproteins would be accomplished by adding α-methyl mannoside to the mobile phase.

An interesting affinity stationary phase has dye molecules as the functional ligand. Most notable is Cibacron Blue-Sepharose (Pharmacia). This dye supposedly interacts with nucleotide cofactor (e.g., NAD^+, nicotine adenine dinucleotide) binding sites on certain proteins. Elution is often effected by using a combination of increasing ethylene glycol and salt. A series of chromatography stationary phases based on different dyes is available from Amicon (Danvers, MA). These dye-affinity columns are evaluated by trial, since their binding properties with any specific protein cannot be predicted.

Thin-Layer and Paper Chromatography

Two noncolumn types of chromatography that occasionally are still used for proteins, peptides, and amino acids are thin-layer chromatography (TLC) and paper chromatography. In these methods, the sample solution is spotted at the origin of a coated plate or thin strip of paper. The plate may be coated with such stationary phases as silica or a gel, as in the separation of larger molecules. In paper chromatography the cellulose is derivatized with one of several ligands such as polyethyleneimine (PEI-cellulose) or diethylaminoethyl (DEAE-cellulose). The plate or paper strip is then placed in a tank with a shallow layer of solvent in the bottom. As the solvent rises up the plate or paper strip, the sample molecules migrate according to their distribution between the mobile and stationary phases.

Principles of Chromatography

The basis of the chromatographic process is illustrated in Figure 10. A sample containing two components, X and Y, are loaded onto a column. Each component will distribute between the stationary phase and the mobile phase according to the following equation:

$$K = \frac{C_s}{C_m}$$

where K is the distribution coefficient and C_s and C_m are the concentrations of the sample components in the stationary and mobile phases, respectively.

Component "X" is migrating through the chromatographic column faster than component "Y," because it interacts less with the stationary phase. It, therefore, has a lower value for K. The retention time (t_r) is related to the degree of interaction with the stationary phase, according to the following equation:

$$t_r = t_o(1 + k')$$

where t_o is the retention time for components that do not interact at all and hence are unretarded and k′ is the capacity factor. This factor is related to the distribution coefficient by the equation:

$$k' = (N_s/N_m)$$

$$K = C_s/C_m = \frac{N_s}{N_m} X \frac{V_m}{V_s} = k' \frac{V_m}{V_s}$$

or

$$k' = K\frac{V_s}{V_m}$$

where N_s and N_m are the number of molecules distributed in the stationary and mobile phases, respectively, and V_s and V_m are the volumes of the stationary and mobile phases, respectively. If $k' = 0$, then the sample compo-

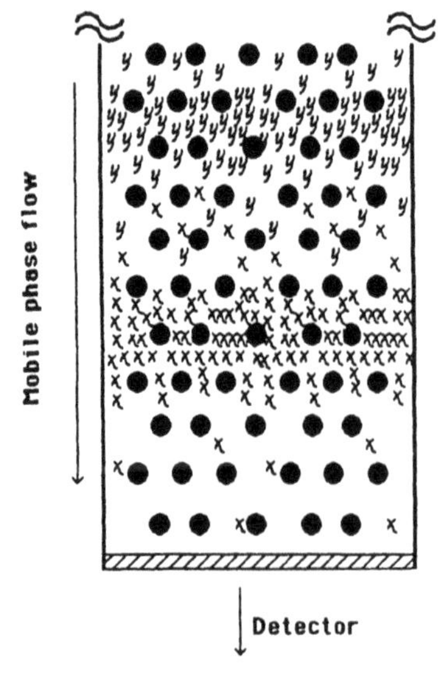

Figure 10 Representation of a chromatographic separation. Two sample components (X and Y) are migrating through a column containing a stationary phase shown as dark circles. Each component is in a separate band and the concentration across each band is essentially a Gaussian distribution. There is some overlap between the bands. The mobile phase flow will carry the components through the detector, as illustrated in Fig. 11.

nent elutes at t_o, which is often referred to as the breakthrough or void volume. If the distribution between the phases is equal (i.e., $k' = 1$), then the sample component elutes at a time equivalent to two column volumes of mobile phase. The k' values for a group of components within a sample mixture may vary widely with a particular mobile phase. In such cases gradient elution allows the k' values to decrease constantly during the chromatographic run by decreasing the distribution coefficient.

As the components elute from the column, their concentrations are monitored with an appropriate detector (Fig. 11). The peak shape approaches that of a Gaussian distribution. The term that describes the degree of separation between two components is resolution. This term is defined according to the following equation:

$$R_s = (t_{r2} - t_{r1})/[(1/2)(t_{w1} + t_{w2})]$$

where t_{r1} and t_{r2} are the retention times of components 1 and 2, and where t_{w1} and t_{w2} are the peak widths at the triangulated base of each component (Fig. 12). The resolution may be improved by either having a greater difference in the retention times or by decreasing the widths of the peaks. To achieve better resolution, one might change the mobile phase, the stationary phase, the temperature, the length of the column, etc.

Column selectivity, α, is measured by the relative separation of two peaks and is given by:

$$\alpha = (t_{r2} - t_o)/(t_{r1} - t_o)$$

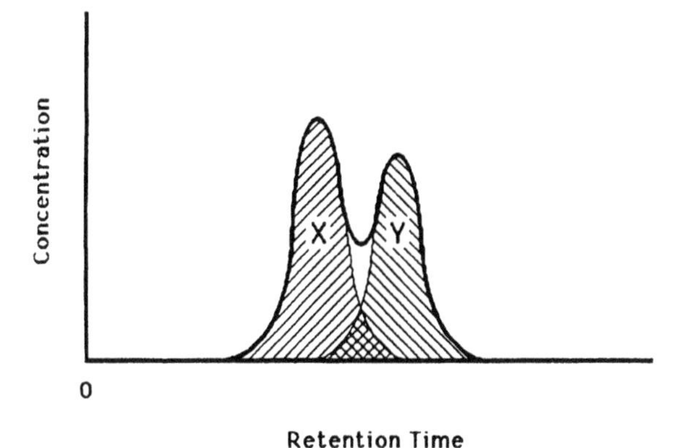

Figure 11 Illustration of the detector response from the separation depicted in Figure 10. Assuming that the detector cannot distinguish between components X and Y, the chromatogram would appear as a double peak. The contribution of each component to the envelope tracing is shown by the hatched lines. Essentially pure component X or pure component Y may be collected at certain retention times.

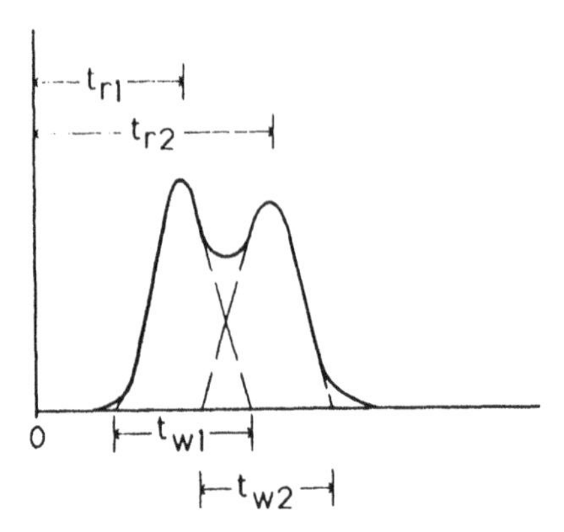

Figure 12 Definition of chromatographic terms. The terms t_{r1} and t_{r2} refer to the retention times of components 1 and 2. The terms t_{w1} and t_{w2} refer to the peak widths of components 1 and 2, as determined from the intercept of each triangulated peak with the baseline.

It is calculated by measuring the retention times of two retained components and subtracting from each the retention time of a nonretained component.

The introduction of HPLC stationary phases provided dramatic improvements in peak sharpness. Thus, components with close values of k′ can be resolved since the peak widths are small. The sharpness of the peaks for an individual column is termed the efficiency. It is defined as:

$$N = 16(t_r/t_w)^2$$

where N refers to the number of theoretical plates (borrowed from distillation terminology), and t_r and t_w are the retention time and peak width for a particular component eluting under isocratic conditions. The value of N is independent of retention time. N calculated for a peak eluting at 5 minutes will be the same as that calculated for a peak eluting in 10 minutes.

The number of theoretical plates is proportional to the column length. Longer columns generally have more plates. In order to obtain a value independent of column length, the height equivalent to a theoretical plate, H, is calculated as follows:

$$H = L/N$$

where L is the column length. It can also be expressed inversely as the number of plates per meter of column length.

The calculated number of plates for each column will depend upon the type of molecule used to determine the retention time and peak width. The calculated value will be constant for a series of similar molecules. It has been observed that the number of plates is always greater when using a small molecule in the calculations rather than a large molecule such as a protein. Typical plate numbers for a small molecule will be in the range of 50,000 to 500,000 per meter. One reason for a smaller number of plates when using a protein in the determination is the lower rate of diffusion of the large molecule and hence the lower rate of reaching equilibrium between the two phases. This is especially apparent when monitoring retention as a function of flow rate. Band spreading, or an increase in peak width, is seen as the flow rate is increased, and the effect becomes more pronounced with increasing molecular weight.

The binding of proteins to a stationary phase appears to be a multipoint interaction. The number of protein–ligand interactions can be determined by measuring the total protein-loading capacity of the column and the number of ligands per area of stationary phase. On a phenyl HIC column the number of interactions varied from 2.7 for lysozyme to 3.8 for Conalbumin. Because the number of interactions exceeds one, it is difficult to resolve proteins isocratically on most stationary phases to which the protein binds. In other words, multipoint binding leads to high values of k'.

Kopaciewicz et al. have proposed a nonmechanistic stoichiometric model to explain the ion exchange process between proteins and the stationary phase. The equations of this model offer a means of measuring the number of interactions (Z) between the protein and the stationary phase. The equilibrium process in ion exchange is given by the expression:

$$P \cdot C_i + Z \cdot D_b \rightleftarrows P_b + Z \cdot aD_o + Z \cdot bC_i$$

where D_b is the concentration of displacing ions associated with the stationary phase. $P \cdot C_i$ is the concentration (moles/liter) of protein in solution with accompanying counter-ion (C_i). P_b represents the amount of protein bound to the stationary phase, while D_o is the displacing ion concentration of the mobile phase (moles/liter). The displacing power of an ion, however, is proportional to its ionic strength, therefore, the constants a and b are used to adjust for this. The term Z is the number of charges associated with the adsorption–desorption process.

Based on the above discussion of chromatographic principles, the approach to optimal elution of proteins differs from that of small peptides and amino acids. In particular, lower flow rates are required for proteins due to mass transfer in and out of the porous matrix. The loss of resolution due to the slower flow rates can be compensated for by the use of shallower gradients. Indeed, protein elution is almost an "all or none" phenomenon as the mobile phase reaches a certain eluting strength. This can be seen from the evidence that short columns provide the same resolution as long columns in protein chromatography.

ELECTROPHORESIS

Electrophoresis refers to the migration of sample components in an electric field. It complements LC for resolving mixtures, especially of proteins. Generally, the proteins migrate through a solid matrix, most commonly polyacrylamide, although other materials such as agarose and cellulose acetate paper are used in certain applications. There are two arrangements for polyacrylamide gel electrophoresis, tube gels and slab gels. Slab gels may be either horizontal or vertical.

Hardware

The hardware for performing electrophoresis consists of a power supply, and an electrophoresis apparatus containing: (1) a support for the matrix and (2) two reservoirs to hold the buffers with electrodes at either end of the matrix. The electrophoresis apparatus can be purchased from one of several manufacturers or can be constructed in a machine shop from plexiglass. The electrodes used are thin platinum wires.

Principles

In electrophoresis, the mobility, u, is defined as the velocity per unit field, or

$$u = v/E$$

where E is the electric field. If a molecule with charge q is placed in an electric field, E, the molecule will move at a constant velocity, v, determined by the balance between the electrical force, Eq, and the viscous drag, fv, in which f is the frictional coefficient.

$$Eq = fv$$

Therefore,

$$u = q/f.$$

The parameter f is a function of some of the physical parameters of the molecule and the media in which the molecule is migrating. From this equation, it can be seen that mobility increases with q, decreases with f, and is zero for uncharged molecules.

Modes of Electrophoresis

Paper Electrophoresis

Paper electrophoresis involves the use of thin strips of cellulose or cellulose acetate as the matrix. The paper is first wet with the electrophoresis buffer and then placed in the electrophoresis tank, such that both ends contact separate buffer tanks containing the electrodes. The sample is spotted at the origin and then current is applied. If a high-voltage gradient is applied to the paper, the paper will be immersed in a nonmiscible liquid for cooling. Once the electrophoresis is complete, the paper is dried and the sample components are identified by a color reaction, fluorescence, etc.

Sodium Dodecyl Sulfate–Polyacrylamide Gel Electrophoresis (SDS-PAGE)

The most popular mode of electrophoresis for proteins is polyacrylamide gel electrophoresis in the presence of sodium dodecyl sulfate (SDS-PAGE). Samples applied to SDS-PAGE gels are first boiled in sample buffer containing 1% SDS and 0.1 M dithiothreitol (or 2-mercaptoethanol) which breaks sulfhydryl bonds and denatures the protein to a random coil configuration. The proteins have a uniform shape and an identical charge-to-mass ratio, since the amount of SDS bound per unit weight of protein is constant. This ionic detergent coats each protein molecule, thereby imparting an overall negative charge. At neutral or slightly alkaline pH, the proteins in the sample mixture migrate toward the anode (positive electrode) according to molecular weight. This phenomenon results from the sieving effect of the polyacrylamide gel. Different concentrations of acrylamide are used for making gels that optimize the sieving effect in different molecular weight ranges. A gradient of acrylamide concentration may also be formed in a single gel to optimize resolution. The direction of the gradient is made in the same direction as migration, i.e., the lower concentration of acrylamide is at the top of the gel. The proteins migrate through the gel until the pores are no longer large enough to permit movement and equilibrium is established. In this case, the separation is based strictly upon size. However, the current may be discontinued before equilibrium is reached, since good resolution can be achieved sooner. In this case, the separation is based upon relative mobility. Usually, a low molecular weight dye like bromophenol blue is added to the sample to indicate the front of migration. This is important, because the proteins are not visible during electrophoresis and the process is terminated when the tracking dye reaches the bottom of the gel.

In certain applications, such as for the identification of disulfide-linked dimers, the dithiothreitol is not included in the sample buffer. This is called a nonreducing gel.

Polyacrylamide Gel Electrophoresis (PAGE)

PAGE may also be done in the absence of SDS at alkaline or acid pH and migration then depends on a combination of the overall charge of the protein and the sieving effect. This is known as a nondenaturing gel since the protein is not exposed to SDS or dithiothreitol.

Usually PAGE (or SDS-PAGE) is performed in a discontinuous gel system. This system consists of a gel with two separate regions, an upper stacking gel and a lower resolving gel (Fig. 13). The stacking gel is less concentrated (lower percentage of acrylamide) than the resolving gel and is prepared in a

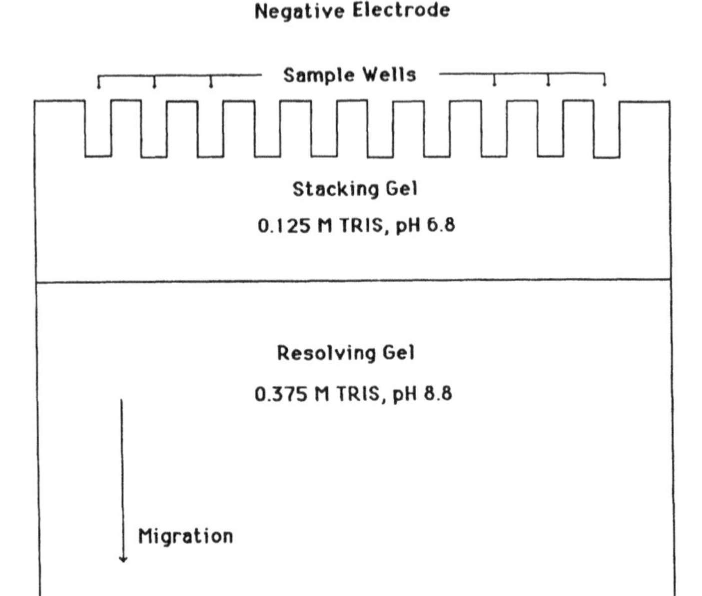

Figure 13 Diagram of a discontinuous gel. The upper or stacking gel is of a lower acrylamide concentration, as well as a lower buffer concentration and pH than the lower resolving gel. Sample is applied to the gel in the wells of the stacking gel.

buffer of lower ionic strength and pH. The larger pore size of the stacking gel allows the molecules to migrate faster. In addition, the lower ionic strength means higher electrical resistance so that the electric field (V/cm) is greater in the upper region and hence makes the molecules migrate faster. As current is applied to an SDS gel, what is called a stacking or dye front forms. This front is due to the difference in mobility of the chloride and glycinate ions in the gel. At the pH at which most SDS-PAGE gels are run, glycinate possesses only a partial negative charge, while chloride has a full negative charge. Therefore, the mobility of the chloride ion is greater. Ahead of the chloride ions the voltage gradient drops, since the resistance to current drops. If a protein molecule migrates ahead of the chloride ions, its mobility is slowed due to this drop in voltage. The voltage gradient behind the glycinate ions increases, because there is higher resistance due to the partial charge of the glycinate ions. If a protein molecule lags behind the glycinate ions, it is forced to catch up due to the greater voltage. This results in a concentration of the molecules at the dye front. Once the interface between the stacking and resolving gels is reached, the sieving effect of the resolving gel then separates the molecules according to size.

Isoelectric Focusing

Another popular mode of electrophoresis is isoelectric focusing. In this mode a pH gradient is established within the electric field. As each protein in the sample mixture migrates toward an electrode it reaches a position in the pH gradient at which its positive and negative charges balance (its isoelectric point) and migration ceases.

The pH gradient is established with a mixture of synthetic, low molecular weight polyampholytes (i.e., molecules containing both positive and negative charges) covering a wide range of isoelectric points in very small increments (0.01 pH units). The mixture of ampholytes is added to the matrix before the current is applied. Therefore, the pH is uniform before the sample is applied. When the electric field is applied to the gel, these small ampholytes begin to migrate, establishing a pH gradient. The ampholytes eventually stop migrating in the pH gradient, each at its own isoelectric point. The proteins also migrate in this pH gradient, however much more slowly, until each one reaches its isoelectric point. This is true as long as the concentration and the buffering capacity of the protein is low enough so as not to disrupt the pH gradient.

Two-Dimensional Separations

In order to resolve molecules of similar size and charge, it is very often effective to combine two modes of chromatography and/or electrophoresis in a so-called two-dimensional separation. The modes of separation most often combined are thin-layer chromatography (chromatography takes place on a thin sheet coated with the matrix) with paper electrophoresis and isoelectric focusing with SDS-PAGE. In a two-dimensional separation the first mode is performed in one direction and then the second mode is performed perpendicular to the first. The two-dimensional separation may be performed on the same media or the first media may be placed in close contact with the second and the sample transferred from one to the other. In combining isoelectric focusing and SDS-PAGE proteins are first run in a tube isoelectric focusing gel for the first dimension. The tube gel is then placed at the top of an SDS-PAGE gel and electrophoresed for the second dimension. Two-dimensional separations are especially important in separating complex mixtures of proteins. A two-dimensional separation of this type is shown in Figure 14.

One application of two-dimensional separations is in peptide mapping of proteolytic digests. The protein of interest is first digested with a specific protease which cleaves the peptide bond after certain amino acids. Generally, the digest is applied to a cellulose-coated thin layer plate and electrophoresed in the first dimension. The plate is then turned 90° and chromatographed in the second dimension. This results in a distribution of spots that is specific for

Figure 14 Two-dimensional gel of a whole *E. coli* cell lysate. The cytoplasmic fraction was first applied to an isoelectric focusing tube gel. After electrophoresis, the tube gel was placed in the well of an SDS-PAGE gel and electrophoresed in the second dimension. N. Stephen Pollitt unpublished data. Used with permission.

the individual protein, since the cleavage sites and distribution of amino acids in the protein is generally unique. This two-dimensional separation is, therefore called a ''fingerprint.''

Capillary Zone Electrophoresis

This technique has recently gained prominence, especially with the introduction of several commercial instruments. Electrophoresis is carried out in a small bore glass (fused silica) capillary tube ($\sim$ 50 μm) which is placed between two electrode/buffer compartments. The sample is typically introduced by dipping one end of the capillary into that solution and allowing a small volume to flow in by gravity. Alternatively, an electric potential may be applied briefly. The capillary is then returned to its buffer compartment.

A high voltage is applied and the individual components will migrate toward the oppositely charged electrode. An important secondary effect is the electroosmotic flow of buffer, generally toward the cathode. Therefore, the migration of a solute is a combination of its electrophoretic mobility and the buffer flow and even neutral molecules can be separated. Interactions of a solute with the capillary wall can also influence migration.

By using a high potential (e.g., 20,000 V) across the length of the capillary (e.g., 50 cm), separation times are typically in tens of minutes, similar to HPLC. The high surface to volume ratio of the capillary permits rapid heat loss; heating would otherwise lead to diffusion and band broadening. The capillary itself can serve as the flow cell for the detector, which must be appropriately designed for this application. Since the effective volume in the detector is so small, detection limits can range from 10^{-13} mole by absorption to 10^{-20} mole by fluorescence. Considering the unique attributes of capillary electrophoresis, as well as the many possible variations on the theme, this technique should become competitive with HPLC for analytical applications.

Detection of Proteins

After electrophoresis is completed, the separated components are visualized by staining or autoradiography (if the sample components are radioactively labeled). The most common staining procedures use Coomassie Brilliant Blue R-250 (in the microgram range) or silver nitrate (in the nanogram range). A typical stained electrophoresis gel and an autoradiogram are shown in Figure 15.

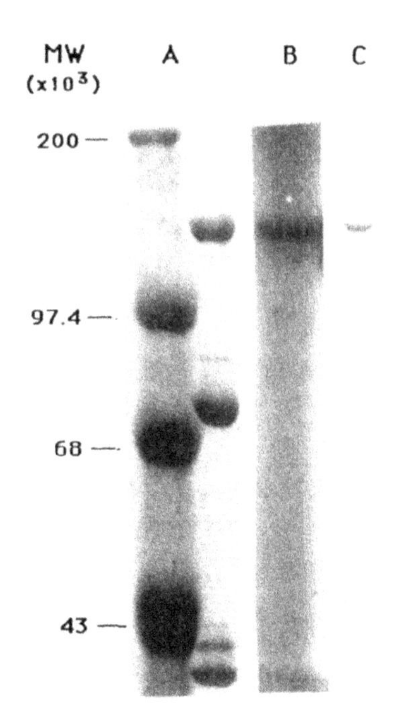

Figure 15 Examples of Coomassie Blue stained gel (A) autoradiogram (B) and Western blot (C) of ^{32}P-labeled Reo III dearing virus. Whole virions were labeled with [α-^{32}P]GTP. The guanylyl transferase enzyme (λ2) found in the virus converts the GTP to GMP which becomes covalently bound to the enzyme. The Western blot is probed with an antibody generated against a peptide comprising amino acid residues 191-206 of the λ2 protein. Thus, only the λ2 protein is visualized by Western blotting or by autoradiography. The additional lane (on left) in panel A contains molecular weight marker proteins.

Alternatively, the separated components may be extracted from the gel or transferred to another medium for other types of analysis, such as determination of biological activity. The protein can be extracted from the gel by one of several methods. In the simplest method the gel slice containing the protein is crushed and then soaked in a buffer. After incubation in the buffer, the solution (now containing the protein) is removed from the crushed gel pieces by centrifugation. This method generally only works for very small proteins or polypeptides. In another method, called electroelution, the gel slice is placed in an electric field. As the protein migrates from the gel slice it is trapped either by a membrane or by a high-salt bridge. The protein may also be transferred to another medium in a process called electroblotting (or electrotransfer). In this procedure, a sheet of a particular membrane (e.g., nitrocellulose or polyvinylidiene difluoride) is sandwiched against the polyacrylamide gel between several sheets of 3 mm Whatman filter paper. When a voltage is applied across the membrane and gel in the proper direction, the protein migrates from the gel to the membrane where it is bound. The membrane, which is physically stronger and more inert than the polyacrylamide gel, may be further processed. Most commonly, particular proteins on the membrane are visualized by staining with a specific antibody in a procedure known as immunoblotting or "Western" blotting or with Coomassie Blue (Fig. 15). Polyvinylidene difluoride has the advantage in that Coomassie blue-stained protein bands can be directly applied to the gas phase protein sequencer. Thus, there are numerous combinations of approaches for analytical and preparative applications.

SUGGESTED READING

Ewing, A. G., Wallingford, R. A., and Olefirowicz, T. M. Capillary electrophoresis. *Anal. Chem.*, 61, 292 (1989).

Freifelder, D. *Physical Biochemistry: Applications to Biochemistry and Molecular Biology*, 2nd Ed., W. H. Freeman and Company, San Francisco, 1982.

Handbook of HPLC for the Separation of Amino Acids, Peptides and Proteins, Vols. 1 and 2. W. S. Hancock (editor), CRC Press, Inc., Boca Raton, FL, 1984.

Johnson, E. L. and Stevenson, R. *Basic Liquid Chromatography*, Varian Associates, Inc., Palo Alto, CA, 1978.

Kopaciewicz, W., Rounds, M. A. Fausnaugh, J., and Regnier, F. E. Retention model for high-performance ion-exchange chromatography. *J. Chromatogr.*, 266, 3 (1983).

Pfannkoch, E., Lu, K. C., Regnier, F. E., and Barth, H. G. Characterization of some commercial high performance size-exclusion chromatography columns for water-soluble polymers. *J. Chromatogr. Sci.*, 18, 430 (1980).

Science, Guide to Scientific Instruments, Vol. 239 Part II (1988).

Snyder, L. R. and Kirkland, J. J. *Introduction to Modern Liquid Chromatography*, 2nd Ed., John Wiley & Sons, Inc., New York, 1979.

5

Proteins as Biological Effectors

Milton T. W. Hearn, Peter G. Stanton, and Joseph Bertolini

Monash University
Clayton, Victoria, Australia

INTRODUCTION

Proteins consist of single or multiple chains of amino acids, linked together by peptide bonds and folded into unique conformations. The linear arrangement or sequence of the amino acids is characteristic of each protein and is referred to as the primary structure. The α-amino acids, as well as other amino acids which are biosynthetically incorporated into proteins, exhibit considerable chemical diversity notably in terms of the structure of side chains which can be either polar (acidic, basic, uncharged polar) or nonpolar. The through-space interactions of these amino acid side chains are thus mediated by hydrogen, hydrophobic, and coulombic bonds and lead to folding of the peptide backbone into a unique topographic hierarchy of secondary, tertiary, and in many cases, quaternary structures. The secondary structure consists of two major structural units—the α-helices and β-sheets which arise from hydrogen bond interactions and result in periodic twisting or pleating of the peptide chain, respectively. The three-dimensional arrangement of the rodlike α-helices and β-pleated sheets into a number of associated small globular or laminar units and domains defines the tertiary structure of the protein. In some proteins there can be further association of these units into dimers, tetramers, etc., and other hierachical arrangements which are referred to as the quaternary structure.

The biological activity of a protein is totally dependent on the integrity of its three-dimensional structure. The folding of the polypeptide chain or the association of folded subunits defines a surface region which serves as the binding or active site through which proteins as well as other biosolutes interact via electrostatic forces, hydrogen bonds, hydrophobic Lipshitz and van der Waals forces, or covalent bonds to effect their biological actions.

Proteins as biological effectors, as distinct from structural proteins, such as collagen and keratin, can be divided into six major groups. These are:

1. Enzymes: Protein catalysts which mediate and accelerate the rate of biological reactions.
2. Regulatory proteins: Hormones, lymphokines and growth factors which interact with membrane receptors to regulate cellular function.
3. Receptors: Proteins on the cell membrane which interact with regulatory proteins and initiate the intracellular biochemical events which account for the biological response.
4. Immune proteins: Proteins synthesized by cells of the immune system to bind to foreign substances or organisms and initiate a series of biochemical and cellular events leading to their destruction and removal.
5. Carrier proteins: Ligand binding proteins which function intracellularly and extracellularly to create reservoirs of biologically important substances.
6. Contractile proteins: Highly specialized proteins which transfer biochemical energy to cellular movement.

In biological systems proteins from these groups are arranged into intricate intercommunicating networks through which the organism identifies and responds to changes in its internal and external environment.

In this chapter we examine in detail some of these protein effector systems. First, the impact of structural conservation and variation on a protein's ability to function as a biological selector and effector will be examined from the viewpoint of the immunoglobulins—the prime defense selector/effector system with modality to control all self- and non-self-recognition phenomena in all vertebrate species. Second, an examination of several of the salient features of the proteins of the blood coagulation or clotting pathway and the immunoglobulin-activated complement pathway is used to illustrate how protein systems can identify threats to the homeostasis of the organism, respond to damage of the blood vessels and capillary walls or overcome the insidious invasion of a foreign substance or organism, and then initiate a cascade of enzymic events which amplifies the initial response and activates the necessary corrective mechanisms. Third, we examine some effector features of the gonadotropins, a family of endocrine glycoprotein hormones secreted by the pituitary gland and which must reach their target cells in the ovary and testis via the circulatory system. These adenohypophyseal glycoproteins exhibit structural heterogeneity associated with variation in polypeptide primary amino acid sequence and the extent of glycosylation, subtle structural variations which correlate with differences in biological potency. These hormones are used to illustrate the relationship of effector–receptor interaction to biological activity and the role subtle structural changes can play as determinants in modulating effector response. Fourth, the discussion on the structure and biochemical mechanisms of growth factor action highlights the diversity of transduction of effector signals into the intracellular compartments by membrane receptors. Finally, the section on antiproteins examines proteins involved in feedback control of response networks, proteins such as the recently discovered antihormone inhibin and antithrombin III, a regulator of unrepentant thrombosis.

IMMUNOGLOBULINS: A HOST DEFENSE SYSTEM

Conservation and Variation of Structure: The Origin of Immunological Specificity

Immunoglobulins, whether cell surface as in the case of T- or B-cell Ig receptors, or as part of the humoral pool of circulating proteins in plasma, are glycoproteins synthesized by lymphocytes in response to an antigenic challenge by a foreign macromolecule, cell, or organism. The primary function of immunoglobulins is to bind foreign, invasive macromolecules as the initial step in their removal from the circulation. The immune response is usually categorized into two systems: (a) The humoral response mediated by B cells which involves the production of circulating antibodies which complex with foreign antigens and facilitate their removal by phagocytic cells and/or enzymatic lysis. (b) The cell-mediated immune response involving specialized lymphocytes designated T cells which recognize invading microorganisms (bacteria, fungi, parasites), virally infected cells, cancer cells, and foreign tissues, and effect their removal by controlling cells with phagocytic and cytotoxic actions.

The basic structural aspects of the immunoglobulin (Ig) molecule, and a description of the five classes of immunoglobulins have been included elsewhere in this book (Chap. 12) and consequently only a brief description of their structure will be given here. In its simplest monomeric form, the Ig molecule is a hinged structure which approximates a 'Y' shaped protein (Fig. 1) with two antigen-binding sites located at the upper extremities of the Y-arms. In fact, monomeric Ig actually consists of four separate peptide chains, two identical heavy (H) chains of approximately 440 amino acids each, and two identical light (L) chains of approximately 220 amino acids each. The H and L chains are spatially arranged such that the amino termini of the four chains are all oriented in the same way, with the antigen binding sites located near the amino termini. The individual chains are held together by both noncovalent and covalent (disulfide bridge) bonds in the 'hinge region' of the Ig structure.

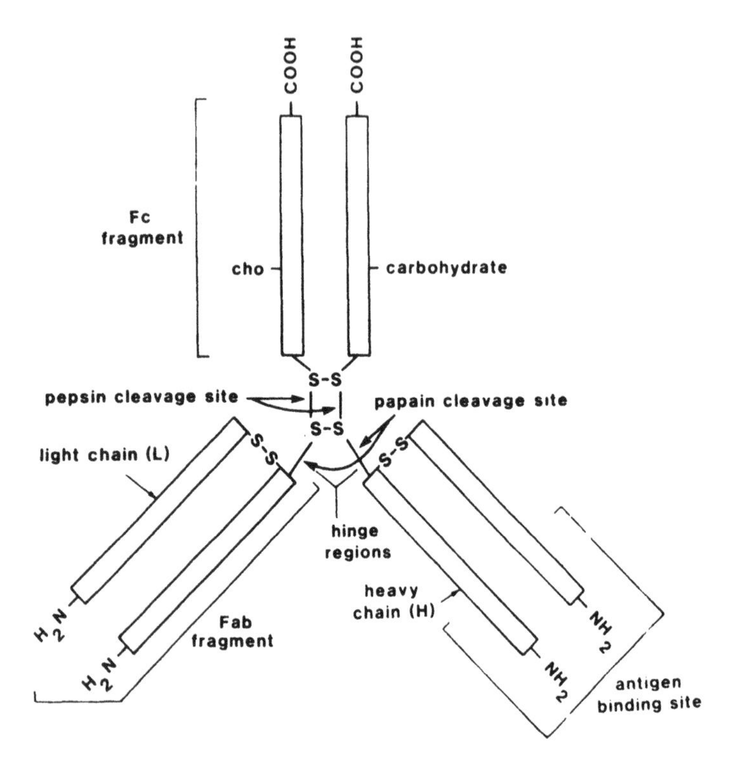

Figure 1 Schematic diagram of immunoglobulin structure. Two light chains and two heavy chains are held together by disulfide bridges in the hinge region of the molecule. Two antigen binding sites are formed by the amino terminal regions of both the H and L chains. Initial structure determination was aided by the presence of specific enzyme cleavage sites located in the hinge region. Cleavage by papain produces an F_c fragment and two F_{ab} fragments, while cleavage with pepsin produces the F_{ab} fragments which are still cross-linked by disulfide bonds.

Amino acid sequence analysis of the H and the L chains of a large variety of Igs has revealed that each chain consists of a constant or C sequence region and a variable or V sequence region, the variable region being located at the amino terminal end of the peptide chain and constituting part of the antigen-binding site. In L chains the variable, or V_L, region consists of approximately half the amino acid backbone (ca $\sim$ 110 amino acid residues) while in H chains the V_H region is also ca $\sim$ 110 amino acid residues in length and therefore one quarter of the total sequence. It is this variability of the amino acid sequence in the V_L and V_H regions that fulfills the structural requirements for the diversity of antigen-binding sites.

When Gerry Edelman and coworkers completed the first amino acid sequence of the immunoglobulin molecule in 1968, they used a protein found in the urine of a hypergammaglobulinemic patient with multiple myeloma, a disease in which a single antibody-producing cell becomes oncogenically transformed, proliferates, and produces a large quantity of a single type of immunoglobulins. From these results it became apparent that the H and L chains were also made up of a number of quite similar regions of amino acid sequence, with the H chain containing three similar regions, or domains, in the constant region (C_H1, C_H2, C_H3) and the one variable domain (V_H) (Fig. 2). The L chain also exhibits this structure with one constant domain (CL) and one variable domain VL. In each case these domains are stabilized by intrachain disulfide bonds (Fig. 2). Each domain is folded separately to form individual units, the variable domains forming the antigen-binding sites, and the constant domains (excepting C_H1) forming the so called F_C fragment, a name

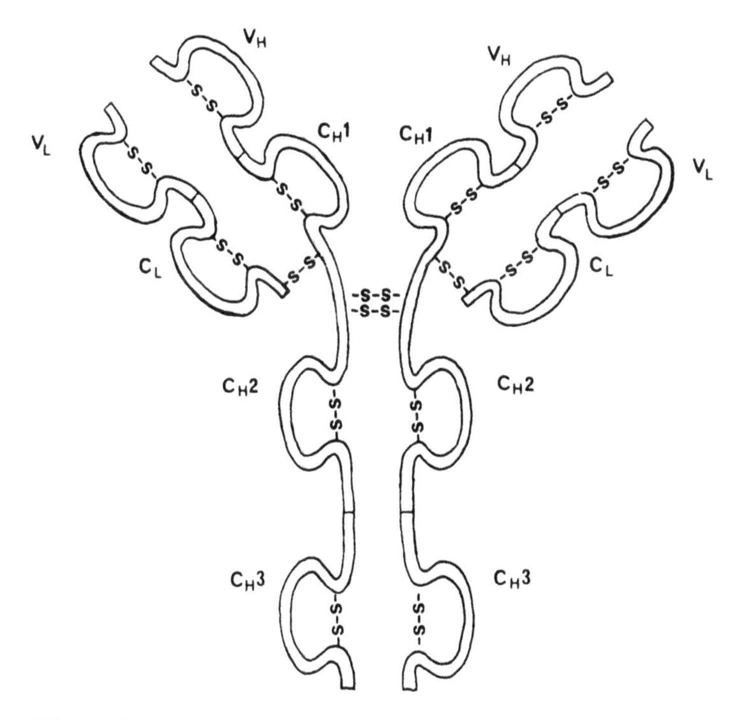

Figure 2 Schematic illustration of the domain structure of the immunoglobulin molecule. Heavy and light chains are folded into repeating domains that are stabilized by disulfide bonds. The antigen binding site is made up of the variable domains (V_H and V_L), and the other biological properties of the molecule are mediated by the C_H3, C_H2, C_H1, and C_L domains.

given to the "crystallizable" fragment derived by papain digestion of intact IgGs by Rodney Porter and colleagues when working at Mill Hill in the late 1950s. To complete the homologies in the structure of the immunoglobulin molecule, each domain is folded into very similar three-dimensional structures consisting of two broad sheets of antiparallel β-strands, stabilized by hydrophobic residues.

Selector and Effector Roles of Immunoglobulins

Enzymatic digestion of immunoglobulin G (IgG) with the protease papain produces three fragments by cleaving the IgG molecule at the hinge region (Fig. 1). Two antigen-binding fragments (Fab) are formed, and one F_C, or crystallizable fragment. Functionally, each Fab fragment retains one binding or selector site for the antigen, while the F_C portion contains the various effector functions of the antibody molecule, such as binding to, and initiation of, the complement cascade. Sufficient evidence has now been accumulated from chemical modification and enzymic studies to demonstrate that particular selector and effector roles of the immunoglobulin molecule may be correlated with individual structural domains, although the precise structural location of all effector binding sites has not yet been fully elucidated. These selector and effector roles have been summarized in Table 1 and include (1) antigen binding, (2) complement fixation, (3) binding to F_C receptors on macrophages and monocytes, and (4) binding to Fc receptors on neutrophils.

Antigen binding is associated with the variable domains on the H and L chains. However, the variability of the amino acids in these domains from different antibodies is not distributed evenly along the peptide chains. Instead, sequence comparisons between immunoglobulin molecules have shown that three short segments exist which exhibit exceptional variability, and these segments have been termed the hypervariable regions. Adjacent to these hypervariable regions are a series of relatively constant residues known as framework regions, which are thought to create a more rigid peptide backbone bringing the three hypervariable segments into close proximity and thus forming a configuration which maximizes the association constant for the antigen binding site (Fig. 3). The immune system of higher vertebrates has been estimated to be able to generate approximately 10 million different antibody binding sites. This amazing number of different proteins is brought about by the immunoglobulins being coded for by three gene pools, each of which contains separate gene segments which can be brought together by specific recombination events during transcription. A schematic for such recombination events is shown in Figure 4.

As has been mentioned, antibody binding of foreign antigens is followed by activation of the complement cascade resulting in the lysis of cells or the ac-

Table 1

Domain	Selector/effector function
V_H and V_L	Antigen binding
C_H1	Binds complement C4b fragment
C_H2	Complement fixation by C1q
C_H3	Binds F_C receptor on monocytes and macrophages
C_H3 and C_H4	Binds F_C receptor on neutrophils

tivation of cellular systems which can dispose of foreign substances. The first step in the activation of the major, or classical, complement pathway is the binding of the immunoglobulin-antigen complex to a subcomponent of the C1 enzyme, termed C1q. Initial studies associated this function with the F_C portion of the Ig molecule, and this has since been further localized to the C_H2 domain. The remainder of the complement cascade is composed predominantly of inactive plasma proteins (zymogens) which are sequentially activated in an enzyme amplification system following the initial triggering by Ig–C1q complex formation. The final product of the complement cascade is the formation of a large protein complex called the lytic complex, which attacks the mem-

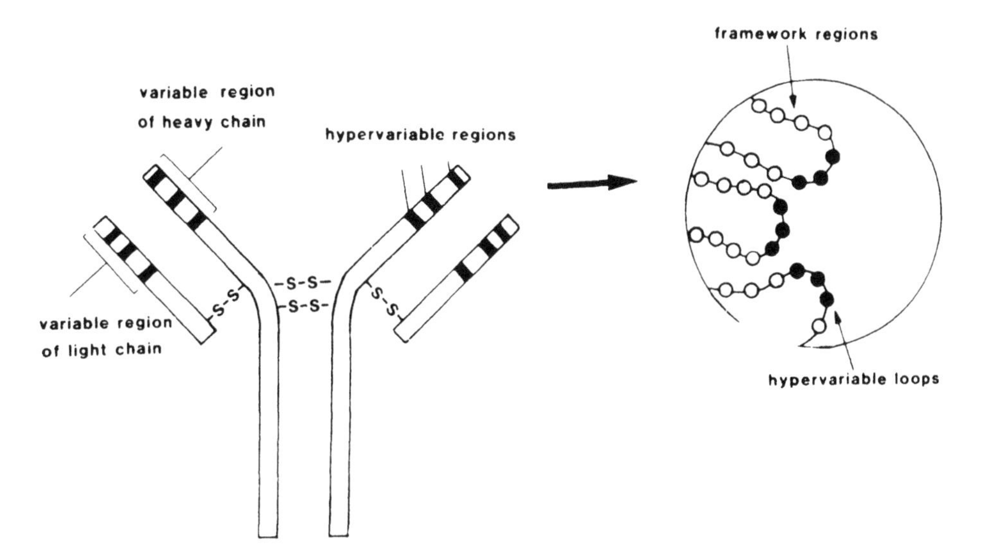

Figure 3 Schematic diagram of the hypervariable structure of the variable domains, V_H and V_L. Three regions of amino acid sequence exhibiting extreme variability are folded to bring them into close proximity with each other, forming the antigen-binding site of the immunoglobulin molecule.

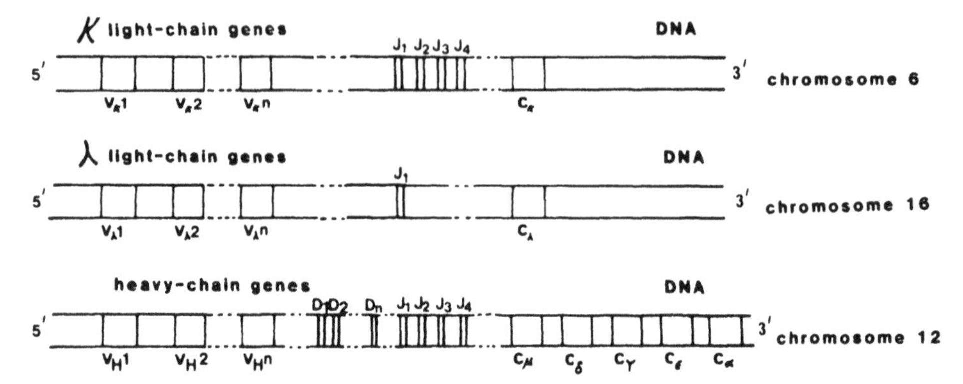

Figure 4 A Multiple genes coding for immunoglobin light and heavy chains. The immunoglobulin light (κ, λ) and heavy chains (μ, δ, γ, ε, α) are coded by separate pools of genes located on various chromosomes. The V genes code for the variable regions and the C genes for the constant regions. There are several hundred V genes. This large number ensures a diversity of immunoglobulins. To construct a light or heavy chain a V gene must be joined to a C gene. With light chain related genes this occurs via a J gene segment, but with respect to heavy chain related genes, an additional D segment is also involved.

branes of foreign cells, disorganizes the plasma membrane, and causes the cells to swell and burst. The complement cascade can also be triggered in the absence of antibody by an alternative pathway following stimulus by the polysaccharide components of the cell wall of foreign microorganisms. In Figure 5 the various stages of both the classical and alternative complement pathway, involving activation of various zymogens, is provided. Complement activation is one method by which Ig molecules may prevent infection. A second method involves the F_C region of IgG molecules (specifically the C_H3 domain) binding to F_C receptors on phagocytic cells such as macrophages and monocytes. The binding of IgG antibody-coated foreign cells to these F_C receptors on the macrophages activates the phagocytic process of the cell resulting in the destruction of the foreign cell (Fig. 6).

A third mode of antibody selector/effector function involves the recognition of antibody-coated organisms by a class of cells known as K cells, or killer cells, which do not phagocytose the foreign cell. K cells bind to a site on the IgG molecule which requires the entire F_C region (C_H2 and C_H3), but as yet the mechanisms by which the K cells kill foreign macroorganisms remains unresolved.

In summary, from a structural point of view, the role of the antibody molecule in host defense is bifunctional. One section of the molecule is involved with selective binding to antigen (the Fab fragment) while a different region

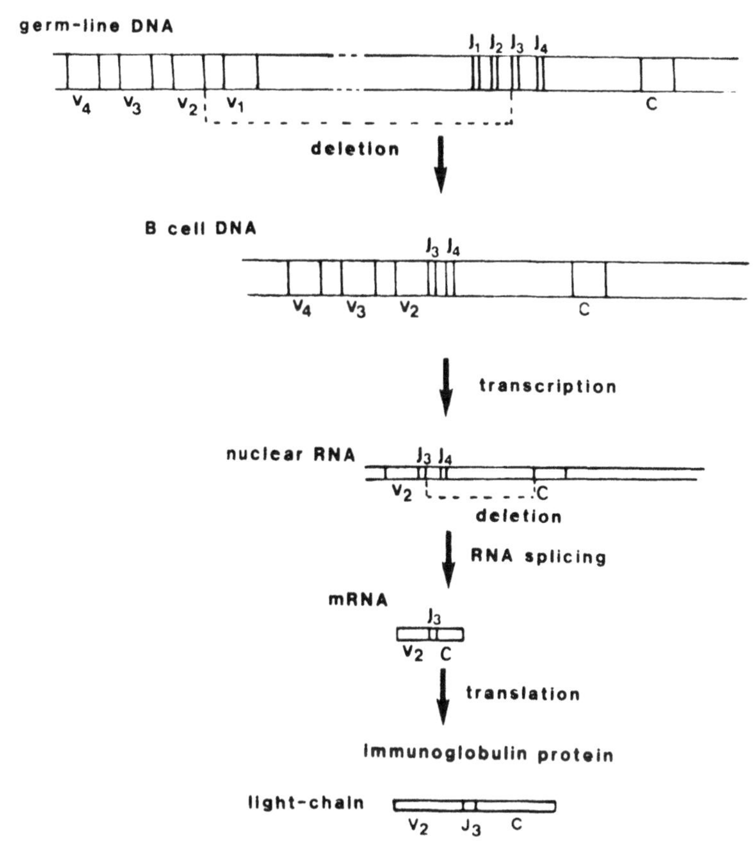

Figure 4 B Assembly and expression of immunoglobulin genes. The assembly (for example) of genes coding for a light-chain occurs as follows. During B-cell differentiation a specific V gene moves adjacent to a J gene segment as a result of somatic recombination events through which intervening DNA is removed. By this process monospecific B cells arise which produce an antibody specific for one antigenic epitope. The V and J genes and other downstream genes are then transcribed. The nuclear RNA primary transcript is processed by the removal of an extra J gene-derived sequence and other intron sequences, with the formation of the mRNA from which the immunoglobulin chain is translated.

mediates effector binding of the immunoglobulin to host tissues, which include killer cells, phagocytic cells and the first component of the classical complement system. This duality permits one protein, albeit a large protein, to act as the prime selector/effector of the host defense system and provide a sophisticated network modality to regulate all self- and non-self-recognition events: the front line trenches of the host's defenses against pathological invasion.

BLOOD CLOTTING: A ZYMOGEN ACTIVATION SYSTEM

The Molecular Basis of Enzymatic Cascades

Many enzymes and hormones are initially synthesized as large inactive precursors and are converted to their biologically active forms in the secretory granules of the cell prior to their release. Alternatively, some proteins, in particular enzymes, are secreted as inactive or zymogen forms, and are activated extracellularly as required by a mechanism involving partial proteolytic cleavage.

Processing of proteins commences intracellularly soon after the polypeptide chain has been assembled by the ribosome. The polypeptides are synthesized

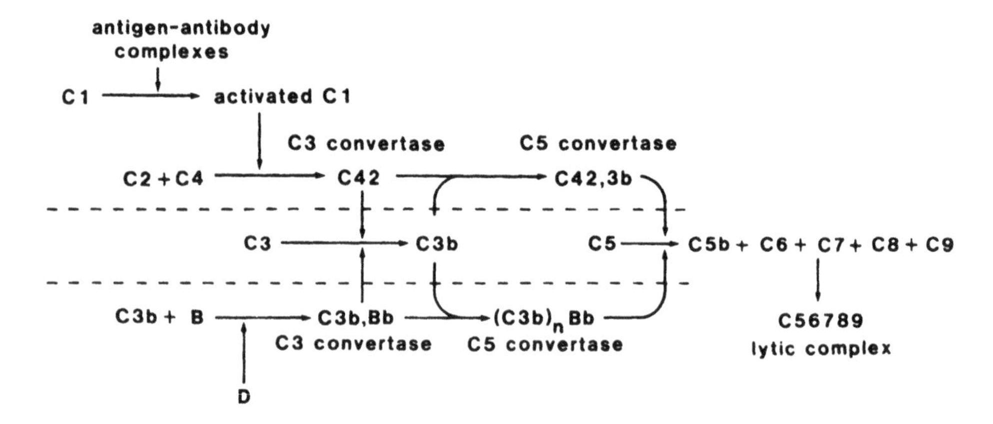

Figure 5 The complement cascade. In the classical pathway, the C1 enzyme complex is activated by the formation of the antigen-antibody complex. Activated C1 causes the production of an activated complex of enzyme C4 and C2. This new complex, named C3 convertase, cleaves C3 to produce two fragments, C3a and C3b, of which C3b binds to the target membrane next to C3 convertase, forming a new complex, C5 convertase. This activated endopeptidase attacks C5, and the smaller released fragments of C5 and C3 act as signals to attract macrophages and to cause the release of histamine from mast cells. The larger fragment of C5 combines with one molecule each of C6, C7, and C8, and six molecules of C9 to form a large molecular weight lytic complex on the plasma membrane of the foreign cell, which causes the membrane to weaken and burst. In the alternative pathway, the trigger is provided both by the presence of C3b and by the polysaccharide components of the cell walls of foreign cells. Several other circulating factors (factors B and D) are also required.

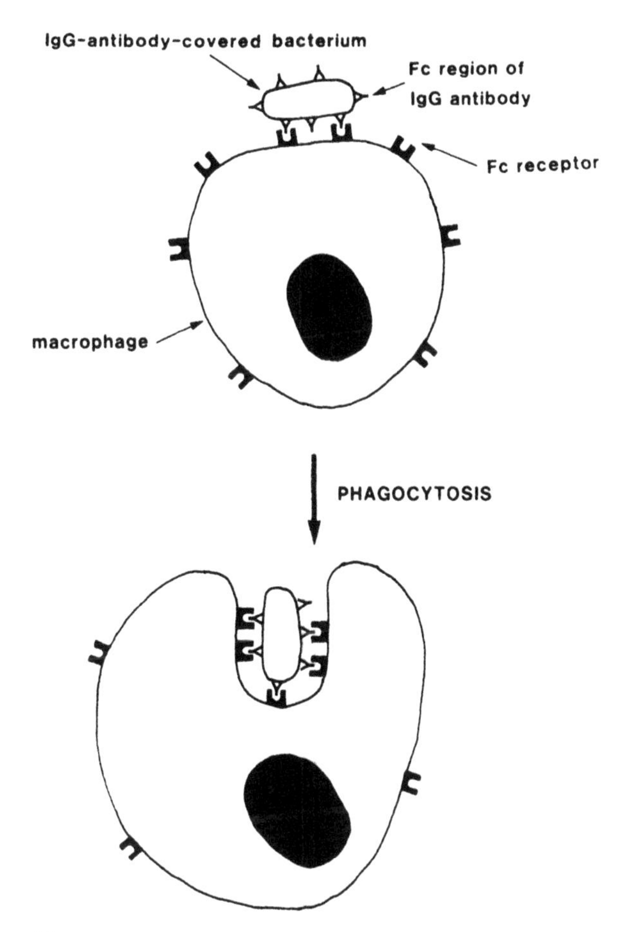

Figure 6 Schematic illustration of the phagocytic process. Antibody molecules coat the outside of foreign cells (bacteria, yeasts etc.), which then bind to Fc receptor on the surface of macrophage, initiating ingestion of the foreign cell by the macrophage.

with a signal sequence which directs them to a specific cellular compartment. Secretory proteins are directed to the endoplasmic reticulum where the first processing step, cleavage of the signal peptide occurs. Further processing may then occur during passage through the Golgi apparatus and secretory granules. At this point the released product may be biologically active or in zymogen form.

The activation of the zymogen could conceivably involve two mechanisms; cleavage of a polypeptide bond allowing access of the substrate to the active site, or a conformational or allosteric change resulting in the formation of an active site. Numerous examples have, however, shown that some zymogens possess weak intrinsic enzymic activity. Binding to the substrate molecule, although at a lower kinetic rate and with lower affinity, suggests that the active site in the zymogen is accessible to the substrate, although the geometry of the fit is far from optimal. Furthermore, it has been shown that activation can be affected without peptide bond cleavage. For example, the blood coagulation protein, Factor XII (Hageman factor) can be activated by two distinct mechanisms; by the proteolytic action of kallikrein or nonenzymatically by contact with a surface such as glass or exposed collagen as results after endothelial rupture, indicating that zymogen activation may require a conformational change and not necessarily only partial proteolytic cleavage for removal of a peptide fragment.

The physiological power of zymogen activation rests on the fact that potentially active compounds can be sequestered in a dormant state but are available for rapid activation when required. The provision of active substances could not be achieved with equal speed by a mechanism involving de novo synthesis.

Some zymogen activation systems are relatively simple. For example, pancreatic trypsinogen is activated by enteropeptidase, which is secreted by the brush border of the small intestine in response to the arrival of the gastric content from the stomach. The activated trypsin in turn regulates a number of other precursors of the digestive enzymes: trypsinogen, chymotrypsinogens, procarboxypeptidases, and proelastase. However, more complex cascade systems exist which serve to rapidly amplify and integrate the initial stimuli and produce a major physiological response. Two examples of this are the complement system and the blood clotting system; both being capable of activation via so called intrinsic and extrinsic pathways.

The key features of all enzyme cascade systems involving sequential activation of various zymogens are:

Signal integration
Signal amplification
Usually covalent modification of the zymogen
Enhanced sensitivity of the zymogen cascade to other stimuli, regulators, or promoters
Lowering of the stimulus concentration for half maximal activation of the cascade system
Control of activation typically requires allosteric ligand binding or regulator effects
Involves a sequence of on/off biological switching mechanisms with the zymogen/active enzyme requiring conformational change from the inactive (or T) precursor form to an active (or R) form or alternatively regulation by protein and antiprotein systems.

Examples of various enzymatic cascade systems are given in Table 2. These examples clearly demonstrate a further key feature of enzymic cascades, namely, they represent a pivotal position in either metabolic control or homeostasis of higher animals.

Intrinsic Versus Extrinsic Pathways

The thrombolytic-fibrolytic pathways of blood coagulation and clot digestion are representative of other pathways involving enzyme cascade activation and control such as milk clotting, hemolymph coagulation in invertebrate animals, the complement pathway, the role of extraneous proteins like the *Staphylococus aureus* staphylocoagulase and snake venom procoagulants, and enzymes involved in nitrogen metabolism and the glycolytic pathways involved with glucose metabolism.

Following damage to a blood vessel a series of biochemical reactions are initiated involving plasma proteins which lead to the formation of a blood clot. Two pathways are involved, one referred to as the intrinsic pathway as all the required components are within the blood, and the other being the extrinsic pathway as extravascular factors are required (Fig. 7).

In the intrinsic pathway the cascade of events is initiated by surface contact of Factor XII or Hageman factor, with the damaged vessel which results in a conformational change and enhancement of proteolytic activity. This converts prekallikrein to kallikrein which then cleaves Factor XII forming an irreversibly activated enzyme. There then follows a series of zymogen activation reactions involving Factors XI, IX, and X (refer Fig. 7). In the extrinsic pathway the initiating event is exposure of a lipoprotein tissue factor from damaged tissue, which activates Factor VII probably by proteolytic cleavage to directly catalyse the activation of Factor X.

Table 2 Enzymatic Cascades

Target enzyme	Signal	Cascade stages	Modification
Glycogen phosphorylase	cAMP	3	Phosphorylation
Glycogen synthase	cAMP others	2	Phosphorylation
Glutamine synthase	Ammonium metabolites	3	Adenylation
Pyruvate dehydrogenase	Metabolites	2	Phosphorylation
Lipase	cAMP	2	Phosphorylation
Prothrombin	Surface contact	>5	Proteolysis

Abbreviation: cAMP:cyclic adenosine 3′, 5′ monophosphate.

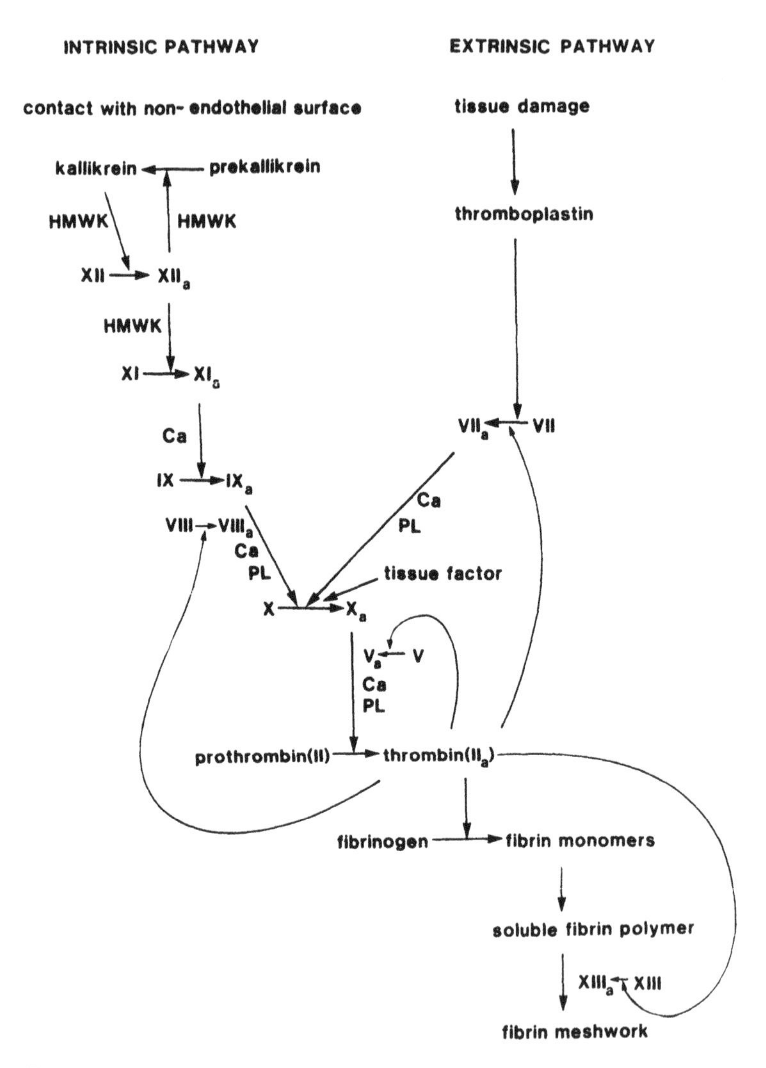

Figure 7 Summary of biochemical events in the blood-clotting process. Blood coagulating enzymes which are in plasma as inactive proenzymes or zymogens are activated by a series of consecutive reactions. The reactions are dependent on a number of cofactors. These include protein (VIII and V), Ca^{+2}, vitamin K, and phospholipid. The vitamin K-dependent coagulation Factors VII, IX, X, and prothrombin (II) reactions proceed at lipid/water interfaces. Blood clotting may be initiated by two pathways as indicated which converge at the activation of Factor X. However, there is an interrelationship between the two pathways; XIIa in the intrinsic pathway is necessary for the activation of VII in the extrinsic pathway, while VIIa of the extrinsic pathway promotes the activation of XII. Note the number of reactions influenced by thrombin. These positive feedback points contribute markedly to the amplification of the initial-coagulation signal. PL—phospholipid, HMWK—high-molecular-weight kininogen.

Activated Factor X proceeds to convert prothrombin to thrombin which acts to convert fibrinogen to fibrin. The conversion of prothrombin to thrombin involves the cleavage by Factor Xa of two sets of peptide bonds, namely an Arg-Thr and an Arg-Ile peptide bond. This conversion liberates the so called Gla- and kringle domains and results in the formation of a two chain-activated thrombin composed of an inert light chain and a catalytic chain held together by a single disulfide bond. The catalytic domain shows considerable amino acid sequence homology with other serine proteases including pancreatic trypsin. The action of thrombin on fibrinogen results in the liberation of amino terminal peptides from the fibrinogen A and B chain and the generation of the fibrin monomer. These fibrin monomers associate in an end-to-end fashion and are cross-linked by glutamyl-lysine bonds formed by Factor XIII, a thrombin-activated transglutaminase.

The properties of the activated proteases are regulated at several stages by the presence of calcium ions, phospholipids, and some protein cofactors. These latter proteins possess no intrinsic catalytic activity themselves, but promote the rates of reactions by influencing the affinity between enzyme and substrate. Two such factors are Factors VIII (the antihemophilic factor) and Factor V which are involved in the activation of Factors X and prothrombin, respectively. Their activation can be elicited by thrombin and therefore could represent sites for positive regulation of the pathway. The importance of these factors, in particular Factor VIII is emphasized by the association of the clotting disorder of hemophilia with a genetic deficiency of this factor.

An interesting feature of the clotting pathway is the role of vitamin K. Prothrombin II and Factors VII, IX, and X contain a number of calcium-binding sites at the amino terminal formed by several γ-carboxyglutamyl residues referred to as the Gla region. Vitamin K has been shown to be essential for the carboxylation of the glutamyl residues in this region. Without these calcium-binding regions, processing to the active forms cannot occur. These reactions occur at lipid/water interfaces and the factors bind to phospholipids through Ca^{+2}-dependent processes mediated by the γ-carboxyglutamate residues. The structural relationship between the vitamin K-dependent proteins is portrayed in the schematic shown in Figure 8. Amino acid sequencing studies have shown that each of the four vitamin K-dependent proteins contain an amino terminal γ-carboxyglutamic acid residue Gla region followed by two potential epidermal growth factor domains, a connecting peptide and a catalytic domain. When these proteins are converted to an activated serine protease either an activated peptide is liberated as in the case of prothrombin conversion to thrombin or, in the case of Factor VII cleavage occurs at a single Arg-Ile peptide bond which results in conversion of the zymogen to its active form (Factor VII → Factor VIIa).

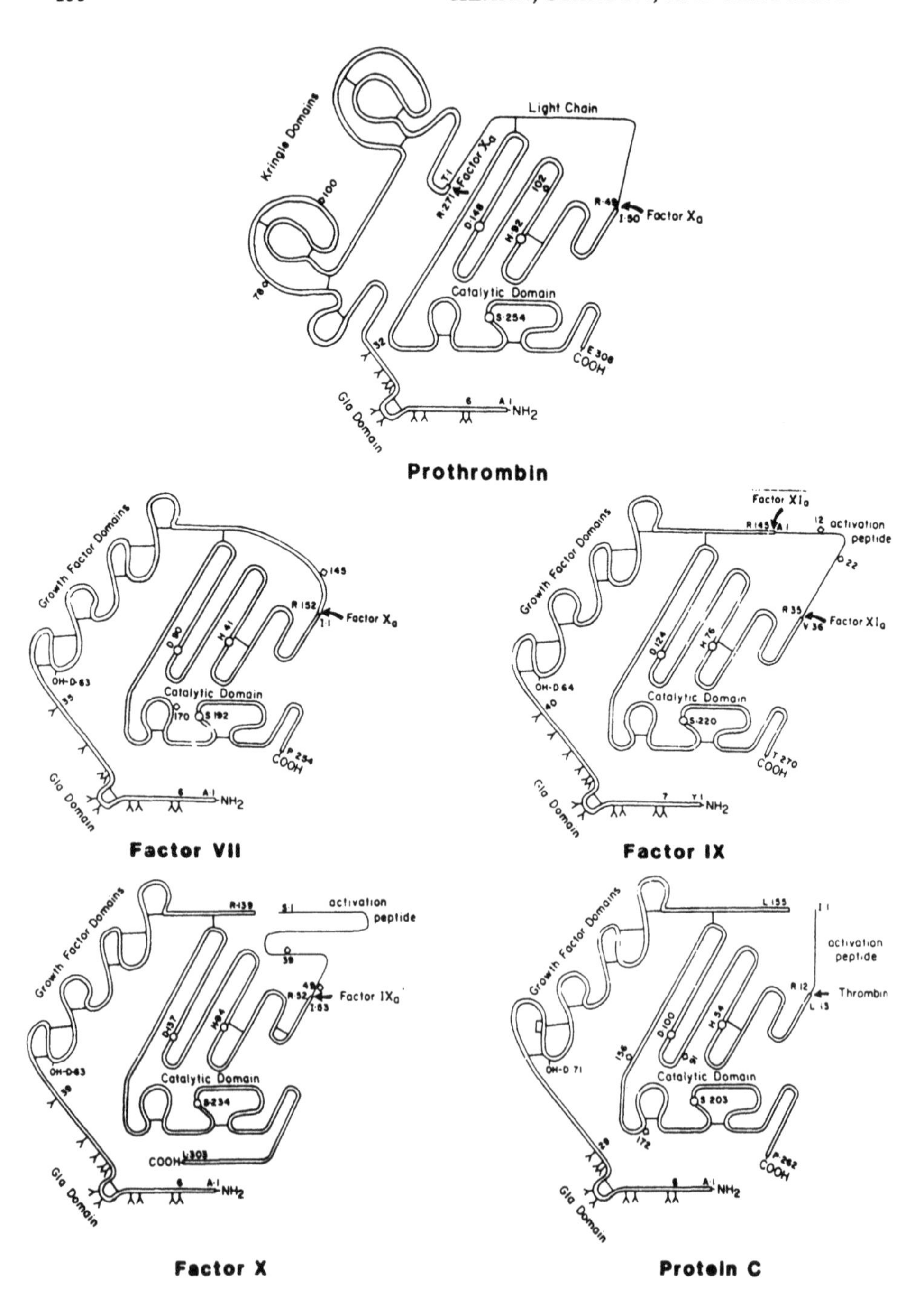

Figure 8 Schematic representation of the structural relationship between the vitamin K-dependent proteins prothrombin, Factor VII, Factor IX, Factor X, and protein C.

It is important that the formation of a clot be specifically limited to the site of injury. A mechanism directly coupled to the clotting process which ensures this has recently been identified. The pivotal protein in the process is protein C. It is activated by thrombin in association with a endothelial cell-bound cofactor called thrombomodulin in a calcium-dependent process. Activated protein C then induces fibrinolysis. In addition, the binding of thrombin to thrombomodulin inhibits its procoagulant functions. Protein C has also been shown to inactivate activated Factor V in association with protein S cofactor. Thus this mechanism ensures that the clot does not spread over intact endothelium.

The blood clotting cascade dramatically illustrates the role that proteins serve as sensors, transducers, and effectors of biological information. These capabilities derive from the ability of proteins to assume altered conformational states consistent with biological activity, allowing interaction with other proteins and affecting a structural change with further functional consequences.

ENDOCRINE PROTEIN HORMONES: MOLECULAR EFFECTORS FROM DISTANT HORIZONS

Classes of Endocrine Effectors

The hormones of the endocrine system can be broadly grouped into two categories based on their structure and mode of action. These categories are the lipid-soluble hormones (steroids, vitamin D and derivatives, and thyroid hormones) and the water-soluble hormones (proteins, peptides, peptide derivatives, and catecholamines). The lipid-soluble hormones are able to cross the barrier presented by the cell plasma membrane without the requirement of cell-surface proteins such as hormone receptors, and once inside the cell initiate their action by binding with intracellular carrier proteins (Fig. 9). Conversely, peptide and protein hormones generally exert their action by binding to specific receptors on the outside of the cell membrane (Fig. 9), and hormone action is initiated through the generation of a "second messenger" on the inner surface of the plasma membrane or internalization of the hormone-receptor complex. Common second messenger enzyme systems recognised to date include the adenylate cyclase complex, which generates cyclic adenosylmonophosphate (cAMP) when activated, the guanyl cyclase complex which acts in a similar fashion to produce cGMP, calcium-dependent protein kinases, which increase the membrane permeability to Ca^{+2}, and the phospholipase C complex, which generates inositol triphosphate, and is coupled to protein kinase C. The second messengers generated by these enzyme complexes are able to diffuse readily throughout the cell, and then activate enzyme cascades or other biochemical events specific for the particular hormone. At present, the detailed molecular interaction of these second messenger systems remains poorly understood, and represents an area of active research.

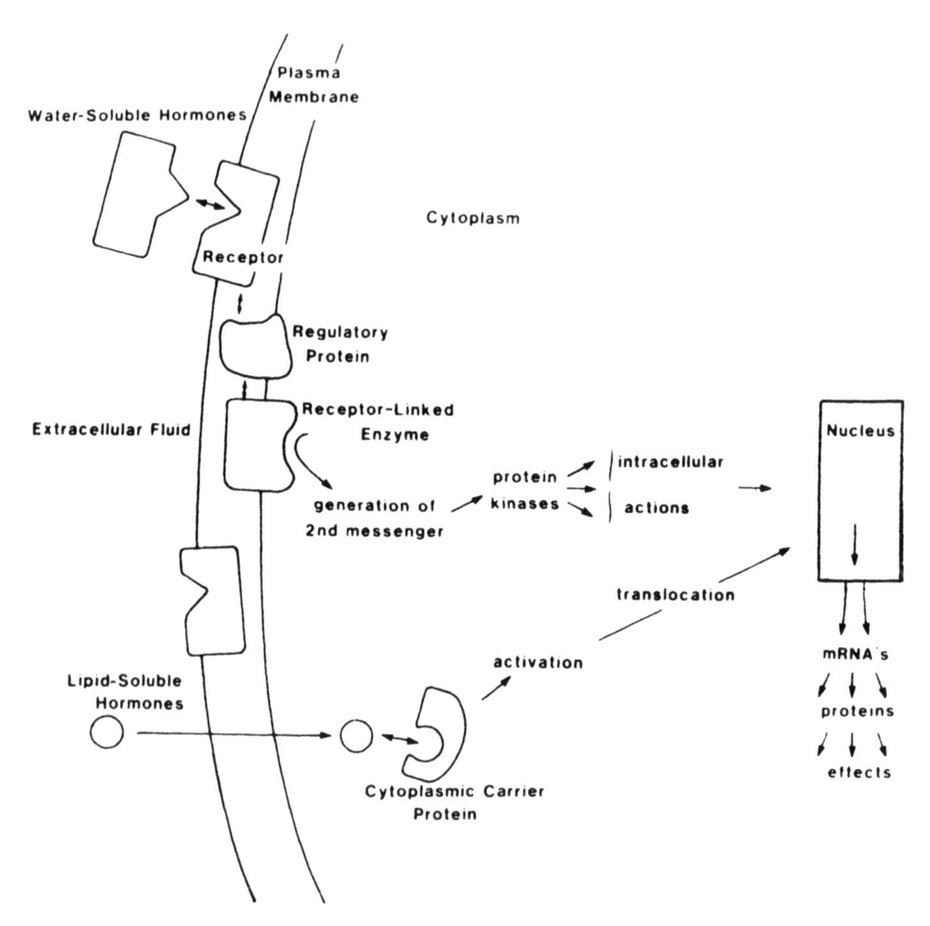

Figure 9 General model for the modes of action of both water-soluble and lipid-soluble hormones: The water-soluble hormones (peptides, proteins, catecholamines) interact reversibly with receptors located on the exterior surface of the target cell membrane. The hormone-receptor complex interacts with membrane-bound enzyme systems to generate the intracellular second messenger, which initiates multiple intracellular events. The lipid-soluble hormones (steroids, vitamin D, and derivatives) cross the cell membrane and bind to cytoplasmic carrier proteins. Binding is followed by an activation process and translocation to the nucleus where the activated carrier proteins bind to chromatin, and stimulate mRNA synthesis. The thyroid hormones operate in a similar manner, although it appears that an intracellular protein carrier is not used.

The action of a hormone on a particular target cell generally produces numerous biochemical responses. For example, the action of thyroid-stimulating hormone on the thyroid cell induces iodide uptake, the biosynthesis of thyroglobulin and release of thyroid hormones, increased biosynthesis of macromolecules, and changes to glucose metabolism and phospholipid synthesis. In a similar fashion, single hormones can have diverse effects on different target tissues as well. A good example of this is the polypeptide glucagon, which acts (a) on the liver to enhance the metabolism of glycogen, (b) on fat cells to stimulate lipolysis, and (c) on the β cells of the pancreas to increase insulin secretion. As a general rule therefore most hormones may affect a wide range of tissues, derived from similar cellular embryology, with the biological response of a particular cell to any hormone determined to a great extent by the differentiated nature of that cell.

The variety of known hormones is immense, as indicated in the summary provided in Table 3. To provide an insight into the structure and function relationships of a particular group of hormones, the following section will deal exclusively with the glycoprotein hormones, and the interested reader is referred to other texts for consideration of the modes of action of other hormone groups. The glycoprotein hormones, as model examples of protein effectors, exhibit substantial structural similarity yet also significant microheterogeneity. They elicit significantly different biological effects and yet are regulated by subtly divergent pathways. At one end of the spectrum, they represent in structural terms the most complex family of protein hormones yet their behavior and properties is symptomatic of all larger polypeptides. They are glycosylated with subunit structure. They have largely resisted all attempts to define their function in terms of precise molecular regions of the primary amino acid sequence due to their convoluted surface topography.

Glycoprotein Hormones: Multimeric Proteins with Model Biological Behavior

As a group, the glycoprotein hormones, chorionic gonadotropin (CG), luteinizing hormone (LH), follicle-stimulating hormone (FSH), and thyroid-stimulating hormone (TSH), have been extensively studied. Structurally, these hormones form a superfamily, the members of which possess quite diverse biological activities (controlling gonadal and thyroidal function, see Table 4) yet exhibit remarkably similar protein structures. To gain insight into this complex family of proteins, it will be useful to initially examine them in terms of their structural features which contribute to specific biological activities and which differentiate each hormonal activity.

Primary Structure

In terms of their primary structure all four hormones are composed of two dissimilar protein chains, or subunits, termed the alpha and the beta, respectively. Each chain consists of approximately 100 amino acids, and the molec-

Table 3 Summary of Polypeptide Hormones

Hormone	Hormone class	Cell source	Principal target tissue	Molecular weight
Chorionic gonadotropin	Glycoprotein	Placenta	Ovary	37,500
Follicle-stimulating hormone	Glycoprotein	Anterior pituitary	Ovary, testis	28,000
Luteinizing hormone	Glycoprotein	Anterior pituitary	Ovary, testis	28,000
Thyroid-stimulating hormone	Glycoprotein	Anterior pituitary	Thyroid	28,000
Growth hormone	Protein	Anterior pituitary	Liver	22,000; 20,000
Placental lactogen	Protein	Placenta	Liver	22,000
Prolactin	Protein	Anterior pituitary	Mammary gland	22,000
Adrenocorticotrophic hormone	Polypeptide	Anterior pituitary	Adrenal cortex, adipose tissue	4,500
β-Endorphin	Polypeptide	Anterior pituitary	Most tissues	4,000
Insulin	Protein	Pancreas	All tissues	5,700
Glucagon	Peptide/protein	Pancreas	Liver, adipose tissue	3,550
Parathyroid hormone	Protein	Parathyroid	Bones, kidney	9,000
Thryoid hormone (and deiodinated derivatives)	Amino acid derived	Thyroid	Most tissues	777
Oxytocin	Peptide	Posterior pituitary	Uterus, mammary gland	1,007
Vasopressin	Peptide	Posterior pituitary	Kidney, arteries	1,084

Angiotensin II	Peptide	Liver	Arterioles, adrenal cortex	1,045
Adrenalin	Amino acid derived	Adrenal medulla	Most cells	183
Nonadrenalin	Amino acid derived	Adrenal medulla	Most cells	198
Testosterone	Steroid	Testes	Most cells	288
Estrogen	Steroid	Ovary	Most cells	272
Progesterone	Steroid	Corpus luteum	Uterus, mammary gland	314
Aldosterone	Steroid	Adrenal cortex	Kidney	360
Corticosterone, cortisol	Steroid	Adrenal cortex	Most cells	346, 362
Prostaglandins	Prostaglandin	Various	Smooth muscle	~ 360
Calcitonin	Polypeptide	Thyroid	Kidney, bones	3,700
Vitamin D	Steroid	Skin	Alimentary tract	385
Inhibin	Protein	Ovary	Gonads, pituitary	58,000; 31,000
Erythropoietin	Glycoprotein	Kidney	Bone marrow	39,000
Serotonin	Amino acid derived	Nerve	Arterioles, CNS	176
Growth hormone-releasing hormone	Polypeptide	Hypothalamus	Pituitary	~ 5,000
Thyrotropin-releasing hormone	Peptide	Hypothalamus	Pituitary	362
Luteinizing hormone Releasing hormone	Peptide	Hypothalamus	Pituitary	1,182

ular weight of the α, β dimer is ~ 28,000 daltons, making these glycoprotein hormones some of the most complex of the circulating protein hormones (Table 3).

The α subunit in all four pituitary glycoprotein hormones from any particular mammalian species is common, having virtually identical amino acid sequence (Fig. 10). Furthermore, the primary structure of the α subunit across various mammalian species remains highly conserved, with only limited amino acid substitutions (Fig. 10).

However, the β subunit has a different primary structure for each hormone (Fig. 10), and is responsible for the hormonal specificity of the intact α, β dimer at the receptor level. This specificity has been confirmed in experiments where hormonal hybrids were formed from, for example, the LH-α subunit with a FSH-β subunit, and the recombined hybrid shown to have activity only in FSH bioassays. Conversely, hybridization of the LH-β subunit with an FSH-α subunit generates a new molecule with only LH activity in various bioassays. Interestingly, the individual isolated β subunits themselves do not possess any intrinsic biological activity, implying that while the β subunit is required in the hormonal complex to confer specificity, the actual regions (or binding determinants) of the native hormone which are recognized by the hormone receptor on the target cell could well reside on both subunits. Two intriguing, and as yet unresolved, hypotheses have been advanced to accommodate the role of the β subunits, namely does the β subunit function as a conformational lock in order for the α subunit to express its function or alternatively does the low affinity binding observed in receptor interaction reflect an induction of preferred conformational fit mediated by β subunit binding to lipids?

Secondary Structure

Intrachain covalent bonds (i.e., bonds between the same protein chain) formed by disulfide bridges between cysteine residues play a major role in the stabilization of the subunit structure of these hormones. The α subunit contains 5 intrachain disulfide bridges, and the β subunits from all hormones contain 6 disulfide links. If the disulfide bonds are even partially cleaved (by reduction), biological activity is lost as the conformation of the subunits is altered. Thus, specific conformational populations of the intact hormone are essential for hormonal activity. In order to elucidate these populations more fully most structural studies have been addressed toward determining the solution structure of the glycoprotein hormones. While this appears to be a simple objective, the practical determination of a finite, and well-characterized quaternary structure has met with many difficulties, and it is not surprising that a complete structural model of the glycoprotein hormones has yet to be achieved. Principal among these difficulties has been the location of all the disulfide

										10										20			
Bovine, ovine	Phe	Pro	Asp	Gly	Glu	Phe	Thr	Met	Gln	Gly	Cys	Pro	Gln	Cys	Lys	Leu	Lys	Glu	Asn	Lys	Tyr	Phe	Ser
Human					Ala	Pro	Asp	Val	-	Asp	-	-	Glu	-	Thr	-	Gln	-	Asp	Pro	Phe	-	-

							30										40						
Bovine, ovine	Lys	Pro	Asp	Ala	Pro	Ile	Tyr	Gln	Cys	Met	Gly	Cys	Cys	Phe	Ser	Arg	Ala	Tyr	Pro	Thr	Pro	Ala	Arg
Human	Gln	-	Gly	-	-	-	Leu	-	-	-	-	-	-	-	-	-	-	-	-	-	-	Leu	-

				50						*				60									
Bovine, ovine	Ser	Lys	Lys	Thr	Met	Leu	Val	Pro	Lys	Asn	Ile	Thr	Ser	Glx	Ala	Tyr	Cys	Cys	Val	Ala	Lys	Ala	Phe
Human	-	-	-	-	-	-	-	Gln	-	-*	Val	-	-	-	Ser	-	-	-	-	-	-	Ser	Tyr

	70										80		*								90		
Bovine,ovine	Thr	Lys	Ala	Thr	Val	Met	Gly	Asn	Val	Arg	Val	Glx	Asn	His	Thr	Glu	Cys	His	Cys	Ser	Thr	Cys	Tyr
Human	Asn	Arg	Val	-	-	-	-	Gly	Phe	Lys	-	-	-*	-	-	Ala	-	-	-	-	-	-	-

Bovine, ovine	Tyr	His	Lys	Ser	COOH
Human	-	-	-	-	

Figure 10 a Amino acid sequences of the alpha subunit of the glycoprotein hormones. The sequence of bovine (b) and ovine (O) LH is used as the reference; dashes indicate residues identical to this sequence in other species. The attachment points of carbohydrate are indicated by CHO.

										10										20					
hTSHβ								Phe-	-	Ile	-	Thr	Glx	Tyr	Met	Thr	His	Val	-	Arg	Arg	Glx	-	Ala	
bTSHβ								Phe	-	Ile	-	Thr	Glu	Tyr	Met	Met	His	Val	-	Arg	Lys	Glu	-	Ala	
pTSHβ								-	-	Ile	-	Thr	Glu	Tyr	Met	Met	His	Val	-	Arg	Lys	Glu	-	Ala	
b,oLHβ	Ser-	Arg-	Gly-	Pro-	Leu-	Arg-	Pro-	Leu-	Cys-	Gln-	Pro-	Ile-	Asn-	Ala-	Thr-	Leu-	Ala-	Ala	Glu	Lys-	Glu-	Ala	Cys-	Pro	
pLHβ	-	-	-	-	-	-	-	-	-	Arg	-	-	-	-	- -	-	-	-	-	Asp	-	-	-	-	
hLHβ	-	-	Glu	-	-	-	-	Trp	-	His	-	-	-	-	Ile	-	-	Val	-	-	-	Gly	-	-	
hCGβ	-	Lys	Glu	-	-	-	-	Arg	-	Arg	-	-	-	-	-	-	-	Val	-	-	-	Gly	-	-	
hFSHβ							Asn	Ser	-	Glu	Leu	Thr	-	Ile	-	Ile	-	Ile	-	-	-	Glu	-	Arg	
pFSHβ									-	Glu	Leu	Thr	-	Ile	-	Ile	Thr	Val	-	Val	Lys		-	Leu	Thr

						30										40								
hTSHβ	Tyr	-	Leu	-	Ile	Asn	-	Thr	-	-	-	-	-	-	Met	Thr	Arg	Asx	Ile	Asx	Gly	Lys	Leu	Phe
bTSHβ	Tyr	-	Leu	-	Ile	Asn	-	Thr	Val	-	-	-	-	-	Met	Thr	Arg	Asx	Val	Asx	Gly	Lys	Leu	Phe
pTSHβ	Tyr	-	Leu	-	Ile	Asn	Ser	Thr	-	-	-	-	-	-	Met	Thr	Arg	Asp	Phe	Asp	Gly	Lys	Leu	Phe
b,oLHβ	Val	Cys	Ile	Thr	Phe	Thr	Thr	Ser	Ile	Cys	Ala	Gly	Tyr	Cys	Pro	Ser	Met	Lys	Arg	Val	Leu	Pro	Val	Ile
pLHβ	-	-	-	-	-	-	-	-	-	-	-	-	-	-	-	-	-	Arg	-	-	-	-	Ala	Ala
hLHβ	-	-	-	-	Val	Asn	-	Thr	-	-	-	-	-	-	-	Thr	-	Arg	-	-	-	Gln	Ala	Val
hCGβ	-	-	-	-	Val	Asn	-	Thr	-	-	-	-	-	-	-	Thr	-	Thr	-	-	-	Gln	Gly	Val
hFSHβ	Phe	-	Leu	-	Ile	Asn	-	Thr	Trp	-	-	-	-	-	Tyr	Thr	Arg	Asp	Leu	-	Tyr	Lys	Asn	Pro
pFSHβ	Phe	-	-	Ser	Ile	Asn	-	Thr	Trp	-	-	-	-	-	Thr	Thr	Gly	Arg	Asx	Leu	Val	Tyr	Lys	Asx

		50										60										70		
hTSHβ	-	-	Lys	Tyr	Ala	Leu	Ser	-	Asx	-	-	-	-	Arg	Asp	Phe	Ile	Tyr	Arg	Thr	-	Glx	Ile	-
bTSHβ	-	-	Lys	Tyr	Ala	Leu	Ser	-	Asp	-	-	-	-	Arg	Asp	Phe	Met	Tyr	Lys	Thr	Ala	Glu	Ile	-
pTSHβ	-	-	Lys	Tyr	Ala	Leu	Ser	-	Asp	-	-	-	-	Arg	Asp	Phe	Met	Tyr	Lys	Thr	-	Glu	Ile	-
b,oLHβ	Leu	Pro	Pro		Met	Pro		Gln	Arg	Val	Cys	Thr	Tyr	His	Glu	Leu	Arg	Phe	Ala	Ser	Val	Arg	Leu	Pro
pLHβ	-	-	-		Val	-		-	Pro	-	-	-	-	Arg	-	-	Ile	-	-	-	Ser	-	-	-
hLHβ	-	-	-		Leu	-		-	-	-	-	-	-	Arg	Asp	Val	-	-	Glu	-	Ile	-	-	-
hCGβ	-	-	Ala		Leu	-		-	Val	-	-	Asn	-	Arg	Asp	Val	-	-	Glu	-	Ile	-	-	-
hFSHβ			Ala	Arg	Pro	Lys	Ile	-	Lys	Thr	-	-	Phe	Lys	-	-	Val	Tyr	Glu	Thr	-	-	Val	-
pFSHβ		-	Ala	Arg	Pro	Asx	Ile	Glx	Lys	Thr	-	-	-	Arg	Glx	-	Val	Tyr	Glx	Thr	-	Lys	Val	-

								80										90						
hTSHβ	-	-	-	Leu	His	-	(Ala	-	Tyr)	Phe	-	Tyr	-	-	-	-	-	-	Lys	-	-	Lys	-	Asx
bTSHβ	-	-	-	Arg	His	-	Thr	-	Tyr	Phe	-	Tyr	-	-	-	Ile	-	-	Lys	-	-	Lys	-	Asx
pTSHβ	-	-	-	His	His	-	Thr	-	Tyr	Phe	-	Tyr	-	-	-	Ile	-	-	Lys	-	-	Lys	-	Asp
b,oLHβ	Gly	Cys	Pro	Pro	Gly	Val	Asp	Pro	Met	Val	Ser	Phe	Pro	Val	Ala	Leu	Ser	Cys	His	Cys	Gly	Pro	Cys	Arg
pLHβ	-	-	-	-	-	-	-	-	Thr	-	-	-	-	-	-	-	-	-	-	-	-	-	-	-
hLHβ	-	-	-	Arg	-	-	-	-	Val	-	-	-	-	-	-	-	-	-	Arg	-	-	-	-	-
hCGβ	-	-	-	Arg	-	-	Asn	-	Val	-	-	Tyr	Ala	-	-	-	-	-	Gln	-	Ala	Leu	-	-
hFSHβ	-	-	Ala	His	His	Ala	-	Ser	Leu	Tyr	Thr	Tyr	-	-	-	Thr	Gln	-	-	-	-	Lys	-	Asp
pFHSβ	-	-	Ala	His	His	Ala	Asx	Ser	Leu	Tyr	Thr	Tyr	-	-	-	Thr	Glu	-	-	-	-	Lys	-	Asx

Figure 10 b (continued)

				100										110										120
hTSH$_\beta$	Thr	Asx	Tyr	Ser	-	-	Ile	His	(Glu	Ala	Ile)	Lys	Thr	Asx	Tyr	-	Thr	Lys	-	Glx	Lys	Ser	Tyr	COOH
bTSH$_\beta$	Thr	Asx	Tyr	Ser	-	-	Ile	His	Glu	Ala	Ile	Lys	Thr	Asn	Tyr	-	Thr	Lys	-	Gln	Lys	Ser	Tyr	Met-COOH
pTSH$_\beta$	Thr	Asp	Tyr	Ser	-	-	Ile	His	Glu	Ala	Ile	Lys	Thr	Asn	Tyr	-	Thr	Lys	-	Glu	Lys	Ser	Tyr	COOH
b,oLH$_\beta$	Leu	Ser	Ser	Thr	Asp	Cys	Gly	Pro	Gly	Arg	Thr	Glx	Pro	Leu	Ala	Cys	Asx	His	Pro	Pro	Leu	Pro	Asp	Ile
pLH$_\beta$	-	-	-	Ser	-	-	-	-	-	-	Ala	-	-	-	-	-	-	Arg	-	-	-	-	Gly	Leu
hLH$_\beta$	Arg	-	Thr	Ser	-	-	-	Gly	Pro	Lys	Asp	His	-	-	Thr	-	Asp	-	-	Gln	$CONH_2$			
hCG$_\beta$	Arg	-	Thr	-	-	-	-	Gly	Pro	Lys	Asp	His	-	-	Thr	-	Asp	Asp	-	Arg	Phe	Gln	-	Ser
hFSH$_\beta$	Ser	Asp	-	-	-	-	Thr	Val	Arg	Gly	Leu	Gly	-	Ser	Tyr	-	Ser	Phe	Gly	Glu	Met	Lys	Gln	Tyr
pFSH$_\beta$	Ser	Asx	-	-	Asx	-	Thr	Val	Arg	Gly	Leu	Gly	-	Ser	Tyr	-	Ser	Phe	Gly	Glu	COOH			

									130										140					
b,o,pLH$_\beta$	Leu	COOH																						
hCG$_\beta$	Ser	Ser	Ser	Lys	Ala	Pro	Pro	Pro	Ser*	Leu	Pro	Ser	Pro	Ser*	Arg	Leu	Pro	Gly	Pro	Ser*	Asp	Thr	Pro	Ile
hFSH$_\beta$	Pro	Thr	Ala	Leu	Ser	Tyr	COOH																	
hCG$_\beta$	Leu	Pro	Gln	$CONH_2$																				

Figure 10 b Amino acid sequences of the beta subunit of the glycoprotein hormones.

Table 4 Principal Functions of the Glycoprotein Hormones

Glycoprotein hormone	Source	Biological action: Male	Biological action: Female
LH	Anterior pituitary	Stimulates testicular androgen biosynthesis	Stimulates ovarian estrogen and progesterone production, oocyte maturation and ovulation
FSH	Anterior pituitary	Control of testicular spermatogenesis	Control of ovarian follicle maturation
CG	Placenta	—	Control of progesterone secretion of corpus luteum
TSH	Anterior pituitary	Controls thyroidal iodine uptake and production of thyroid hormones	

bridges which form the intrachain bonds, as many of the cysteine residues are located very close to each other in the primary sequence, and uncontrollable disulfide interchange has been shown to occur during isolation.

Despite the high degree of intrachain disulfide bond formation in the glycoprotein hormones, the alpha and beta subunits are held together by noncovalent bonds (hydrogen bonding, Van Der Waals forces, hydrophobic interactions). This means that the native hormones are stable under physiological conditions of pH and temperature, but dissociate under acidic conditions to form the biologically inactive subunits. Measurements made using spectroscopic methods such as circular dichroism and fluorescence detection, and also by probing the surface structure with specific antibodies, have clearly demonstrated that the tertiary structure of the subunits changes during dissociation of the intact hormone to the α and β forms. This structural change also contributes to the lack of biological activity found for the individual subunits.

Tertiary Structure

From a conceptual viewpoint, one would expect the four different glycoprotein hormones to exhibit a similar overall tertiary structure, based on the ability of recombined hybrid subunits (e.g., FSH-α and LH-β) to express biological activity. This concept is supported by epitope (or the antigenic-binding segment) mapping studies with monoclonal antibodies (MAbs) raised against determinants present on the α subunit of the α-β complex of hCG. For example, these epitopes were also present in the same topographical relationship on hFSH, hTSH, and hLH, suggesting that the α subunit (at least) maintains a similar overall conformation irrespective of the type of β subunit to which it is complexed. The presence of a number of areas on all the β subunits which exhibit considerable sequence homology (e.g., regions cysteine^{34}-cysteine^{38}, leucine^{49}-proline^{50}, valine^{58}-tyrosine^{61}-leucine^{71}-proline^{75}, serine^{83}-cysteine^{90}, cysteine^{92}-cysteine^{95}, see Fig. 10), also implies a series of common domains, or contact points, on the four β sequences and strengthens the argument for a common basic conformation.

Glycoprotein Hormone Function

Various models have been proposed to account for the mechanism by which the glycoprotein hormones bind to their specific cell surface receptors and elucidate hormone action. Experimental observations have now clearly shown that the hormone-receptor binding process is a biphasic phenomenon involving two hormone-receptor binding affinity constants. As the hormones are composed of two subunits, a popular model (Fig. 11) suggests that the hormone-specific β subunit initially binds to the receptor, activating a secondary binding process involving the binding of the α-subunit, the push-pull hypothesis. This model also provides an explanation for the function of the common α-subunit, which may be intercalated within the cell membrane to stimulate the adenylate cyclase enzyme complex. However, there is as yet no evidence to support the direct interaction of the alpha subunit with either the regulatory or the catalytic subunits of the adenylate-cyclase complex, although the glycoprotein hormones do stimulate the formation of cAMP in appropriate bioassays.

A number of other possible roles have been suggested for the α subunit. As has already been discussed, the α, β interaction is essential to maintain the correct conformation of the native hormone for proper receptor binding. Hence the alpha subunit may function primarily to enable the β subunit to assume the proper conformation for interaction of the β subunit with the receptor. Alternatively, and more probably, receptor-binding domains may be located on both subunits.

In considering these aspects of the structure and function of the glycoprotein hormones, a number of intriguing questions must be asked, namely (a) which functional groups on the α-β complex are important in expressing hormone function at the receptor level?, (b) what is the role of the carbohydrate moieties in hormone binding?, and (c) why are there such a high number of disulfide bridges in the hormones?

Wherever possible in studies on protein structure–function relationships the trend has emerged to obtain part of the information required to answer these questions by x-ray crystallographic methods. However, this approach presumes the ability to grow suitable crystals of the proteins being studied. Due to the structural microheterogeneity this has not been possible to date with the glycoprotein hormones. A number of other methods are available to

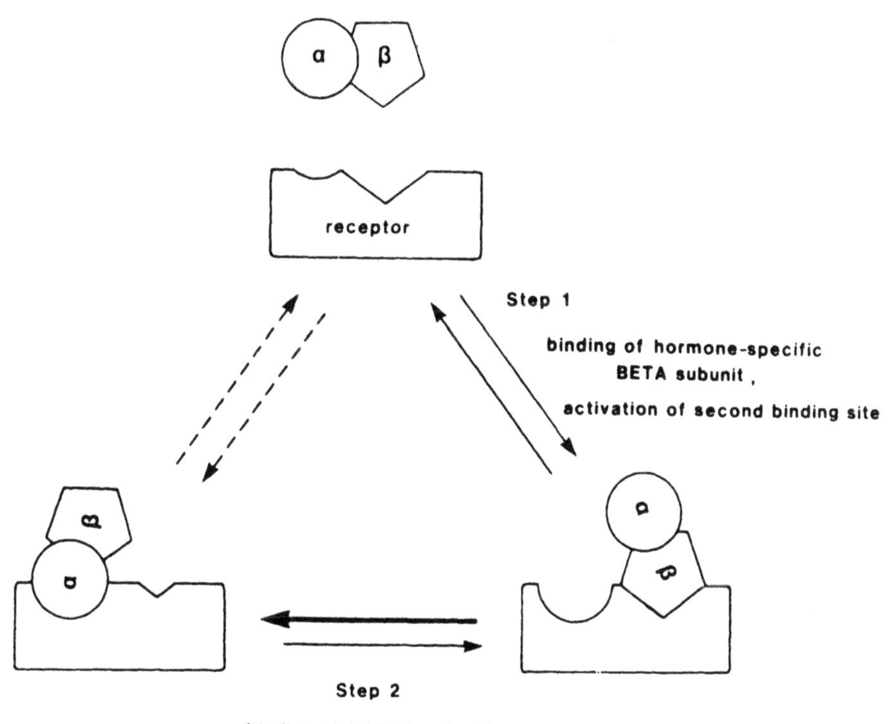

Figure 11 Schematic model for the role of glycoprotein hormones subunits in hormone action. This model proposes a two step mechanism for hormone binding, with the first step being the specific binding of the beta subunit to the hormone receptor. Binding of the beta subunit induces a conformational change in the receptor and activates a second binding site for the (common) alpha subunit, which stabilizes the hormone-receptor complex and initiates postreceptor mechanisms such as second-messenger activation.

probe protein tertiary structure, including spectroscopy, chemical and/or enzymatic modification or removal of selected groups or regions, and epitope mapping through the use of monoclonal antibodies. None of these methods alone provide a complete and self-consistent structural picture *per se*, but when considered together yield considerable insight into the tertiary structure and function of the glycoprotein hormones. For example, the use of hCG-specific monoclonal antibodies (MAbs) has, in recent studies, allowed the characterization of the spatial relationships of epitopes on the α and β subunits, both pre- and postreceptor binding. Current data suggest that both subunits contain receptor-binding domains which can be blocked by preincubation with MAbs. Similarly, MAbs which bind to the β subunit of hCG and hLH but do not prevent receptor binding have been characterized, indicating that at least part of the β subunit is exposed following receptor binding. However, it appears that antigenic determinants on the α subunit are masked when the intact hormone is receptor bound as indicated in Figure 12. These findings correlate well with the hormone-receptor binding model discussed earlier (Fig. 11). Furthermore, independent evidence has shown that for hormone preparations in which the subunits have been covalently cross-linked together the modified hormone retains 50–100% of both binding and biological activities, and hence the activation of postreceptor events must proceed via a process which does not require the physical dissociation of the hormone within the membrane.

Role of the β Subunit

As described earlier, the hormone-specific β subunits of the four glycoprotein hormones confer hormonal specificity, as well as displaying a considerable degree of sequence homology. An obvious question is therefore, which regions, or functional groups are responsible within the subunit sequence to elicit, for example, FSH and LH behavior at the receptor level?

To date a definitive answer to this question remains elusive despite the considerable effort which has been expended in this area. Thus, studies have shown that various areas of the β sequence are essential as receptor-contact regions in that their selective removal or chemical modification alters or destroys receptor-binding activity. In particular, an octapeptide loop close to the carboxyl terminus of all the β subunits has been proposed as a determinant in the expression of hormonal specificity (Fig. 13). This loop is stabilized by a disulfide bond, and examination of sequence data reveals that a number of nonconservative substitutions of amino acids in the loop allows differing net charges to be introduced. Hence the proposal suggests that a neutral or net positive change results in LH-like activity, and a negative change is required for FSH/TSH activity, while TSH specificity is dictated by a crucial aromatic amino acid substitution in the centre of the loop (Fig. 13). This model has proved technically difficult to assess, but with the availability of newer biochemical methods such as site-directed mutagenesis, this interesting hypothesis can now be further evaluated. As is evident from the above discussion, indirect evidence is no substitute for direct, molecular definition when evaluating the effector roles of proteins!

Role of Carbohydrate in Hormone Function

An interesting facet of the function of the glycoprotein hormones is the role of the carbohydrate moieties, which differ between the four hormones. There are

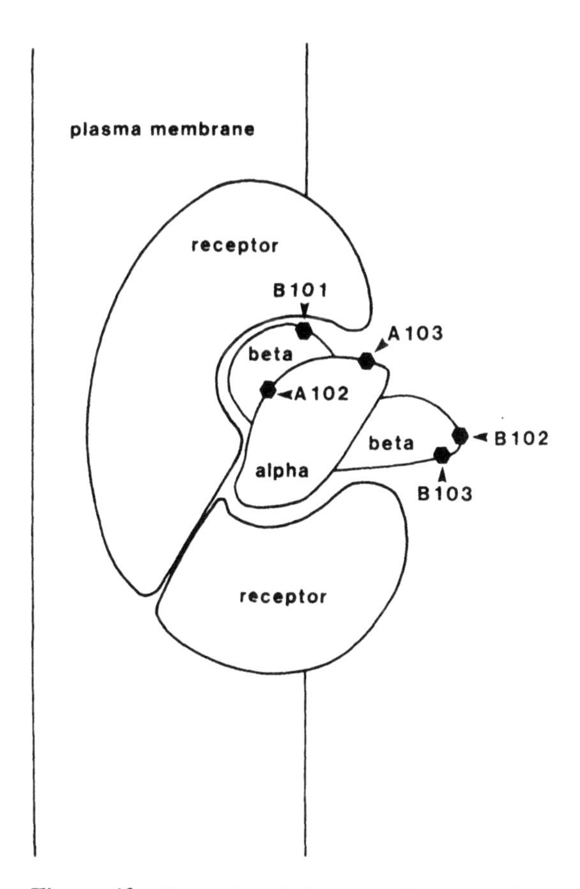

Figure 12 Illustration of the antigenic sites on hCG relative to the receptor binding site. The antigenic surface structure of hCG was probed with monoclonal antibodies specific for binding determinants located on the alpha and beta subunits. This probing was carried out both pre- and postreceptor binding. Monoclonal antibodies B102 and B103 bind to similar sites on the β subunit, and these sites are still exposed after formation of the hormone–receptor complex. MAbs A102 and B101 block the binding of hCG to receptor, and are not exposed in the hormone–receptor complex, indicating that they are located in the receptor-binding region of hCG. MAb A103 does not prevent formation of the hCG–receptor complex. However the binding site corresponding to MAb A103 is not exposed to the solvent after complex formation, and is thus located in a region where receptor, hormone, and solvent are near each other. [Adapted from Moyle, W. R., Ehrlich, P. H., and Canfield, R. E. *Proc. Natl. Acad. Sci. (USA)* 79, 2245–2249] (1982).

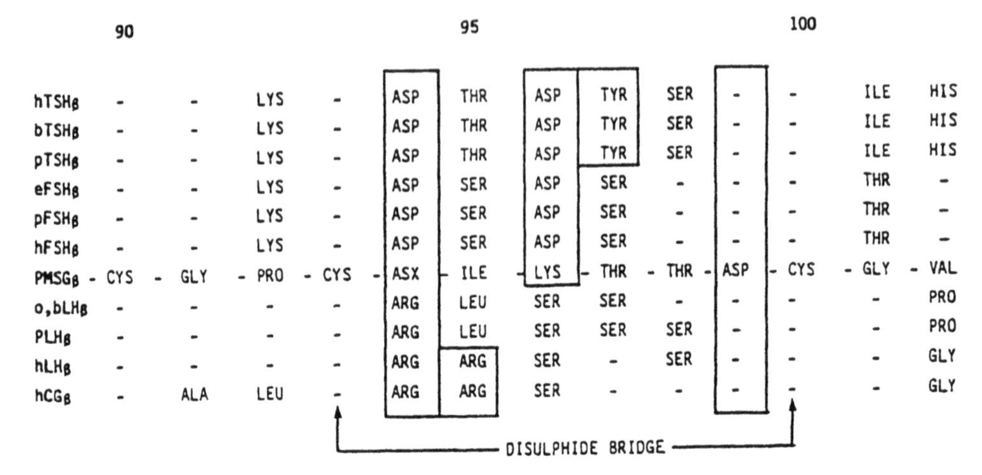

Figure 13 Octapeptide loop model:Comparison of the amino acid sequences of the β subunits of the glycoprotein hormones in the region of the determinant loop (amino acids 93–100). The sequence of pregnant mare serum gonadotrophin β subunit is used as the reference sequence, and dashes indicate residues identical to this sequence (h:human, p:porcine, o:ovine, b:bovine, e:equine).

three or four carbohydrate attachment sites for the pituitary-derived hormones, with two sites on the alpha subunit and either one or two on the β subunit. On the β subunit carboxy terminal extension of chorionic gonadotropin there are a further four attachment sites. Glycoprotein hormones of human origin contain variable amounts of sialic acid, a negatively charged carbohydrate group, which interestingly is absent in the hormones isolated from bovine pituitaries. Conversely, recent studies have shown that the terminal oligosaccharides of bovine hormones are sulfated, which is absent from hCG and hFSH, but present in hLH and hTSH. In addition, purified preparations of the glycoprotein hormones show considerable microheterogeneity, in that numerous isoforms of each hormone may be purified by ion-exchange chromatography, gel electrophoresis, and isoelectric focusing. This microheterogeneity arises principally as a result of changes in the carbohydrate contents of each hormone, although variants with minor alterations in amino acid sequence due to protease activity have been described. Why do these hormones exhibit microheterogeneity and what physiological function role(s) may be ascribed to the carbohydrate moieties? Successive removal of sialic acid residues from hCG results in a drastically reduced circulatory half life of the hormone and increases its biological activity, suggesting that sialic acid is present both in a 'protective' role and also as a modulator of hormonal biopotency. Similar trends have been observed with the other gonadotropins. The level at which the sialic acid groups on the hormone affect its biological potency is at present unclear, as experimental desialylation of gonadotropins does not significantly alter their receptor binding characteristics, implying control at a postreceptor level.

Evidence is now accumulating to demonstrate that the assembly of the carbohydrate moieties on the hormones at the posttranslational stage can be independently controlled by both metabolic (e.g., by glucose starvation) and endocrine factors (e.g., LHRH, TRF, steroids). For example, removal of gonadal steroids from circulation by castration can result in an increase in the sialylation of FSH, increasing its circulatory half life and thereby changing its biological effectiveness in vivo. Similarly the distribution of basic isoforms of FSH is greater in young women than in men or elderly women. Hence the pituitary gland is capable of responding to changes in external stimuli by altering the type of glycoprotein hormone to be released. Clearly, the modulation of hormonal biopotency by selective alteration of sialic acid groups is not a universal mechanism, as the bovine hormones do not contain significant amounts of this negatively charged carbohydrate. It is tempting to speculate that the sulfation process may be a mechanism controlling the clearance rates and potencies of the bovine hormones in a similar manner as does sialic acid, but further evidence is needed to corroborate this. However, the recent demonstration that human hormones contain both sialic acid and sulfated hexosamines (hLH, hTSH) complicates this issue, and it is apparent that our understanding of the role of the carbohydrate moieties in glycoprotein hormone action requires considerable refinement if the full picture of the effector roles of these complex proteins is to be achieved.

GROWTH FACTORS: THE MEPHISTOPHELES OF THE PROTEIN EFFECTOR WORLD

Biological Role of Growth Factors

Growth factors are polypeptides that stimulate proliferation and pleiotypic responses (nutrient uptake, increase in cellular metabolism, ion fluxes, cytoskeletal changes) within cells and maintain their differentiated character. These polypeptide mitogens differ from endocrine hormones in that they are not secreted by specialized glands into the circulation and transported to target organs, but are potentially secreted by all cells into their surrounding extracellular space or microenvironment and reach their target cells by diffusion. Growth factors may act on the secreting cells or adjacent cells. This mode of action is referred to as autocrine and paracrine, respectively (Fig. 14). The growth factors have a pivotal role in regulating tissue growth during normal development in the fetus, neonate, and adult. In addition they play a

Table 5

Growth factors	Protein size	Cell source	Target cells
EGF	6kD	Submaxillary gland, kidney (mouse); submandibular glands, intestinal mucosa (humans)	Mesenchymal and epithelial cells (fibroblasts, kidney, intestinal epithelial, glial, granulosa, endothelia)
TGF-α	5.6kD	Transformed cells (murine sarcoma virus), placenta, embryos	As above
TGF-β	25kD (homodimer)	Platelets, liver, brain, kidney, lung, heart, submandibular gland, placenta, embryos, transformed cells (feline sarcoma virus)	Hepatocytes, fibroblasts, epithelial cells, keratinocytes
IGF-I	7kD	Liver, kidney, smooth muscle, mesenchymal tissue (fibroblasts, chondrocytes, osteoblasts, adipose cells)	Mesenchymal and epithelial cells (chondrocytes, adipose cells, muscle, fibroblasts), granulosa cells
IGF-II	7kD	Liver, kidney, colon, brain, placenta	Mesenchymal and epithelial cells (chondrocytes, fibroblasts, adipose cells)
PDGF	Human:30–32 kD Heterodimer, glycoprotein A Chain 14–18 kD B Chain 16 kD Porcine: homodimer (B-B) osteosarcoma cells: homodimer (A-A) Simian sarcoma transformed cells: p28^{sis}-homologous to B	Platelets, monocytes, macrophages, endothelial cells Smooth muscle cells, osteosarcoma cells, transformed cells (simian sarcoma virus)	Mesenchymal cells (fibroblasts, smooth muscle, glia, chondrocytes, placental cytotrophoblasts)

Table 5 *(Continued)*

Growth factors	Protein size	Cell source	Target cells
HBGF	Basic pI 18 kD Acidic pI 14 kD-exhibit considerable. N-terminal heterogeneity	Pituitary, brain, ovary, prostate adrenal, muscle, kidney, testis, cartilage, plasma, liver, lung, rat chondrosarcoma, hepatoma, endothelial cells, tumor cell lines.	Endothelial cells, fibroblasts, chondrocytes, granuloma cells
NGF	26 kD (homodimer)	Submaxillary, gland, (mouse), peripheral nerves	Sympathetic and sensory neurons
IL-1	Human:15 kD (3 charged forms) Mouse:16, 17, 18 kD	Monocytes, macrophages	Thymocytes, T lymphocytes, B lymophocytes, neutrophils, chondrocytes, hepatocytes, muscle cells, epithelial cells
IL-2	15 kD (glycoprotein)	T-helper cells	Cytotoxic T lymphocytes.
IL-3 (Multi-CSF)	23–28 kD (glycoprotein)	T lymphocytes	Eosinophils, mast cells, T lymphocytes, granulocyte and macrophage progenitors
CSF-I (M-CSF)	70 kD (homodimer) glycoprotein-60% carbohydrate	Mouse-L cells	Macrophage progenitors
CSF-2	23 kD glycoprotein-50% carbohydrate	Endotoxin-induced mouse lung, T lymphocytes endothelial cells	Macrophage and granulocyte progenitors

(Abbreviations:) EGF: Epidermal growth factor, TGF-α: transforming growth factor-α, TGF-β: Transforming growth factor β, IGF-I: insulin-like growth factor I, IGF-II: insulin-like growth factor II, PDGF: platelet-derived growth factor, HBGF: heparin binding growth factor, NGF: nerve growth factor, IL-1: interleukin-I, IL-2: interleukin-2, IL-3: interleukin-3, CSF-I: colony-stimulating factor-1, CSF-2: colony-stimulating factor-2.

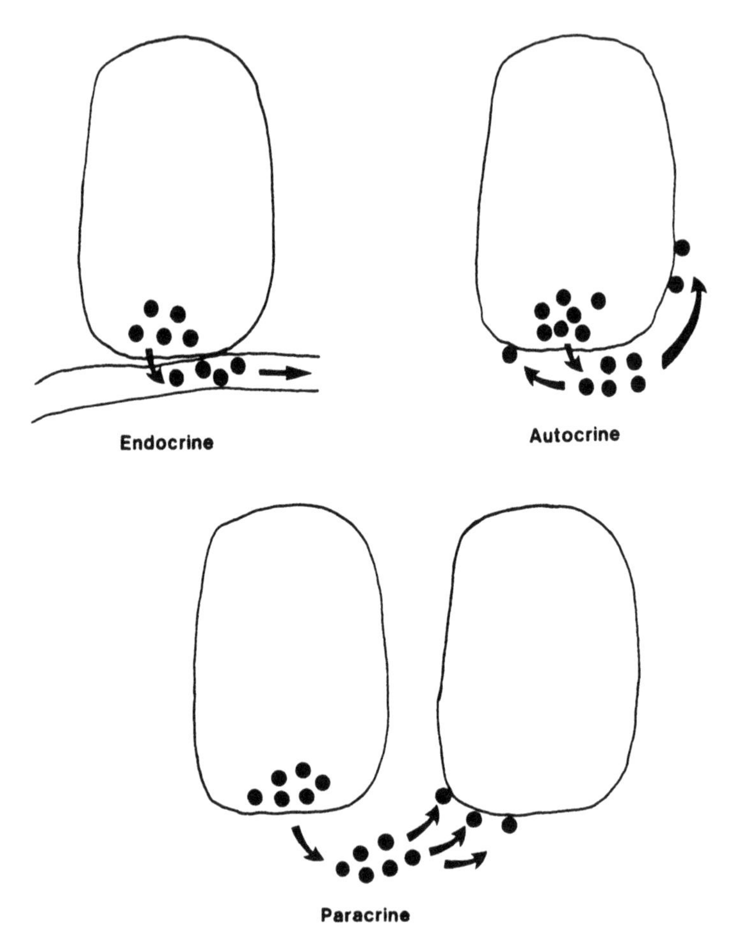

Figure 14 Illustration of autocrine, paracrine and endocrine modes of action by polypeptide hormones. Hormones secreted by specialized endocrine tissues (e.g., pituitary) are carried to target cells by the blood stream. In autocrine and paracrine systems the polypeptides (e.g., growth factors) are secreted into the cellular microenvironment and act on the secreting cell or diffuse to a nearby target cell. The mode of regulation of autocrine and paracrine factors is not well understood, but it appears that cell contact is an important regulator. Disruption as occurs in injury may trigger secretion and initiate a cascade, as growth factors stimulate the secretion of other factors.

key role in wound healing and are a factor in the etiology of tumors. Table 5 summarizes the structure and cellular or tissue source of selected examples of growth factors and their respective target cells.

Growth factors control cell proliferation by regulating the events associated with DNA synthesis and cell division. They exhibit, in some cases, complementary functions, acting at different stages of the cell cycle. Platelet-derived growth factor and heparin-binding growth factors are termed competence factors as they prime the cell for DNA synthesis by stimulating cellular events necessary for the transition from the quiescent G_0 phase to the G_1 prereplicative phase of the cell cycle (Fig. 15). Insulin-like growth factor 1 is termed a progression factor as it is required in order for the cell to traverse the G_1 phase, enter the S phase and commence DNA synthesis. This concept of competence and progression factors has arisen from numerous experiments largely with mouse fibroblast cells. Whether this concept is totally applicable to other

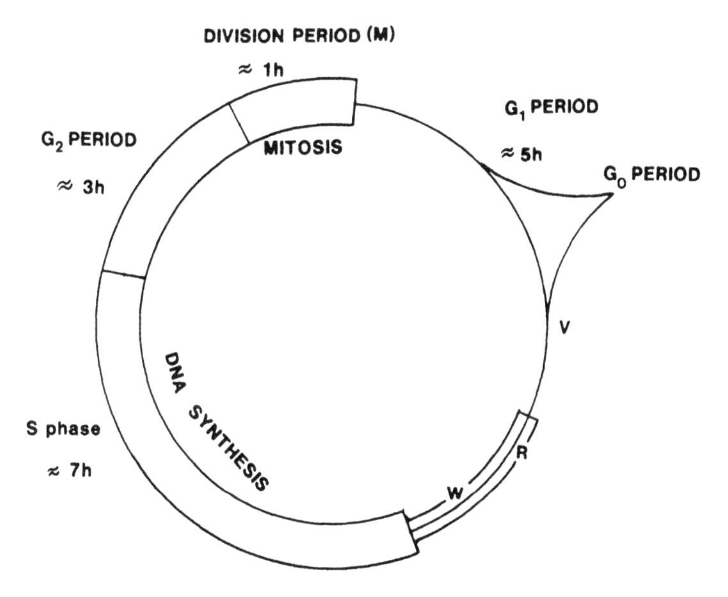

Figure 15 Schematic representation of the events of the cell cycle. The cell proliferation cycle consists of a number of phases regulated by growth factors. A quiesent cell in G_0 enters the cycle at G_1 when stimulated by a competence factor (e.g., platelet-derived growth factor). Cells progress to the V point under the action of other growth factors (e.g., epidermal growth factor, insulin-like growth factor I). Progression beyond this point is strictly dependent on growth factors (e.g., insulin-like growth factor I). The cycle reaches the restriction or R point where cells are committed to enter the S phase. A further regulatory point W is encountered immediately before the S phase where cells can be arrested when deprived of growth factors. DNA synthesis occurs during the S phase and after a short gap phase (G_2) mitosis and cell division occurs (M). The scheme illustrated pertains mainly to fibroblasts in culture. The time course and nature of regulatory growth factors for other cells may vary. Generally, however, the S, G_2, and mitotic periods are relatively constant in the cells of the same organism. The G phase is most variable, with arrested cells remaining in the G_0 state of the G phase for days, months, or years.

cell types and whether all growth factors can be divided into these two functional groups remains to be determined. It is an important question, as knowledge of the cooperativity of action of growth factors on cellular proliferation and function for particular target cells would aid in the formulation of serum-free defined synthetic culture media, which would greatly enhance the reproducibility of experiments utilizing cultured cells. The concept also suggests that cellular transformation may require the activation of complementary cellular processes arising either from increased growth factor secretion, or mechanisms involving membrane and intracellular signal transduction. Thus knowledge of the molecular controls of the cell cycle is central to understanding the mechanisms of cellular transformation and hence the etiology of tumors.

Mode of Action

The action of growth factors on cellular function is mediated through membrane receptors. The receptors for epidermal growth factor, platelet-derived growth factor, insulin-like growth factor I, colony-stimulating factor I, have internal domains with phosphotyrosine kinase activity which are activated by the binding of the growth factor to the extracellular ligand binding domain (Fig. 16). The phosphorylation of tyrosyl residues of specific substrates resulting in their activation is one mechanism by which the extracellular signal is transduced and the intracellular response effected. At present the molecular nature of the physiological substrates is not known.

The tyrosine kinase domain of growth factor receptors has close homology to proteins coded for by cellular proto-oncogenes (there are currently 15 known proto-oncogenes that code for protein kinases) and viral oncogenes. It is evident, therefore, that inappropriate phosphorylation of intracellular proteins that mediate growth factor action on the cell cycle could lead to cellular transformation and unregulated proliferation. The identification of these proteins would therefore be of great interest. Experiments have been performed where the phosphorylation pattern has been examined in virus transformed cells, however the results have not been conclusive. Phosphorylated proteins found in Rous sarcoma virus-transfected cells in which the V-*src* oncogene is expressed include three structural proteins (vinculin, p81,p36), three glycolytic enzymes (enolase, lactate dehydrogenase, phosphoglycerate mutase), and two proteins of unknown function (p50 and p42). However, the possible importance of these proteins for the transduction of the mitogenic signal is undermined by the observation that in cells infected by the McDonough and UR2 sarcoma virus, respectively, viruses which express protein kinases coded for by the v-*fms* and v-*ros* genes do not exhibit phosphorylation of these substrates. Furthermore, with the exception of vinculin, the proteins phosphorylated by virally coded enzymes are not phosphorylated in normal cells. It is

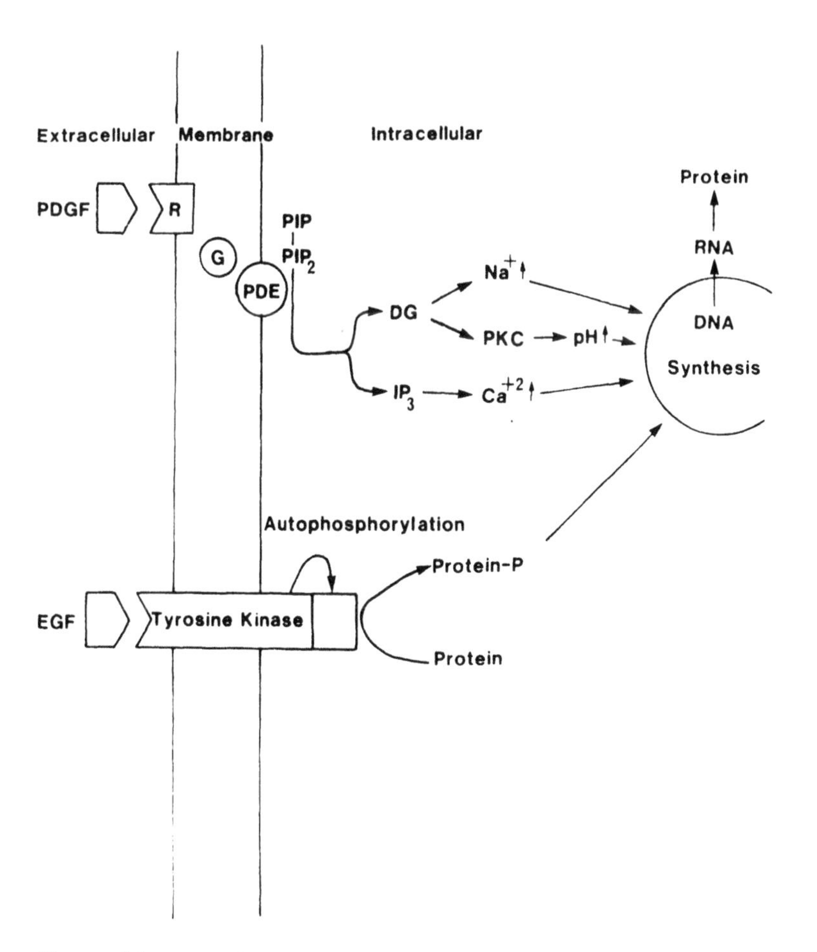

Figure 16 Intracellular events associated with growth factor action. The growth factor or effector molecule interacts with specific receptors on the cell membrane. The epidermal growth factor (EGF) receptor has tyrosine kinase activity which is activated by autophosphorylation. It in turn activates intracellular substrates by phosphorylation which mediate the intracellular response. Platelet-derived growth factor (PDGF) acts via an alternative mechanism. Following binding to a receptor the extracellular signal is transferred via the G protein, which requires guanosine triphosphate (GTP) to function, to a phosphodiesterase enzyme (PDE) which amplifies the signal. The enzyme converts phosphatidyl inositol 4,5-biphosphate (PIP_2) to the second messenger diacylglycerol (DG) and inositol triphosphate (IP3). DG increases intracellular sodium and pH which is part of an as yet unknown mechanism contributing to DNA synthesis and other cellular responses. IP_3 mobilizes calcium ions which is another intracellular messenger.

likely that differences exist in the specificity of tyrosine protein kinases coded by various viral oncogenes and proto-oncogenes resulting in some cases in the indiscriminant phosphorylation of proteins. It is an important caveat to the temptation of extrapolating mechanisms of cellular transformation in eukaryotic cells from viral transfection studies. Most probably, low abundance, as yet uncharacterized, proteins that have a regulatory role in key cellular events are the substrates for growth factor-activated tyrosine kinases.

The intracellular action of growth factors may also be mediated through an alternative pathway (Fig. 16). Ligand binding to the receptor results in activation of phosphoinositadase C and the cleavage of diacylglycerol inositolphosphate to two intracellular messengers, diacylglycerol and inositol triphosphate. Diacylglycerol activates protein kinase C which phosphorylates a number of as yet poorly characterized substrates at their serine and threonine residues and in addition activates the $Na^{+}-H^{+}$ exchange system leading to an increase in cellular pH. These mechanisms have an important but as yet poorly understood role in regulating intracellular function. It has been found, however, that ion fluxes influence the expression of the fos gene and high pH promotes DNA synthesis. Inositol triphosphate serves to increase cytosolic calcium concentrations by mobilizing stores in the endoplasmic reticulum. It is an important cofactor in a number of metabolic processes and represents a further link in the chain of signal transduction within the cell.

The mechanism by which growth factor activation of the receptor-associated phosphotyrosine kinase activity results in activation of phosphoinositide turnover is believed to be mediated by the G protein, the α subunit of which is related to the p21 protein coded by the ras proto-oncogene. The G protein regulates adenylate cyclase which converts adenosine triphosphate to cyclic adenosine 3′, 5′ monophosphate also an intracellular mediator for some metabolic processes. Binding to guanosine triphosphate is required for activation of the G protein. This interaction is self-limiting as the G protein hydrolyses the nucleotide. The mechanism by which the G protein then regulates phospholinositase C is currently thought to occur indirectly via promotion of cyclic adenosine monophosphate formation, or may involve direct coupling via the α subunit or *ras* protein. The critical role of the G protein in the regulation of cellular function is suggested by the fact that in many transformed cells increased production of the *ras* protein or site-specific mutagenesis which inhibits its hydrolytic activity has been observed which can lead to an amplification of the extracellular signal and cellular transformation.

The above scheme for the transduction of signals initiated by growth factors into the cell is based largely on work performed with platelet-derived growth factor. Work is required to elucidate the mechanisms that apply to other factors. It appears that similar mechanisms are operative for insulin-like growth factor I and heparin-binding growth factor. Epidermal growth factor, however, does not appear to activate phospholipase C, and the increase in intracellular calcium induced by binding to the receptor unlike platelet-derived growth factor is of extracellular origin. The receptors for insulin-like growth factor II and transforming growth factor β do not have intracellular tyrosine kinase domains. The mechanism of transmembrane signal transduction with these two factors is not known.

Other mediators of growth factor action are proteins coded for by the c-*myc* and c-*fos* genes. Growth factors have been shown to increase transcription of these proto-oncogenes and produce nuclear acting proteins. Their exact function is not known, but they may regulate other genes necessary for cell division. The importance of these genes in the regulation of cellular function is suggested by the observed increase in expression of these genes in transformed neoplastic cells.

Proto-oncogene Growth Factors and the Unrepentent Cell

The interest in growth factors is based on the apparent relationship between growth factors, proto-oncogenes and tumor formation (Table 6). Thus some oncogenes express proteins that are homologous to a growth factor, growth factor receptor of intracellular mediators of growth factor action. Malignant cells are characterized by their ability to multiply in an autonomous manner. As growth factors promote cell proliferation it is thought that transformation could result from (1) increased production of autocrine acting growth factors; (2) increased response of the cell due to an increase in membrane growth factor receptors; (3) amplification of the intracellular signal due to changes in the postreceptor signal transduction pathway (Fig. 17). Examples of cellular transformation resulting by each of the above mechanisms are listed in Table 6.

Transformation of cells by the Simian sarcoma virus is thought to occur by the autocrine mechanism. The putative transforming protein p28-*sis* has been identified and shown to have significant homology to platelet-derived growth factor. Antiplatelet-derived growth factor antibodies block DNA synthesis in these cells and tumor size correlates with the amount of immunoreactive "pseudo"-platelet-derived growth factor secreted by the transformed cells. A number of natural tumors (glioblastomas, fibrosarcomas, osteosarcomas) have also been described where expression of c-*sis* and the production of platelet-derived growth factor like molecules occurs and therefore presumably autocrine stimulation of cellular proliferation. Wilm's tumor of the kidney has also been shown to be associated with greatly increased production of insulin-like growth factor II.

Increased growth factor receptor number is thought to contribute to the transformation of cells in squamous cell carcinoma, and brain tumors of glial origin, and possibly in some breast cancers and bladder carcinomas. The

Table 6 Proto-oncogenes Involved in the Cellular Growth Factor-Receptor-Response Pathway

Oncogene	Product
c-*erb* b	Epidermal growth factor receptor-like protein containing cytoplasmic phosphotyrosine kinase domain but not extracellular ligand-binding domain and thus may be constitutively activated.
c-*fms*	Codes for colony-stimulating factor I receptor.
c-*sis*	Platelet-derived growth factor-like protein secreted extracellularly which results in autocrine stimulation of the cell.
c-*ras*	Located on the inner face of the plasma membrane and is involved in the transduction of signals from certain receptors. Binding to guanine nucleotide is necessary for its activity. Intrinsic guanosine triphosphatase activity ensures transient nature of this activation. Mutation can impair this activity resulting in continuous activation.
c-*hst* and c-*int*-2	Gene sequences indicate that they code for proteins homologous with heparin binding growth factor and therefore could be involved in autocrine stimulation of the cell.
c-*myc* and c-*fos*	Code for DNA binding proteins localized in the nucleus. The expression of these genes correlates with proliferative activity and is induced by growth factors.

c-*erb*-B gene product is homologous to the epidermal growth factor receptor, but lacks the external ligand-binding domain. Therefore, it may function as a constitutively activated kinase resulting in consistent and unregulated stimulation of cells. Tumors associated with amplified expression c-*erb*-B have not yet been reported. The transformation of macrophage cells by histiocytosis sarcoma virus is thought to be due to the associated increase in multicolony-stimulating factor receptors mediating the action of interleukin-3.

Increased expression of proto-oncogenes, in particular c-*ras*, c-*myc*, and c-*fos* whose products normally mediate the intracellular events initiated by growth factor binding to the membrane receptor, has been observed in a wide variety of neoplastic cells. This increase occurs as a result of gene amplification due to genetic translocation, alternatively the proto-oncogene product may be expressed in an excessively activated form due to accumulated mutations in its structure.

The growth factor oncogene concept has in the last several years provided a useful model for explaining the biochemical basis of cellular transformation

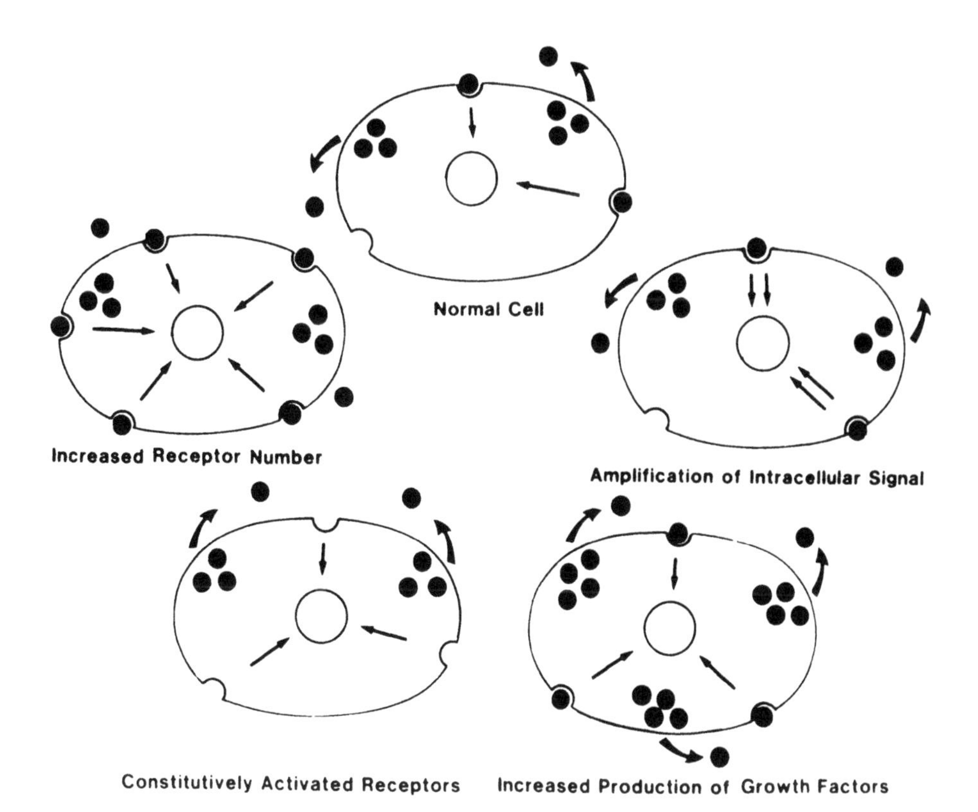

Figure 17 Mechanisms for increased or unregulated stimulation of cells by growth factors.

and proliferation. It is clear that many points of potential clinical intervention exist in particular for the development of growth factor inhibitors.

The chemotactic and proliferative action of growth factors on cells suggests a role for their use in promoting wound healing. Their use could result not only in accelerated healing, but also regulated healing, therefore diminishing the extent of scaring. This application requires that large amounts of growth factor be available. Therefore, there is the challenge of developing sources by recombinant genetic engineering and establishing large scale, high throughput purification systems.

The future application of one growth factor, heparin-binding growth factor, is particularly interesting. This factor has been shown to have possibly the greatest tissue distribution of all the factors. This is probably understandable in light of its ability to promote capillary formation or angiogenesis. Vascularization is required in all tissues. Excessive capillary infiltration is characteristic of a number of clinical conditions, such as diabetic retinopathy, arthritis, hemangioma, retrolental fibroplasia, and hemarthrosis. In addition, neovascularization is essential for tumor growth, so that nutrients can be supplied to the proliferating cells and waste products removed. Promotion of angiogenesis would, however, be desirable in scleroderma as decreased vascularization of tissues is thought to be the primary event which leads to the associated fibrosis. Enhanced vascularization of the infarcted heart, of wounds and burns would contribute to improved recovery from these conditions.

Thus the production of large amounts of heparin-binding growth factor or antagonists and the devising of appropriate treatment or administration regimens would be of great clinical and commercial significance.

ANTIPROTEINS: EXEMPLARY CANDIDATES FOR FEEDBACK CONTROL BY PROTEIN EFFECTORS

The production of small and large molecules by biochemical processes is frequently controlled by feedback pathways. For example, the activity of the first enzyme in a multistep series of enzymic reactions may be actively controlled (either in a negative or positive control manner) by the concentration of the final product of that pathway. Some examples of biological systems controlled by such feedback loops have already been alluded to in this chapter, such as the complement cascade and blood coagulation. The synthesis of many hormones is also under feedback regulation, and a classic example of this is the negative feedback by gonadally produced steroids on the secretion of follicle-stimulating and luteinizing hormones (FSH and LH) by the pituitary.

Two broad categories of feedback loops can be identified based on the mode of action of their regulating components: (1) allosterically modulated binding systems and (2) covalently modulated allosteric systems. Allosteric binding modulation of controlling enzymes refers to the reversible noncovalent interaction of effectors (or inhibitors) to binding sites other than the enzyme active site resulting in protein conformational changes leading either to an active structure with high affinity for the substrate or an inactive or inhibited structure with a low substrate affinity. Examples of noncovalent allosteric regulation include phosphofructokinase while covalently modulated allosteric regulation, a feature of many enzymatic cascades involved in metabolic pathways, include phosphorylation of glycogen phosphorylase, glycogen synthase, and pyruvate dehydrogenase or adenylation of glutamine synthase. Other examples of feedback control affected by the covalent modification of the controlling enzyme can involve cleavage of a peptide bond, and conversion between the inactive and active forms (e.g. zymogen activation). Examples of both categories of control or regulatory proteins are numerous. In general allosteric binding modulations tend to involve small molecules and ions, such as the end products of a biological pathway, whilst covalent modulation invariably involves enzymes themselves often, but not necessarily, with proteolytic or bond fusion activities. A further subset of protein modulators are the antiproteins, which function as specific inhibitors of the regulating proteins or activated zymogens. Examples of antiproteins which will be considered here include the plasma protease inhibitors which control the function of biologically important proteases such as thrombin, plasminogen activator and plasmin, and the gonadal superfamily of protein hormones represented by the inhibin family which selectively suppress or attenuate the pituitary secretion of FSH.

Antiproteins to Cascade Zymogens

As a group, the plasma protease inhibitors represent nearly 10% of the total protein in blood plasma, and control a number of critical events associated with connective tissue turnover, coagulation, fibrinolysis, complement activation, and inflammatory reactions. Specific proteinase inhibitors which have been extensively studied include α_2-antiplasmin (α_2-AP), plasminogen activator inhibitor (PAI), and antithrombin III(AT-III). PAI and α_2-AP are both essential for the control of fibrinolysis, or the proteolytic degradation of fibrin clots. Fibrinolysis is mediated by the enzyme plasmin, which is formed in the circulation from the inactive precursor plasminogen through the action of plasminogen activators (Fig. 18). The protease inhibitors PAI and α_2-AP inhibit the activities of plasminogen activator and plasmin, respectively, by the rapid formation of very stable 1:1 protease/inhibitor complexes. The inhibition is competitive, in that the inhibitors bind to their target enzymes through a substratelike region contained in the inhibitor's amino acid sequence. Mechanistic studies with α_2-AP have demonstrated that this process occurs in two steps: (a) through the formation of a reversible complex with plasmin by noncovalent bonding, particularly involving the lysine residues on the A chain of plasmin, and (b) irreversible complex formation accompanied by a complete loss of plasmin activity, and the cleavage of a peptide bond in α_2-AP (Fig. 18). The inhibitor–protease complex is so stable that it resists dissociation by protein denaturants such as urea or detergents, with the complex being stabilized by the formation of a proteinase inhibitor covalent bond. The protease inhibitor antithrombin III also plays a major role in the regulation of serine proteases in the formation of blood clots in the coagulation cascade (thrombolytic pathway) system. Antithrombin III (AT-III) is a glycoprotein of molecular weight $\sim$ 58,000, and directly inactivates thrombin with 1:1 stoichiometry. This effect is significantly enhanced in terms of kinetics of the inhibition by the presence of heparin although the 1:1 stoichiometry is unaffected. In addition to

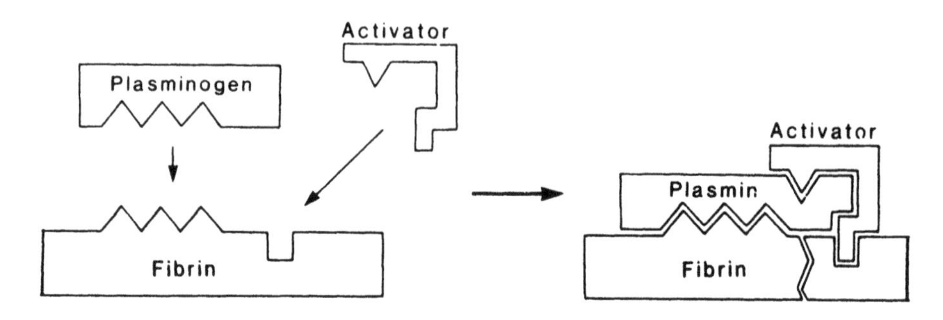

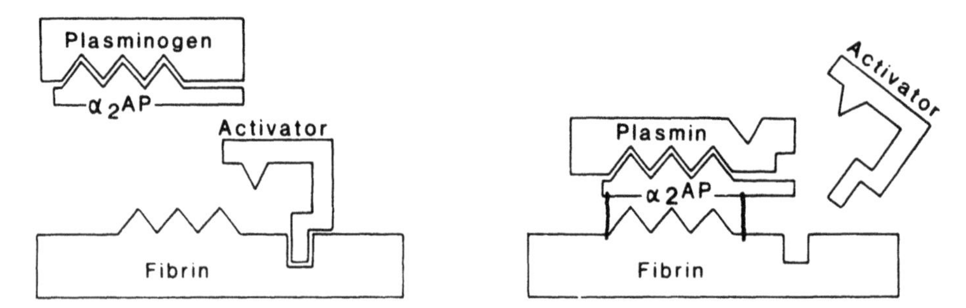

Figure 18 Mechanism of fibrinolysis and its inhibition by α_2-antiplasmin. Fibrinolysis requires the activation of plasminogen to plasmin by plasminogen activators, with subsequent lysis of the fibrin molecule. Inhibition of this process occurs through (a) binding of α_2-antiplasmin with plasminogen, decreasing the amount of plasminogen available to bind with fibrin, and (b) direct inhibition of fibrin-bound plasmin.

thrombin, AT-III inactivates a number of other serine proteases (e.g., plasmin, plasma kallikrein, trypsin), although these processes are unlikely to be of physiological importance as the association rate constants for these interactions are much slower than for thrombin. Heparin accelerates the rate of AT-III inactivation of thrombin by binding with or near an N-terminallylocated tryptophan residue on the protein, causing a conformational change (which buries the tryptophan) and activates the AT-III

Recent research has demonstrated that the protease inhibitors have other important roles in cellular metabolism. Several as yet structurally uncharacterized protease inhibitors have the ability to suppress cell transformation in vitro and carcinogenesis in vivo, presumably by inhibiting one or more intracellular proteases involved in the activation of proteins crucial for the production or expression of the transformed phenotype. Candidates for the substrate proteins include growth factors, many of which are posttranslationally cleared before becoming active. Hence increased levels of growth factor(s) may occur in carcinogenic tissues, leading to altered growth patterns. The direct link between protease inhibitors and cancer has also been demonstrated for α_2-macroglobulin inhibitor, which is absent in all neoplastic lesions, and inter-α_1-proteinase inhibitor (also called α_1-antitrypsin), where with individuals manifesting to congenital deficiency exhibit a greater susceptibility to the early onset of lung disease (familial emphysema). This finding implies that α_1-PI plays an important role in the defense mechanisms of the lung toward proteolytic attack.

Protein Inhibitors and Synergists to Hormonal Effectors

As discussed in the earlier section on the effector roles of the pituitary glycoprotein hormones, follicle-stimulating hormone (follitropin, FSH) is the key glycoprotein hormone controlling folliculogensis and spermatogenesis in the ovary and testis, respectively. Feedback control of the biosynthesis and secretion of FSH and LH by the anterior pituitary has long been known to be exerted by gonadally secreted steroids (estradiol and testosterone). However, more than 50 years ago, McCullagh postulated that the gonads contained a protein which was involved in the nonsteroidal regulation of FSH secretion. Knowledge of the structure and function of this antihormonal gonadal protein—aptly called inhibin—languished until quite recently. The considerable progress in the characterization of inhibin-related proteins over the past two years or so is due to the emergence of powerful new high-resolution protein purification strategies, improved microsequencing capabilities for amino acid sequence determination and more stringent design and synthesis of minimum degeneracy oligonucleotide probes for mRNA–cDNA hybridization studies. These studies have resulted in the characterization of the amino acid and cDNA sequences of bovine, ovine, porcine, and human inhibins reported from these laboratories at Monash University, and the porcine and human follicular fluid inhibin by the Genentech-Salk Institute collaborating group.

Structure/function studies have demonstrated that inhibin exists in several posttranslational forms (58 and 31 kD forms) based on disulfide linked α and β subunits, which are coded for by different genes at different genomic loci. The 31 kD forms, inhibins A and B, possess the common α subunit, whereas the sequences of the β chains differ near the N terminus, giving rise to β_A and β_B chains, respectively. A schematic representation of the structure and feedback control of gonadal inhibin on pituitary FSH secretion is shown in Figure 19. The FSH inhibitory nature of inhibin requires the disulfide bonding of the α and β subunits to remain intact. However, in recent developments, two inhibin-related proteins which are potent and selective stimulators of pituitary FSH secretion have been described; one, follitropin-releasing protein (FRP), has been shown to consist of disulfide-linked dimers of the β_A chains, while the other, activin, is a heterodimer of β_A and β_B cross-linked chains (Fig. 19). Both act as stimulators of FSH release from the pituitary gland, and hence their activities are distinct from luteinizing hormone-releasing hormone (LHRH), which stimulates the pituitary to secrete both FSH and LH. In addition, the time scale of action of FRP and activin is 50–100-fold slower than LHRH and both are unaffected by LHRH antagonists. It is clear therefore that these stimulators act via a controlling pathway different from LHRH. While the presence of FSH stimulators (FRP and activin) and FSH inhibitors (inhibin) in the gonads suggests a complex control mechanism, experiments with both inhibin and the β dimers present in pituitary cell cultures have shown that

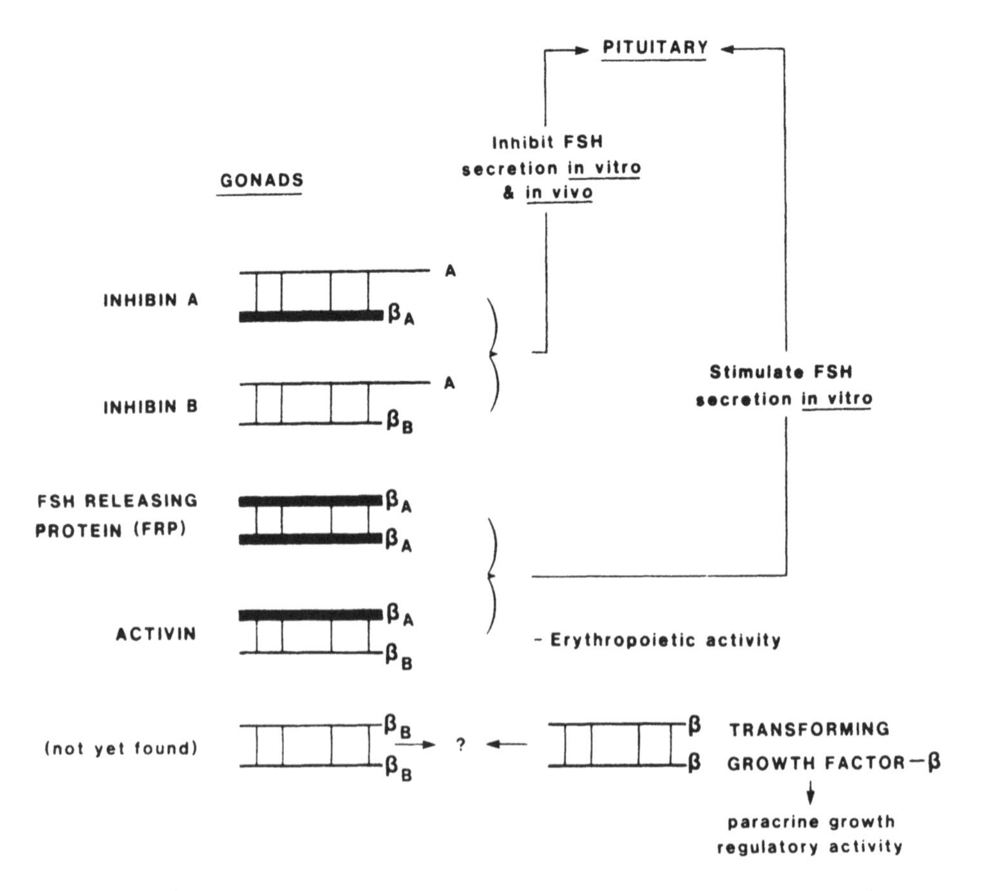

Figure 19 Schematic illustration of the subunit structure of inhibin and inhibin-related proteins.

the inhibitory effect of inhibin predominates. In fact, the role of activin as an erythroid differentiating factor in erythropoiesis may prove to be its dominant function in vivo.

It is now clear that inhibin, activin, FRP, and the structurally related mullerian inhibitory factor, the *Drosophila* decapentaplegic gene complex factor, T-cell-suppressing factor, *Xenopus* differentiating factor and β-transforming growth factor (TGF-β) all belong to a superfamily of hormonal regulators with similar structural characteristics. Their synthesis, through the reorganization of individual subunits either at the genomic or posttranslational levels, represents a potent physiological mechanism to extend the diversity of available biologically active proteins from the gene pool.

SUMMARY AND CONCLUSIONS

This chapter presented an overview of the role of proteins as biological effectors. From a simplistic point of view, based solely on comparison of the structural diversity of immunoglobulins, blood coagulation proteins, gonadotropins, growth factors, and antiproteins, it could be concluded that the functional mechanisms of these protein families bear little or no relationship. Despite this enormous divergency in structure–function relationship, there are in fact elements of commonality in their effector roles arising as a direct consequence of the ability of these classes of protein effectors to act as exquisite examples of the processes of biorecognition. All these case histories, and the numerous other familial case studies of protein effectors which could have been employed to illustrate the different functional roles of proteins, owe their biological properties to their primordial protein antecedents which have traversed the harsh wilderness of evolution in biorecognition phenomena and survived to elicit specific effector roles. Dictated by underlying physicochemical constraints, deceived at times by the lulling tones of the siren entropy, and constantly vulnerable to the vagaries of other more pervasive forms of biological networking and information transfer encoded in the genes of virus and invading microorganisms, protein biorecognition in higher life forms, and particularly in mammals, represents the finely tuned molecular avenues for the genome to transfer its information to the next generation. The examples summarized in this chapter illustrate the complex, and in disease states imperfect, functional potential of proteins to be manifested in the jigsaw of biorecognition and be realized in the network of nature's biological effectors.

Proteins thus represent a diverse range of effector molecules whose properties are totally dependent on their conformational or topographic status. The three-dimensional structure defines active sites on the molecule through which intermolecular interaction and biorecognition phenomena can occur. The cipher for this surface topography is, of course, coded in the primary amino acid sequence.

Much experimental work is being directed in these and other laboratories at elucidating the principles governing the folding of unique peptide sequences into three-dimensional structures. Further advances in the theoretical understanding of the thermodynamics of protein folding as observed by x-ray crystallography, nuclear magnetic resonance and other spectroscopic techniques will greatly aid this quest. In addition, more comprehensive computer-aided algorithms for structure simulation, improved models of protein conformational behavior, and greater insight into the molecular forces which control sequence nucleation will also be required.

The ultimate application of this knowledge is the specific production of proteins with the required biological properties which could encompass biological activity, specificity, biological half-life, and tolerance to extreme reaction or (biological) environmental.

As the structure of a protein is directly related to the encoding gene, the techniques of molecular biology such as in vitro mutagenesis, gene cloning and gene expression will provide abundant sources of engineered protein analogues which will initially allow structure–function studies to be undertaken more fully, and subsequently provide a source of new protein structures, appropriately engineered, for clinical and industrial applications. This revolution in protein chemistry will thus drive biotechnology well into the twenty-first century just as the revolution in molecular biology in the 1960s and early 1970s provided the basis for the spectacular advances in genetic engineering. Since proteins, as effectors, represent the commercially and intellectually dominant end-point of these developments, can there be any doubt left in the public perception that the next decades will represent a great renaissance in protein chemistry filled with exciting challenges, research satisfaction, and exuberant vitality?

ACKNOWLEDGMENTS

The support of the National Health and Medical Research Council of Australia, the Australian Research Grants Commission, the Buckland Foundation, the Potter Foundation, and Monash University Research Foundation in enabling different aspects of research summarized in this chapter is gratefully acknowledged.

SUGGESTED READING

Antiproteins

Kennedy, A. R. In *Mechanisms of Tumor Promotion*, vol. 3. CRC Press, Boca Raton, FL, 1984, pp. 13–55.

Robertson, D. M., Hearn, M. T. W., Forage, R. G., and DeKretser, D. M. In *Molecular and Cellular Endocrinology of the Testis*. Edited by M. Stefanini. Elsevier Science Publ. B. V. Amsterdam, 1986, pp. 267–271.

Troll, W., Wiesner, R., and Frenkel, K., *Adv. Cancer Res.* 49, 265–283, (1987).

Immunoglobulin Structure and Function

Clark, W. R. *The Experimental Foundations of Modern Immunology*, 2nd ed. John Wiley and Sons, New York, 1983.

Roitt, I., Brostoff, J., and Male, D. *Immunology*, Gower Medical Publishing, London/New York (1985).

Zymogen Activation Systems

Jackson, C. M. and Nemerson, Y. Blood coagulation. *Ann. Rev. Biochem.* 49, 765–811 (1980).

Neurath, H. and Walsh, K. A. Role of proteolytic enzymes in biological regulation (a review). *Proc. Natl. Acad. Sci. (USA)* 73, 3825–3832 (1976).

Reid, K. B. M. and Porter, R. R. The proteolytic activation systems of complement. *Ann. Rev. Biochem.* 50, 433–64 (1981).

Endocrine Protein Hormones

Ascoli, M. (Ed.) *Luteinizing Hormone Action and Receptors*. CRC Press, Inc., Boca Raton, FL, 1985.

Pierce, J. G. and Parson, T. F. *Ann. Rev. Biochem.* 50, 465–495 (1981).

Wilson, J. D. and Foster, D. W. *Williams Textbook of Endocrinology*, 7th Ed. W. B. Saunders Co., Philadelphia, 1985.

Growth Factors

Bradshaw, R. A. and Prentis, S. *Oncogenes and Growth Factors*. Elsevier Science Publishers, Amsterdam, 1987.

Goustin, A. S., Leof, E. B., Shipley, G. D., and Moses, H. L. Growth factors and cancer. *Cancer Res.* 46, 1015–1029 (1986).

6

Isolation of Natural Proteins

Stanley Stein

Center for Advanced Biotechnology and Medicine
Piscataway, New Jersey

INTRODUCTION

Proteins and peptides control most of the processes of a living organism. This chapter describes the concepts and procedures used in the isolation of a protein which is involved in a specific biological mechanism. The isolated protein may be used as a tool, for example, to study that biological response. Eventually, the particular protein must be prepared in a homogeneous form so that its chemical structure can be elucidated. Rather than preparing the purified protein from natural sources, it is becoming increasingly more common to produce the protein by alternate means. If the polypeptide is sufficiently small (ca. < 40 residues), it may be chemically synthesized. Otherwise, it may be prepared using recombinant DNA technology. Other chapters in this book describe these alternate methods for producing peptides and proteins. This chapter deals with the isolation of proteins and peptides, mainly for the purpose of structural analysis.

There have been major advances in the methodology for analyzing proteins. A sample of a homogeneous protein or peptide containing as little as 100 picomoles is often sufficient to obtain some sequence data. This is fortunate, because many of the proteins of biomedical interest are present in trace quantities in natural sources. Only a limited amount of protein sequence data is typically necessary, since the complete primary structure of the protein can more readily be obtained by sequencing the gene for that protein. Indeed, there are many examples, such as the one of gamma-interferon given below, in which the natural protein is isolated and analyzed to confirm the structure already predicted by the gene sequence.

Many approaches are possible in an isolation project and choices must be made. These depend on factors, such as the facilities and techniques available, time constraints for completion of the project, previous experience of the investigators, funding, etc. The concepts involved and the actual procedures employed in an isolation project will be illustrated with an interesting group of proteins, the interferons. Additional information will be provided through another interesting group called the opioid peptides. The author of this chapter has participated in research programs related to the isolation of opioid peptides and interferons from natural sources, which is one of the reasons for using these illustrations. Although these specific examples cannot provide a thorough understanding of the entire subject, it is hoped that they can offer the reader a broad perspective on the topic.

THE COMPLEXITY OF NATURAL PROTEINS

The interferons and the opioid peptides both comprise groups of biologically active polypeptides. There is an assortment of factors responsible for this diversity. From these descriptions, the reader should appreciate the efforts of the research investigator in deciphering such puzzling situations.

Background on the Interferons (1)

The name for this group of proteins is based on their ability to "interfere" with the progress of a viral infection in an animal. Interferons are secreted by cells infected with a virus. The secreted protein molecules bind to specific receptors on neighboring, uninfected cells and induce these cells to enter into a protected, antiviral state. An unusual feature of the interferon proteins is the multiplicity of forms that exist. This situation deserves some discussion, since it can arise in an isolation project and lead to confusion. In humans, three separates types of interferon, called alpha, beta, and gamma, are known. The alpha-interferon (IFN-α) group actually encompasses a family of structurally similar proteins; each member of this family is encoded by a different gene (Fig. 1). There is a single structural gene for beta-interferon (IFN-β) and another one for gamma-interferon (IFN-γ). Generally, multiple active forms represent the same protein, but with slight modifications. This situation is observed in the case of IFN-γ. There is a single protein that has two possible carbohydrate attachment sites. Each of these forms has been identified and the monoglycosylated and diglycosylated forms have been isolated. Only one form of IFN-β has been identified so far.

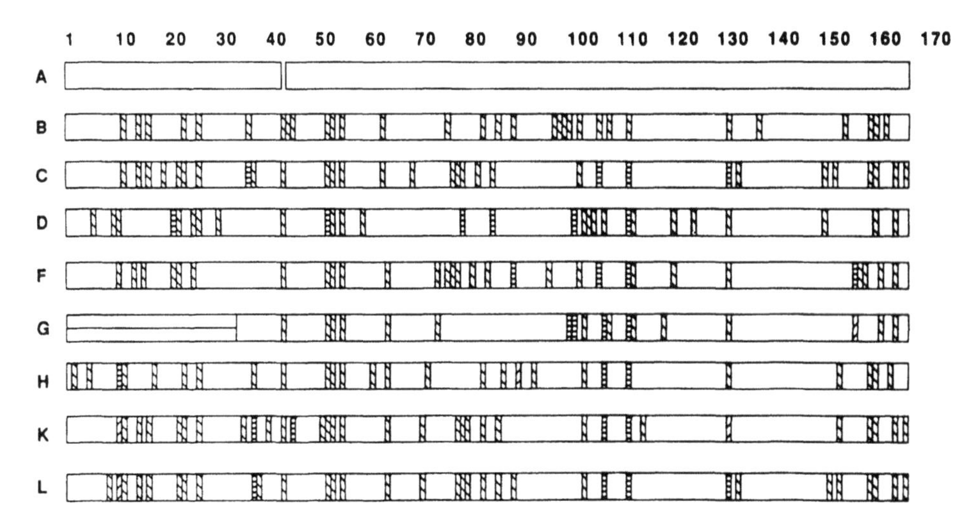

Figure 1 Descriptive representation of the family of human alpha-interferons. The structural similarities among nine subtypes of alpha-interferon are presented in a comparative format. Subtype A consists of one less amino acid than the other eight subtypes and a gap is inserted at this apparent amino acid deletion. Subtype B is then compared with A and an amino acid difference at a particular position is shown as a cross-hatched bar. There are 31 differences (including the gap) out of 166 amino acid residues between subtypes A and B, corresponding to a homology of 81%. Subtype C is then compared with A and B. At position 101, for example, the amino acid is the same in subtypes A and C, but different from that in B. At position 103, B and C have the same amino acid, which differs from that in A. At position 107, A, B, and C each has a different amino acid, as indicated by the new cross-hatching pattern in C. Such comparisons with each further subtype are then made. At any single position for the nine subtypes, there is never more than four different amino acids. Indeed, all the subtypes have the same amino acid at a majority of the positions.

Human beta-interferon and gamma-interferon show little, if any, structural homology to each other or to alpha-interferon (their sequences are not shown in this figure). Only a single primary sequence has been found for each of these interferons, in contrast to the multiplicity of sequences in the alpha-interferon family. Different subtypes have been found for gamma-interferon, but these are due to carbohydrate moieties attached at positions 25 and 97 of the protein chain.

Background on the Opioid Peptides (2)

Opiates, such as morphine, have profound physiological effects, including desensitization to pain and induction of a state of euphoria. An opiate substance binds to specific receptors on nerve cells to cause these phenomena. It was postulated that there must be a natural mechanism in which endogenous substances interact with these receptors to regulate pain perception, certain aspects of mood and some other physiological responses. What has emerged after several years of research in many laboratories is that there is a family of related opioid receptors and three distinct groups of structurally related opioid peptides (Fig. 2). The common feature of opioid peptides is the amino acid sequence Tyr-Gly-Gly-Phe-Met (called Met-enkephalin) or Tyr-Gly-Gly-Phe-Leu (called Leu-enkephalin).

In continuation of the discussion above, another explanation for the existence of multiple forms, which is often observed with small peptides such as the enkephalins, is the presence of processing intermediates. The precursor, proenkephalin, contains 8 copies of the enkephalin sequence within its structure. It has been demonstrated that proenkephalin is processed by proteolytic enzymes into small fragments. In the brain, one finds, essentially, some proenkephalin and some enkephalin; once processing of the precursor molecule begins, it continues until the enkephalins are generated. In the adrenal gland, on the other hand, one finds a mixture of enkephalin-containing polypeptides of various sizes and sequences arising from the very same precursor, proenkephalin. That is, processing is less complete. Each of these larger enkephalin-containing polypeptides may, in its own right, have a distinct hormonal action when secreted by the adrenal gland into the blood stream.

To make matters more complicated, there are two other precursor proteins that contain the enkephalin sequence and release peptides that are active at the opiate receptor. One of these precursors, pro-opiomelanocortin, contains but one copy of the Met-enkephalin sequence. Processing of this precursor leads to the opioid peptide beta-endorphin, which is the enkephalin sequence extended by an additional 26 amino acids at the carboxy terminal. Shorter forms, known as alpha and gamma-endorphin have been identified. Anther type of processing mechanism has been demonstrated in which the amino terminal of endorphin is acetylated. Each of these forms has been shown to possess a distinct biological activity and each may, therefore, have a unique physiological function.

The third opioid precursor is prodynorphin, which contains two copies of the Leu-enkephalin sequence. The biologically active enkephalin-containing peptides released from prodynorphin are called neoendorphin and dynorphin.

The message is that there are many, unpredictable possibilities with peptides and proteins in natural sources. The investigator must be cognizant of such situations when working on an isolation project. Eventually, when a bio-

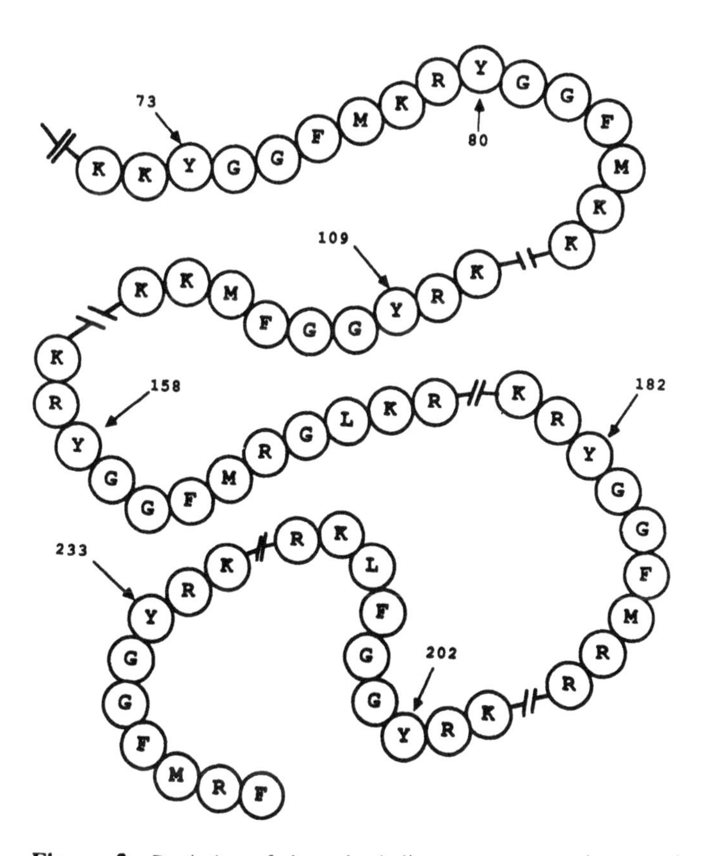

Figure 2 Depiction of the enkephalin precursor protein, proenkephalin. The term Met-enkephalin refers to the polypeptide tyrosine (Y)-glycine (G)-glycine (G)-phenylalanine (F)-methionine (M), whereas Leu-enkephalin has a leucine (L) replacing the methionine. Proenkephalin contains eight copies of enkephalin sequences, five being Met-enkephalin, one being Leu-enkephalin, one being the octapeptide, Met-enkephalin-Arg^6-Gly^7-Leu^8 and one being the heptapeptide, Met-enkephalin-Arg^6-Phe^7. Each enkephalin sequence is bounded by double basic amino acids, either lysine (K) or arginine (R), which serve as the enzymatic cleavage sites for release of the enkephalins from the precursor. Each number in the figure refers to the position in proenkephalin relative to the N-terminal.

Two other proteins are known which serve as enkephalin precursors (not shown in this figure). Pro-opiomelanocortin has one copy of the Met-enkephalin sequence. The biologically active peptide, called beta-endorphin, comprises the C-terminal 31 amino acids of pro-opiomelanocortin, with the Met-enkephalin sequence at the amino-terminal of beta-endorphin. Prodynorphin, the other precursor, includes two copies of the Leu-enkephalin sequence. The active peptides released from this precursor are called neoendorphin and dynorphin. Furthermore, various other forms of endorphin, neoendorphin and dynorphin have been identified. Thus, there are three distinct precursor proteins that release biologically active enkephalins or enkephalin-containing polypeptides.

logically active peptide or protein is to be used as a drug, an understanding of the physiological forms of that substance may be critical.

PRELIMINARY CONSIDERATIONS

The forethought and the continual evaluation of an isolation project can be essential to its eventual successful completion. There are many factors to be considered even before the first step of purification is undertaken. Careful attention to the points cited below is recommended.

The Assay

In an isolation project one starts with a crude mixture with the intent of purifying a specific component. Each step of the purification process separates the mixture of proteins into various fractions and the fraction containing the protein of interest must be located. An assay for specifically monitoring the presence of the particular protein in the presence of an excess of extraneous proteins is thereby required. This is generally done by some type of bioassay.

Interferon Bioassay

The most convenient procedure for measuring interferon activity is called the cytopathic effect inhibition assay (CPE). Essentially, the interferon makes cells resistant to destruction by viruses, as diagramatically shown in Figure 3. A photograph of a microtiter plate containing an actual interferon assay is shown in Figure 4. The assay is performed as follows: A test sample of interferon is placed in the first well of one of the rows of the microtiter plate. The sample solution is then serially diluted across the row; that is, the second well is a 1:2

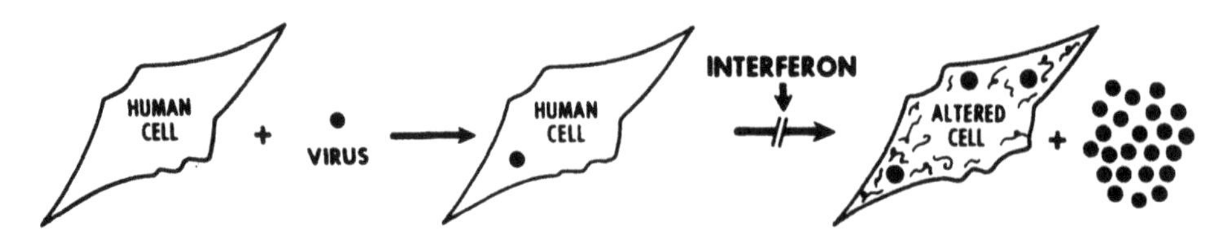

Figure 3 Antiviral activity of interferon. In the process of infection, the virus injects its nucleic acid, either DNA or RNA into a host cell. The virus alters the host cell and reproduces severalfold. The progeny virus destroy or lyse the host cell. When interferon binds to a specific receptor on the cell, it initiates a series of intracellular events that prevent the replication of the virus. The CPE assay is based on this ability to prevent cellular destruction. The cells are first incubated for several hours to induce the antiviral state and they are then challenged with virus. After an overnight incubation, the cells are treated with a vital stain and checked for survival.

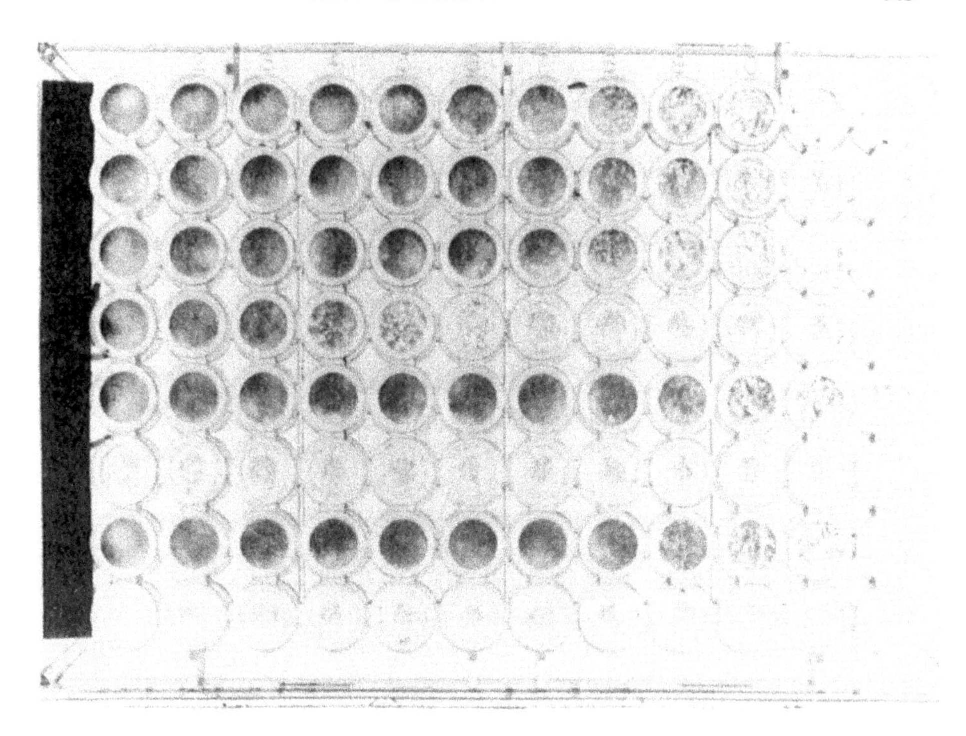

Figure 4 Photograph of a CPE microtiter plate. Rows 1, 2, 3, 4, 5, and 7 contain serial dilutions of various samples of interferon (across). Row 4 represents the standard and contains serial dilutions of a calibrated interferon solution. The endpoint of the standard row is well 5, in which there is about a 50% protection of the host cells. The endpoint for row 1, for example, is well 9, indicating an interferon concentration 16-fold higher (2^4) than the standard. Rows 6 and 8 are cell controls, illustrating the absence of interferon activity.

dilution, the third well is a 1:4 dilution, etc. A different test sample may be added to each row. Cells are then added to each well, followed by the challenge virus. The cells attach to the bottom of each well. After the overnight incubation, the wells are rinsed and the cells are stained with crystal violet. Viable cells take up the dye, whereas cells lysed by the virus appear colorless. Thus, the appearance of color means that interferon was present in the test sample at a sufficient concentration to protect the cells. When reading the plate, one determines the greatest dilution of the test sample that protects the cells. This endpoint is compared with a laboratory interferon standard and converted into units.

The interferon assay illustrates the attributes of an assay that are important to the eventual success of an isolation project. These are specificity, quantitation, rapidity, sensitivity, and capacity for handling large numbers of samples. Other attributes such as expense, reliability, and difficulty of the assay may also be factors to consider.

Specificity concerns the property that the assay will measure only the substance of interest; in other words, there will be no false positive or negative results. In this assay, one is measuring the protection of cells which would otherwise be killed by a virus. This is a relatively unusual property that most proteins or other natural substances would not be expected to have. Although there is great specificity, it may still be possible to have false positive or false negative results.

To illustrate a possible false negative result, one may conceive that some component in the sample being tested is acting as a poison. Even if interferon were present in the test sample, the cells would be killed, not by the virus, but by the poison. For this reason, a control assay should always be done with the sample diluent. For example, if the purification step involves reverse-phase chromatography using acetonitrile mixed with aqueous trifluoroactic acid, then these solutions should be checked in the assay. At this point, the sensitivity of the assay plays a helping role. Of course, if one applies concentrated acid or solvent to cells in culture, they will not survive. However, as is the case with the interferon assay, if the sensitivity is so great that the test sample may be diluted a million-fold into a physiological buffered solution, then the potential poisons are also diluted out. In the highly sensitive interferon assay these circumstances exist and the interferon samples are readily measured. One must also guard against other potential sources of false negative responses. Interferon, which has antiproliferative activity as well as antiviral activity, may be toxic to the cells at very high doses. Analysis of highly concentrated interferon samples may actually give negative results unless a thorough dilution analysis is done.

False positive results must also be guarded against, since they could lead the investigator to isolate the wrong substance. This would generally be more of a problem with crude extracts rather than with partially purified samples. In the highly specific interferon assay system this situation is uncommon.

Besides contributing to the specificity factor, a highly sensitive assay also preserves precious active material. Less than 1 picogram of interferon is readily measured by the CPE. It will be seen below that an isolation process may yield only microgram amounts of purified interferon. If the CPE were to require, for instance, microgram amounts of interferon, then most of the material would have been expended just to monitor the purification steps, leaving little interferon at the end for structural analysis.

Opioid Peptide Bioassay and Radioreceptor Assay

The assay used for the original isolation of Met-enkephalin and Leu-enkephalin was based on the inhibitory effects of opiates on the contraction of an electrically stimulated intestinal muscle strip. A section of the ileum, freshly removed from a guinea pig, was mounted in a bath of saline solution. Contractions of the muscle preparation, induced by regular pulses of electricity, were measured by an attached transducer. Addition of sample of an opiate to the bath caused the amplitude of the contractions to diminish. This assay, the basis of which corresponds to a well known side effect of opiates on intestinal motility, has the attributes of specificity and sensitivity, as well as being reasonably quantitative. Unfortunately, only a few samples could be analyzed each day by this assay.

A particularly convenient assay, used in some later studies, involved competitive binding to opiate receptors on cells (3). Most biologically active peptides and proteins manifest their effects by binding to their corresponding receptors on target cells. Receptor assays, such as the one described below, are particularly useful for screening compounds as potential agonists or antagonists.

In this assay, a tumor cell line (NG-108) having a high number of opiate receptors on the surface of each cell was derived from mouse brain tissue. Just like the cell lines used for the interferon assays, the NG-108 cells could be readily grown by standard cell culturing techniques. An excess of a radiolabeled opiate and the test sample was added to a suspension of the cells. After collecting and washing the incubated cells by centrifugation, bound radioactivity was measured. If a test sample contained an opiate, some of the receptor sites on the cells would become occupied and would not be available for binding the radiolabeled ligand. The decrease in bound radioactivity would be proportional to the concentration of opiate in the test sample, as shown in Figure 5. This radioreceptor assay was sufficiently selective to be useful for monitoring opioid peptides in impure extracts. It had a relatively broad specificity, allowing it to detect alkaloid type opiates as well as opioid peptides.

Protein Stability

Denaturation refers to the loss of biological activity due to a conformational change. Peptides are generally more stable toward denaturation than are proteins. By virtue of their small size, peptides tend to be flexible and can readily revert to their active conformation. Proteins with many disulfide bonds tend to be more stable, since they are held in the correct conformation by these covalent bridges between different sections of the polypeptide chain. At an early stage of the purification project, it is advisable to test the stability of the biological activity. For example, it may be necessary to work in the cold if the protein loses activity at room temperature. The protein may be more stable at certain values of pH. It is particularly interesting to determine if the activity is preserved in the presence of organic solvents, which would then allow the use of reverse-phase high-performance liquid chromatography (HPLC) for purification. If the protein is stable in the presence of sodium dodecyl sulfate (SDS), it may be possible to use preparative gel electrophoresis as a final purification step.

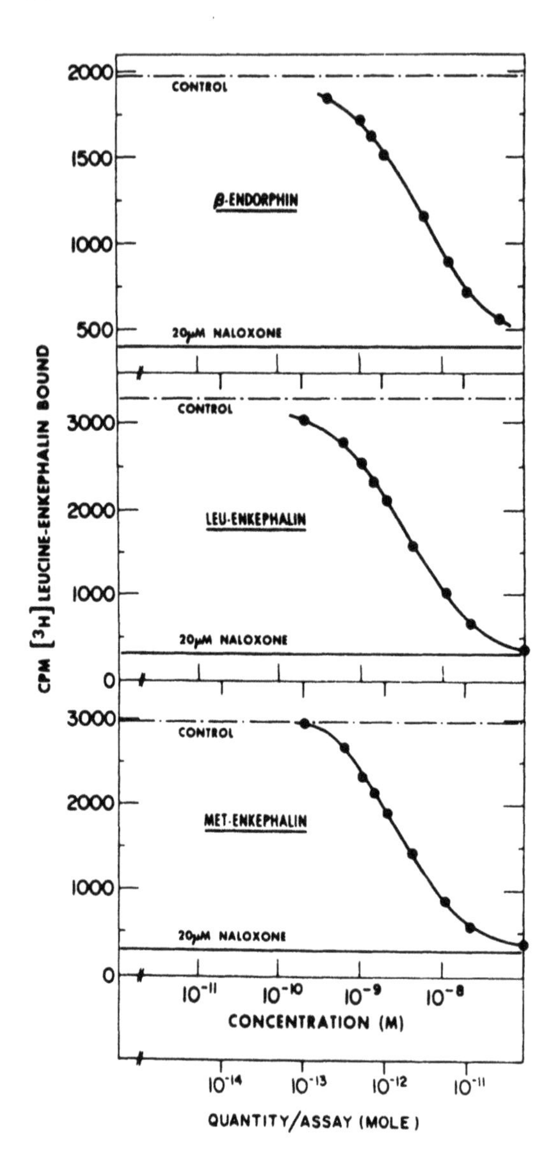

Figure 5 Radioreceptor assay for opiates: An equal number of NG-108 cells are placed into individual assay microcentrifuge tubes. A fixed concentration of radiolabeled enkephalin and a sample of the standard (or the test sample) is added to each tube. There is a competition between the radiolabeled enkephalin and the opioid peptide in the standard (or test sample) for the limited number of receptors on the cells. When there is little, if any opioid peptide present in the sample ($< 10^{-10}$ M), then the maximal amount of radioactive enkephalin can bind to the cells. To measure this binding, the cells are spun down, washed with buffer and then measured by scintillation

Abundant quantities of proteolytic enzymes are present in all living cells. These proteases are usually released during the initial extraction step. Accordingly, a general practice is to add protease inhibitors, at least during the early steps of the purification process. Phenylmethylsulfonyl fluoride (PMSF) is often added to inhibit the so-called serine-proteases (i.e., a serine residue is at the active site of the protease), such as trypsin and chymotrypsin. The polypeptide, pepstatin is added to inhibit proteases, such as pepsin, which are active at acidic pH. Chelating agents, such as ethylenediamine tetraacetic acid (EDTA) are added to inhibit metalloproteases.

Since small peptides are generally resistant to harsh treatment, it is often possible to inhibit the proteases under conditions which do not affect the biologically active peptides. In a procedure (4) for the isolation of peptides from the thymus gland, the tissue is first frozen and then homogenized over dry ice. The tissue is then extracted with 6 M guanidine hydrochloride. Prior to using this procedure, many of the naturally occurring thymic peptides were only previously found in degraded form. In the case of the protein, platelet-derived growth factor (PDGF), platelets were boiled prior to extraction in order to inactivate the proteases which are so rich in these cells. Even so, the protein was found to have many nicks in the polypeptide chain. However, the PDGF protein was held together by a high number of disulfide bonds and remained biologically active.

Source of the Starting Material

Enkephalins

The levels in the brain of the enkephalin peptides are on the order of 100 nanograms per gram of tissue (i.e., about 1 part per 10,000,000). The extraordinary work of Hughes, Kosterlitz, and co-workers led to the isolation of microgram amounts of a mixture of two similar pentapeptides, called Met-enkephalin (Tyr-Gly-Gly-Phe-Met) and Leu-enkephalin (Tyr-Gly-Gly-Phe-Leu), from kilograms of brain (5). Small, biologically active peptides, such as these, are typically synthesized as part of a larger precursor protein, which is then proteolytically processed to yield the final peptide products. The isolation

---

counting. As more nonradioactive enkephalin or endorphin is added, there is a decreased binding of the radiolabel to the cells, due to competition for the receptors. When assaying an unknown, the bound counts are measured and the concentration of the opioid peptide is read from the curve.

Nonpeptide opiates, such as the morphine analog, naloxone, also bind to the same receptor on NG-108 cells. The minimal, nonspecific binding, determined in the presence of a huge excess of unlabeled opiate is shown in this figure. A high ratio of specific to nonspecific binding, which is the case in this assay, is a desirable feature.

of the enkephalin precursor protein seemed a formidable task. Besides having only trace levels of the precursor in brain tissue, a purification project for a protein tends to be more difficult than for a peptide.

A breakthrough in this project was the discovery that the adrenal glands have unusually high levels of the enkephalins. Anyone who has experienced going to the slaughterhouse to obtain fresh animal tissues will appreciate the difference in obtaining 10–20 adrenal glands rather than hundreds of brains. Working with adrenal glands, from the first step of removal from the animal through the extraction and processing steps, is immensely easier and less costly than working with brain tissue. To further simplify matters, the cortex portion of the adrenal gland could be removed, the remaining medulla could be gently homogenized, and the secretory granules, which contain the opioid peptides, could be collected by a centrifugation procedure. A highly enriched preparation was thus obtained even before a single chromatographic step. It then became possible to isolate the precursor protein, called proenkephalin, and elucidate the biosynthetic pathway leading from this precursor to the pentapeptides. Although such a fortunate situation is not typically encountered in a project in which a protein must be isolated from animal tissues, some forethought in screening different tissues or alternate sources, such as tumor tissues or cell cultures, can often be well worth the effort. There are numerous examples of this in the scientific literature.

Interferons

The various members of this group of proteins are of intermediate size (about 20,000 daltons). A common feature is that they have potent immunomodulatory activities and are hence found in trace quantities in animal tissues. The isolation of these proteins was accomplished by producing them in cells maintained in tissue culture. In the case of alpha- and gamma-interferons, lymphocytes (white blood cells) served as the source, whereas beta-interferon was obtained from cultured human foreskin fibroblasts. To induce the cells to synthesize and secrete their respective interferons, a stimulus, perhaps resembling the natural occurrence, was developed. For alpha- and beta-interferons, this involved challenge with an infectious virus. In the case of gamma-interferon, it was found that a combination of chemicals which act as a mitogen and a carcinogen was an effective inducer.

Maintaining cells in a viable and productive state in culture is not a trivial matter. Generally, one adds calf serum (obtained from coagulated blood) to supply the myriad of hormones and other factors required by living cells. Unfortunately, serum contains a massive amount of extraneous proteins, a circumstance which adds considerably to the difficulty of trying to purify the particular protein of interest. Preferentially, one should use serum-free medium for maintaining the cells in culture. There have been great advances in the development of defined media for maintaining different types of cells in culture over prolonged periods of time. Since the induced cells must survive for only one day, the requirements are less stringent. In the case of alpha-interferon, the culture medium was supplemented with casein (milk protein) instead of serum. The rationale was that most of the casein could be readily removed by acid precipitation. With beta- and gamma-interferons, neither serum nor additional protein had to be added to the basic culture medium in order to obtain a high titer of the interferon.

It should be noted that the opioid peptide research was done with tissues from laboratory and slaughterhouse animals. The interferons, however, were of human origin. Alpha- and gamma-interferons were produced from white blood cells, which are readily available from hospitals or blood banks, whereas the fibroblast cells used for the production of beta-interferon were derived from secondary cultures of foreskin fibroblasts. Even though the homologous proteins are typically present in animals, it is important to obtain the human protein. In the case of the interferons, there is considerable species specificity. If the protein is to be eventually used for medical purposes, as is the case with interferons, the human version of the protein is absolutely required.

THE ISOLATION PROCESS

Alpha-Interferon

This family of proteins was originally known as leukocyte interferon. At the time this research was done, the subtypes of leukocyte interferon were referred to as alpha, beta, and gamma. The reader is cautioned to distinguish between the naming of the subtypes of leukocyte interferon in the next few paragraphs with the later adopted nomenclature in which the entire family of leukocyte interferons is known as alpha-interferon.

As described above, prior to beginning any attempts at purification, a convenient bioassay and a procedure for production of the protein were put into place. The concept of using casein rather than serum to supplement the culture medium was based on the combination of circumstances that leukocyte interferon is stable in dilute acid and that casein is insoluble in dilute acid. Therefore, acid precipitation steps were utilized to enrich and concentrate the crude interferon preparations. As can be seen in the purification table (Table 1), adjustment of the incubation (or conditioned) medium from pH 7 to pH 4 caused most of the casein to precipitate, while the interferon remained in solution. Besides the trace amounts of interferon, the remaining 20% of the soluble protein included some casein, other proteins secreted by the cells, and debris from cells that had expired during the overnight incubation. The trichloroacetic acid precipitation step served to concentrate the remaining pro-

Table 1 Purification of Human Leukocyte Interferon

Step	Units recovered $\times 10^{-6}$	Protein recovered (mg)	Relative specific activity (U / mg)	Degree of purification	Recovery range per step (%)
1. Incubation	50	10,000	5×10^3	1	—
2. pH 4 supernatant	50	2,000	2.5×10^4	5	100
3. 1.5% trichloroacetic acid precipitate	40	1,000	4×10^4	8	80–100
4. Triton X–100 / acetic acid supernatant	40	250	1.6×10^5	32	70–100
5. 4% Trichloroacetic acid precipitate	35	175	2×10^5	40	80–90
6. Sephadex G–100	32	57	5.6×10^5	112	70–90
7. Lichrosorb RP-8 (pH 7.5)	28	11	2.5×10^6	500	80–100
8. Lichrosorb diol					
Peak α	11	1.1	1×10^7	5000}	
Peak β	2.5	ND	ND	ND}	70–90
Peak γ	12.5	0.21	6×10^7	12,000}	
9. Lichrosorb RP-8 (pH 4) (Peak γ)	1.6	0.0064	3×10^8	60,000	40–60
10. Lichrosorb RP-8 (pH 4) (Peak γ)	8.2	0.021	4×10^8	80,000	40–60

For determination of protein recovered in each fraction, bovine serum albumin was used as a standard. The absolute specific activity determined by amino acid analysis of the homogeneous peak of step 10 was found to be 2–4 $\times 10^8$ units / mg (see text). Step 10 was performed on pooled material from several preparations. ND, not determined.

Source: From Ref. 6.

tein from liters of solution into a pellet which could be redissolved in a small, more manageable volume. A precipitation reagent, more typically ammonium sulfate, is often utilized for concentrating and partially purifying a crude protein extract. Two more rounds of acid precipitation, the first in which the interferon was soluble and the second in which it was not, yielded a 40-fold purified, as well as concentrated product.

The recovery of interferon, as measured by the bioassay, was also high. This latter parameter must be considered along with the degree of purification when evaluating the usefulness of a purification step. The quantitative nature of the bioassay is essential to appropriately evaluate each purification step.

A series of chromatographic steps was then employed to obtain a purified leukocyte interferon subtype. Separation by size was done on a column of Sephadex G-100. The mobile phase buffer contained 4 M urea. Without this denaturant, it was found that the interferon protein formed a series of aggregates with other proteins that distributed throughout the column eluate. Only a threefold enrichment was obtained, but the removal of large proteins, which tend to form aggregates, was an additional benefit of this step.

This particular project was one of the first in which a protein was successfully purified by high performance liquid chromatography (HPLC). The fractions from the Sephadex column, that contained the interferon, were applied to a reverse-phase HPLC column. This column had been pre-equilibrated with an aqueous buffer at pH 7.5. The proteins in the sample mixture, having concentrated on the column, were eluted by pumping a gradient of increasing concentration of propanol through the column. It can be seen from Figure 6 that most of the extraneous proteins eluted before the interferon, resulting in a further enrichment of about fivefold.

A normal phase HPLC column was used for the next purification step. This type of column, which owes its hydrophilic nature to the presence of hydrogen bond-forming diol groups, is rarely used for protein chromatography. In the case of leukocyte interferon, it was able to resolve three major subtypes and provide a 24-fold purification of the gamma subtype. At the time of this research, it was not known what mechanism was responsible for the multiple peaks on the diol column. In retrospect, it is known that there is a family of closely related proteins, arising from different genes, produced by leukocytes. The gamma subtype of alpha-interferon was purified to homogeneity by additional steps of reverse-phase HPLC. In preparation for the previous diol step, propanol was added to the sample to bring it to a final concentration of 80% propanol. For the upcoming reverse-phase step, it was necessary to remove the propanol from the diol fractions prior to loading. This was conveniently accomplished by extracting the propanol into hexane. The general rule is that a sample must be in the appropriate milieu so that the proteins will concentrate on the column during loading. Reverse-phase chromatography was now done at pH 4, again using a gradient of propanol to obtain the homogeneous gamma subtype of the alpha (or leukocyte) interferon.

This example illustrates the potential of reverse-phase HPLC for purifying proteins. Besides providing high resolving power in an individual chromatographic run, it is possible to change the selectivity of reverse-phase chromatography in different ways. In this example, the selectivity was changed by a shift in the pH of the mobile phase. By adjusting the pH one changes many of the charges on the side chains of the amino acid residues in the proteins. The cumulative changes are different for each protein in the mixture to be resolved and, hence, their relative retention on the column will differ. Simply changing

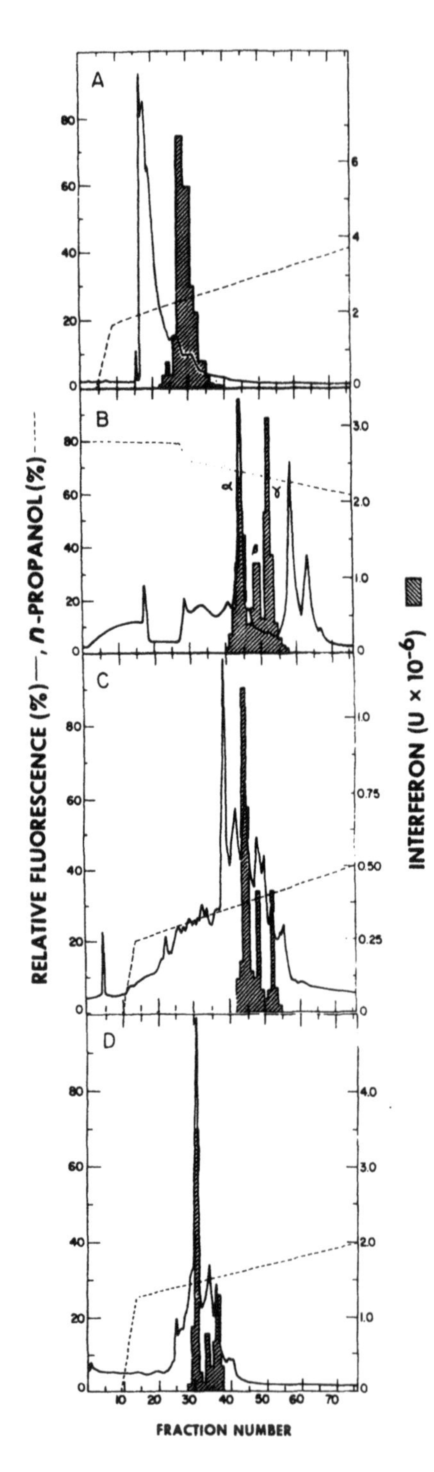

Figure 6 High-performance liquid chromatographic purification of human alpha-interferon. Partially purified interferon (after precipitation steps and Sephadex chromatography) was applied to a reverse-phase column of octyl-silica in aqueous buffer at pH 7.5. The column was then eluted with a gradient of increasing concentration of *n*-propanol (Panel A). Protein elution was monitored by an automated fluorescence technique (not described in this chapter) using the amine-specific reagent, fluorescamine. Aliquots from the collected fractions were measured in the interferon CPE assay. The interferon-containing fractions were pooled, propanol was added to a final concentration of 80% and the sample was applied to a normal-phase column of diol-silica (Panel B). Elution of the column with a gradient of decreasing propanol concentration resulted in discrimination of three different subtypes of the alpha-interferon. Fractions corresponding to the gamma-subtype were pooled and the propanol was extracted into hexane. The aqueous solution was then applied to the same reverse-phase column at pH 4.0 and eluted with a gradient of increasing propanol (Panel C). The major peak of activity was collected and rechromatographed on the same column, but with a more shallow gradient (Panel D). The final chromatography step represented a pool of material from several preparations carried through the step in Panel C. The fraction with the major peak of activity contained essentially homogeneous interferon.

the organic modifier (e.g., propanol to acetonitrile) or the hydrophobic ligand on the column can also provide a change in selectivity. Unfortunately, some proteins cannot tolerate the organic solvents used for reverse-phase chromatography. Otherwise, reverse-phase HPLC should be preferentially considered for use in a purification program.

Beta-Interferon

This interferon is secreted by fibroblasts into the culture medium after challenge with an infectious virus. It was not necessary to add serum or a protein supplement to the culture medium during the overnight induction period. Beta-interferon was purified to homogeneity by a two-step procedure.

The first step of this procedure illustrates the potential of affinity chromatography. One class of affinity chromatography is based on the use of dyes. These organic molecules tend to resemble important biological cofactors such as NADH (nicotine-adenine dinucleotide) and proteins often have domains which recognize and bind them; it is not known whether there is any physiological significance to these binding sites. One can purchase kits containing a variety of individual dyes covalently attached to chromatographic supports. The procedure for evaluating this method is to pass a small sample of the crude extract through a column of each dye to see if the biological activity is bound. If so, then attempts are made to elute the biological activity from the column. Generally, a high concentration of salt or a combination of high salt and a weak organic modifier is used for elution.

In the case of beta-interferon, it had been determined that Cibacron blue (used in blue jeans) was especially useful. Accordingly, several liters of conditioned medium were passed through a 25-ml column of the commercial product, Blue-Sepharose, which selectively bound the beta-interferon. The column was eluted with 1 M sodium chloride containing 30% ethylene glycol, which removed some additional protein, and finally with 1 M sodium chloride containing 50% ethylene glycol, which removed the beta-interferon (Fig. 7). This simple affinity chromatography step had removed most of the contaminating proteins and yielded a concentrated solution of about 10% purity.

Reverse-phase chromatography was used for the second step of purification. The ethylene glycol was diluted out and the sample was applied to the reverse-phase column. Proteins were eluted with a gradient of increasing propanol at pH 4.2 (Fig. 8). The peak of protein, that corresponded to the fractions of biological activity, was homogeneous beta-interferon.

Gamma-Interferon

This protein is also referred to as immune interferon, because its synthesis in cultured lymphocytes is induced with a mitogen rather than by an infectious

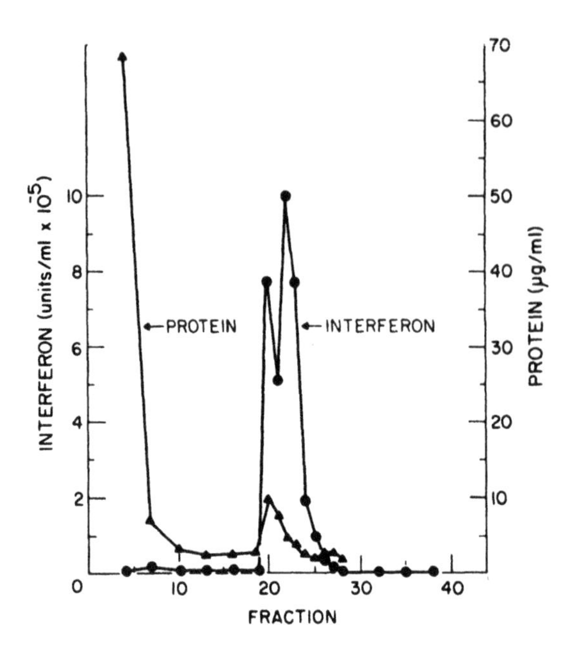

Figure 7 Affinity chromatography of human beta-interferon: Several liters of conditioned medium were passed through a small column (25 ml) of Cibacron blue-Sepharose. Beta-interferon was accumulated on the column, while most extraneous proteins were not adsorbed. The column was washed with nine bed volumes of 30% ethylene glycol in 1 M sodium chloride-buffered solution (fractions 1–9), followed by elution with 50% ethylene glycol in the same buffered solution.

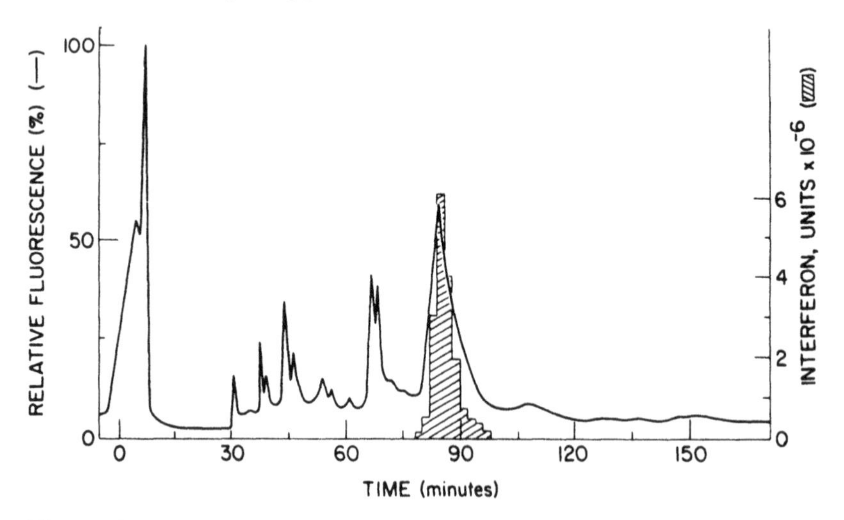

Figure 8 High-performance liquid chromatography purification of human beta-interferon. The fractions from the Cibacron blue-Sepharose column (see Fig. 7), which contained interferon activity, were applied to an octyl-silica column in aqueous buffer at pH 4. The column was eluted with a gradient of increasing concentration of *n*-propanol (c.f. Fig. 6). Essentially homogeneous interferon was obtained.

virus. It does, however, have antiviral activity and is measured by an antiviral assay. Conditioned medium can be prepared under serum-free conditions.

Two similar schemes for purification were reported (7, 8). Both utilized adsorption onto controlled-pore glass, a type of silica, as the first step. Silica has an acidic surface that can form hydrogen bonds with proteins (similar to the diol column used in the alpha-interferon purification). Gamma-interferon is an extremely basic protein and adsorbs strongly to the silica surface. In one process, the protein-loaded silica was washed with 1 M sodium chloride and the gamma-interferon was then eluted with 1 M sodium chloride containing 30% ethylene glycol. In the other process, elution was accomplished with 0.5 M tetramethylammonium chloride, a chemical which possesses both a positive charge and a hydrophobic character.

In one process, the purification steps were then affinity, permeation, and reverse-phase chromatography. Gamma-interferon had previously been shown to be a glycoprotein and its particular carbohydrate moieties could be bound by the protein concanavalin A. The gamma-interferon was enriched on an affinity column of concanavalin A-Sepharose and specifically eluted with 0.3 M alpha-methyl-D-mannoside. The eluate was concentrated under pressure on a filter membrane (10,000 dalton cutoff) and run on a permeation column. The final purification step was by reverse-phase chromatography. It was generally known that gamma-interferon is denatured and losses biological activity by exposure to organic solvents and acidic conditions. After much experimentation, these investigators found a suitable solution to the problem. They ran the reverse-phase column at pH 7.5 with a gradient of dioxane and then immediately removed the dioxane from the collected fractions on small permeation columns.

In the other process, the eluate from the controlled-pore glass was desalted and concentrated under pressure through a filter membrane. It was then applied to an HPLC cation-exchange column. Elution was done with a gradient of increasing concentration of sodium chloride in a constant background of 20% ethylene glycol. This particular HPLC column, the Mono-S, has a polymer matrix, rather than the more typical silica. The presence of ethylene glycol served to prevent hydrophobic interactions and allowed the chromatography to be based solely on ionic interactions.

The course of events of the gamma-interferon research was one in which the protein was identified first by recombinant DNA techniques. A recombinant plasmid containing a cDNA sequence for human gamma-interferon was prepared from induced lymphocytes. Expression of this sequence in both eukaryote and prokaryote cells yielded a protein with the biological properties attributed to gamma-interferon. The sequence of the protein was deduced from the sequence of this structural gene. Accordingly, the natural protein was isolated in order to confirm the structure predicted from the gene. As it turned out, posttranslational processing events yielded a significantly modified protein.

Both isolation processes described above yielded two major forms of gamma-interferon of apparent molecular weights of 25,000 and 20,000. Careful structural analysis established that both forms were constituted from the same protein having a calculated molecular weight of about 17,000. The 20,000 dalton form was shown to have a single carbohydrate moiety, and the 25,000 dalton form to have two carbohydrate moieties attached to the protein. The amino terminal sequence predicted from the gene was Cys-Tyr-Cys-Gln-Asp-Pro-etc. In actuality, the natural, mature protein had Gln at the amino terminal; that is, the Cys-Tyr-Cys residues were not present. Furthermore, the residue of glutamine at the amino terminal had formed the cyclic structure known as pyroglutamate. The amino terminal sequence of the mature protein was, therefore pGlu-Asp-Pro-etc. Recombinant DNA techniques can often offer a more rapid pathway to the characterization of a protein corresponding to a biological activity. As revealed in this illustration, the protein must eventually be isolated from natural sources to elucidate the actual structure of the mature protein.

Enkephalin Polypeptides

The preliminary steps in this project were the development of a convenient radioreceptor assay and the identification of the adrenal gland as a rich source

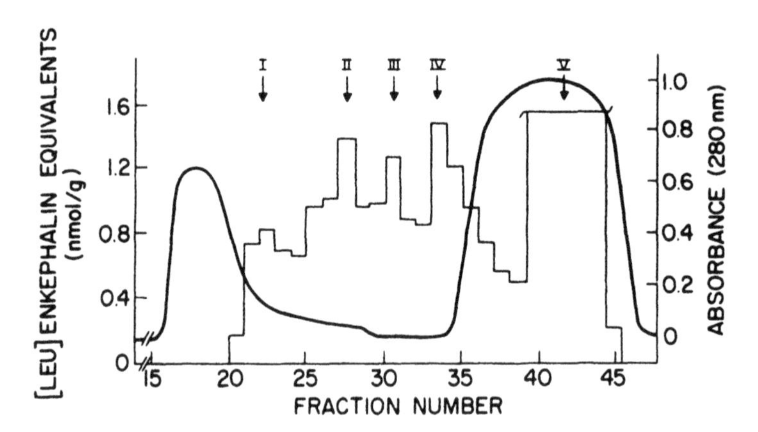

Figure 9 First chromatographic purification step for enkephalin-containing peptides. About 50 g of adrenal medulla tissue was used to prepare chromaffin granules, which were extracted with dilute acid and chromatographed on a Sephadex G-75 column. Proteins and peptides in the column effluent were monitored by measuring the absorbance at 280 nm, whereas enkephalin-containing peptides were determined by the radioreceptor assay. An aliquot from each fraction was digested with the proteolytic enzyme, trypsin, to release enkephalins from the precursor and intermediates prior to the radioreceptor assay. Five pools of fractions were made, as indicated in the figure, and each was processed further, separately.

of material. The isolation of chromaffin granules from bovine adrenal medulla provided an even more enriched source of material. The first chromatographic step was a size separation on a Sephadex column (Fig. 9). The collected fractions were tested in the radioreceptor assay and five distinct size groups were discerned. Each of these groups was carried forward separately. The fractions in peak IV, representing peptides in the size range of about 2,000–5,000 Daltons, were pooled and applied to a reverse-phase HPLC column (Fig. 10). Fractions from this chromatographic run were then rechromatographed by reverse-phase HPLC, until homogeneous enkephalin-containing polypeptides were obtained.

Each of the purified peptides was eventually identified by sequence analysis. Peptide E, for example, is a peptide comprising the sequence between residues 80 and 113 (inclusive) of proenkephalin (see Fig. 2) and contains two copies of the Met-enkephalin sequence within its structure. These peptides may be more than just intermediates in the biosynthetic pathway to the smaller enkephalins. On a molar basis, Peptide E has an unusually high activity in the ileum contraction opiate bioassay, implying a unique physiological function. Various biochemical studies have indicated that these enkephalin-containing polypeptides are released into the blood stream.

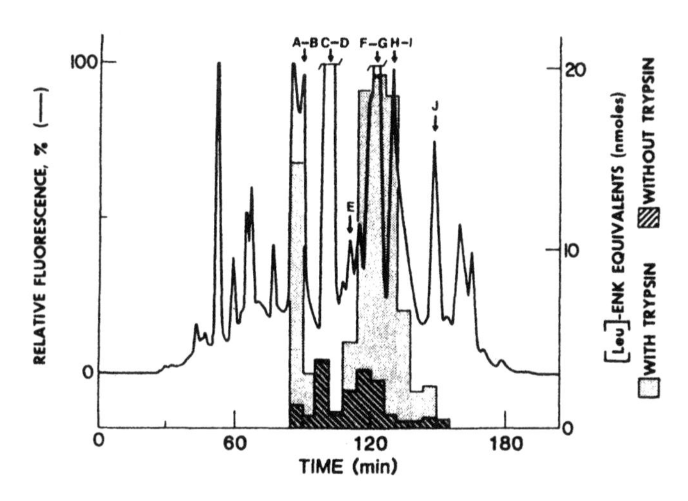

Figure 10 Rechromatography of peak IV of the enkephalin-containing peptides. The pool of peak IV (see Fig. 9) was applied to a octadecyl-silica, reverse-phase HPLC column. Peptide concentration was automatically determined in a continuous fashion using a fluorescence detection system. Enkephalin-containing peptides were monitored by the radioreceptor assay. Fractions were rechromatographed on a cyanopropyl-silica, reverse-phase HPLC column (not shown) to obtain the individual, homogeneous enkephalin-containing peptides.

PROOF OF PURITY

Attempts at Further Purification

When one is dealing with a previously uncharacterized protein, it is difficult to determine that all contaminating proteins have been removed. A protein may be judged to be pure if it gives a single symmetrical peak on an HPLC column. Attempts may be made to resolve the sample into components by chromatographic procedures not previously used in the purification process, such as a modified reverse-phase step. In each instance, the biological activity should be found in the fractions corresponding to the peak of protein. When no further increase in the specific activity (bioassay units / unit weight of protein) can be attained, the protein may be considered to be pure.

Electrophoresis

Another procedure providing high resolution of protein mixtures is electrophoresis. In the most popular version, a sample of the protein is applied to the top of a polyacrylamide gel. The anionic detergent, sodium dodecyl sulfate, is added to impart a negative charge to all proteins in the sample so that migration is toward the anode when an electrical potential is applied across the gel. After a suitable running time, the proteins in the gel are visualized by a staining procedure.

The use of electrophoresis for the determination of purity is illustrated in Figure 11. A single band is observed on the stained gel for purified beta-interferon (9). In many cases, the biological activity of the protein is preserved through the electrophoresis process. It may be observed in Figure 11 that this was true with beta-interferon. An unstained companion track on the gel was cut into sections. Each section was extracted and the extracts were tested for antiviral activity. The correspondence of biological activity with the single band of stained protein is strong evidence, but not totally conclusive, of homogeneity.

The high resolving power of electrophoresis should be considered as a prime technique for protein purification. Indeed, the first successful purification of small amounts of beta-interferon, which yielded amino terminal sequence data, was accomplished with gel electrophoresis as the final step. Likewise, the residual activity after gel electrophoresis of gamma-interferon allowed the initial observation of the 20,000 and 25,000 dalton forms. One should not overlook the potential of electrophoresis as a short-cut in a purification project. With the recent advances in microsequencing technology, it is often possible to extract a previously uncharacterized protein from a gel and obtain some sequence data. This data can be useful for other approaches aimed at characterizing the unknown protein, such as the preparation of DNA probes

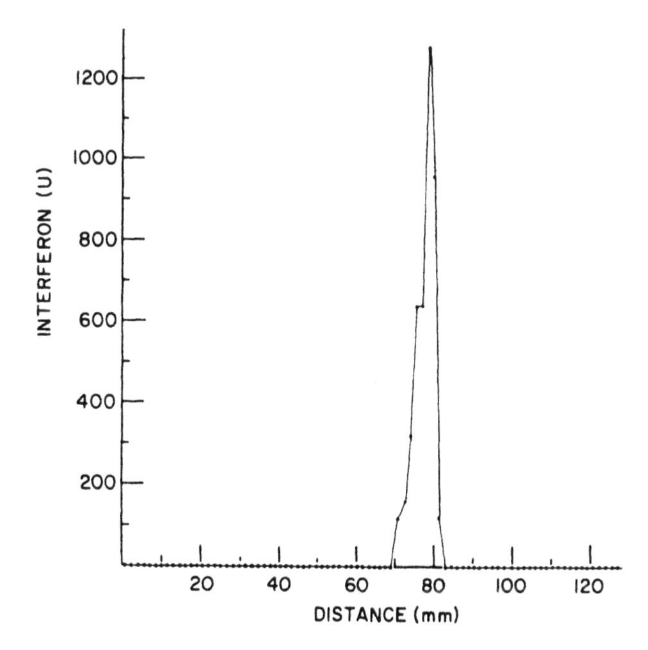

Figure 11 Polyacrylamide gel electrophoresis of homogeneous human beta-interferon. Equivalent samples of interferon were placed in two parallel tracks of a 5–15% gradient gel. Application of an electrical potential caused the proteins to migrate, in the presence of sodium dodecyl sulfate, toward the anode. After electrophoresis, the gel was sliced between the tracks. One track was stained with Coomassie blue, which revealed the presence of only one detectable protein. The other track was cut into sections, which were then extracted with buffer for testing in the CPE assay. The sections corresponding in migration distance to the stained band contained the interferon activity. This provided proof that the purified protein was interferon. Had the interferon in the sample been only a trace component in the presence of a large excess of an extraneous protein, then the correspondence between the two tracks, most likely, would not have occurred.

or the synthesis of a peptide for generation of antibodies cross-reacting with-the protein.

CONCLUDING REMARKS

The few examples presented above should give the reader an appreciation of the types of procedures that can be applied to a purification project. It is again stressed to be attentive to the preliminary details of the bioassay and the source of the crude material. Careful investigation of the properties of the protein and attempts to find short-cuts, such as some type of affinity step, can greatly contribute to the successful outcome of the purification project.

REFERENCES

1. Pestka, S. The human interferons—From protein purification and sequence to cloning and expression in bacteria: before, between and beyond. *Arch. Biochem. Biophys.* 221, 1–37 (1983).
2. Udenfriend, S. and Kilpatrick, D. L. Biochemistry of the enkephalins and enkephalin-containing peptides. *Arch. Biochem. Biophys.* 221, 309–323 (1983).
3. Gerber, L. D., Stein, S., Rubinstein, M., Wideman, J., and Udenfriend, S. Binding assay for opioid peptides with neuroblastoma X glioma hybrid cells: Specificity of the receptor site. *Brain Res.* 151, 117–216 (1978).
4. Haritos, A. A., Goodall, G. J., and Horecker, B. L. Prothymosin alfa: Isolation and properties of the major immunoreactive form of thymosin-alfa in rat thymus. *Proc. Natl. Acad. Sci.* (*USA*) 81, 1008–1011 (1984).
5. Hughes, J., Smith, T. W., Kosterlitz, H. W., Fothergill, L. A., Morgan, B. A., and Morris, H. R. Identification of two related pentapeptides from the brain with potent opiate agonist activity. *Nature* (*London*) 258, 577–579 (1975).
6. Rubinstein etal. *Proc. Natl. Acad. Sci.* (*USA*) 76, 640–644 (1979).
7. Rinderknecht, E., O'Connor, B. H., and Rodriguez, H. Natural human interferon-gamma. *J. Biol. Chem.* 259, 6790–6797 (1984).
8. Friedlander, J., Fischer, D. G., and Rubinstein, M. Isolation of two discrete human interferon-gamma subtypes by high performance liquid chromatography. *Anal. Biochem.* 137, 115–119 (1983).
9. Stein, S., Kenny, C., Friesen, H. J., Shively, J., Del Valle, U., and Pestka, S. Amino-terminal amino acid sequence of human fibroblast interferon. *Proc. Natl. Acad. Sci.* (*USA*). 77, 5716–5719 (1980).

7

Large-Scale Production of Recombinant Proteins: Human Leukocyte Interferon

Fazal R. Khan and Vishva R. Rai

Hoffmann-La Roche, Inc.
Nutley, New Jersey

INTRODUCTION

Many proteins have been identified that may have potential use as therapeutic products. In most cases these proteins are highly potent and are, therefore, normally present in minute quantities in biological tissues. The advent of the recombinant DNA revolution now makes it possible to prepare these proteins in large amounts and at reasonable cost. Human leukocyte interferon represents one of the most prominent recombinant protein products of the biotechnology era to have gone through comprehensive clinical trials.

Interferon was discovered in 1957 by Isaacs and Lindemann (1). It was first described as an agent that is released by virus-infected animal cells, and that upon contact with uninfected cells, makes them resistant to viral infection. A further biological property of interferon is its antiproliferative activity, especially against transformed (cancer) cells. It was subsequently found that there are different types of interferon, termed α, β, γ. In humans there are numerous subtypes of α-interferon and scientists at Hoffmann-La Roche have produced recombinant human leukocyte interferon α-2a (rIFNα-2a) for clinical use.

A description of the process used for the purification of clinical quality rIFNα-2a is presented in this chapter. It should be noted that many factors must be taken into account when a particular isolation scheme is developed. One critical factor concerns the properties of the protein to be purified. In this instance, leukocyte interferon is particularly insensitive to denaturation by chaotropes (i.e., reagents that disrupt hydrogen bonding), detergents (e.g., Triton X-100 and Tween-20) and dilute acid. Although there are many proteins of current biomedical interest having similar stability properties (e.g., interleukin-2), most proteins would generally not be amenable to some of the harsh conditions employed in the process described in this chapter.

CLONING THE LEUKOCYTE INTERFERON GENE

Much of the pioneering work on the isolation and structural analysis of the protein itself was done under the direction of Sidney Pestka at the Roche Institute of Molecular Biology. Further details on the isolation of interferon from human leukocytes can be found in Chapter 6 by Stanley Stein, and the cloning of the interferon structural gene is described in Chapter 11 by Sidney Pestka. The successful cloning in *Escherichia coli* of human leukocyte interferon was a collaborative effort by scientists at Genentech and Hoffmann-La Roche (2). This recombinant organism expresses rIFNα-2α at high levels. The fully active protein could then be isolated from *E. coli* grown in large quantities by fermentation. This purification process is summarized in Table 1, and is described below.

ANALYSIS OF THE PRODUCT

The goal of the large-scale purification process is to obtain a product suitable for clinical use. Numerous quality control criteria are required to ensure purity such as, presence of antibody, DNA, endotoxin, and other chemicals used in the process. In addition, high-performance liquid chromatography (HPLC) analysis and polyacrylamide gel electrophoresis (PAGE) are used to monitor the purity of the product. In the PAGE procedure, the protein sample is ap-

Table 1 Flow Chart of the Recovery Process for rIFNα-2a

Step 1:	Extract *E. coli* cells
Step 2:	Concentrate by ultrafiltration
Step 3:	Immobilized antibody affinity chromatography
Step 4:	Copper chelate chromatography
Step 5:	Carboxymethyl cellulose cation-exchange chromatography
Step 6:	Gel filtration chromatography
Step 7:	Concentrate and sterile filter
Step 8:	Bulk rIFNα-2a for storage or formulation

plied to the gel and, in the presence of sodium dodecyl sulfate, the proteins migrate in an electric field toward the anode, and separate according to molecular weight. The gel is removed from the electrophoresis tank and the protein bands are visualized by staining, such as with the dye Coomassie Blue. This technique has excellent resolving power and, hence, is useful in monitoring the purification process (see Chap. 4). The electrophoretogram in Figure 1 will be referred to in the description of the purification process.

PURIFICATION OF rIFNα-2a

Extraction

The *E. coli* cell paste is stored frozen until required. Extraction is carried out in cold at 2–8°C. The frozen cells are suspended in four volumes of an extraction solution which is composed of 2.0 M guanidine hydrochloride, 2% Triton

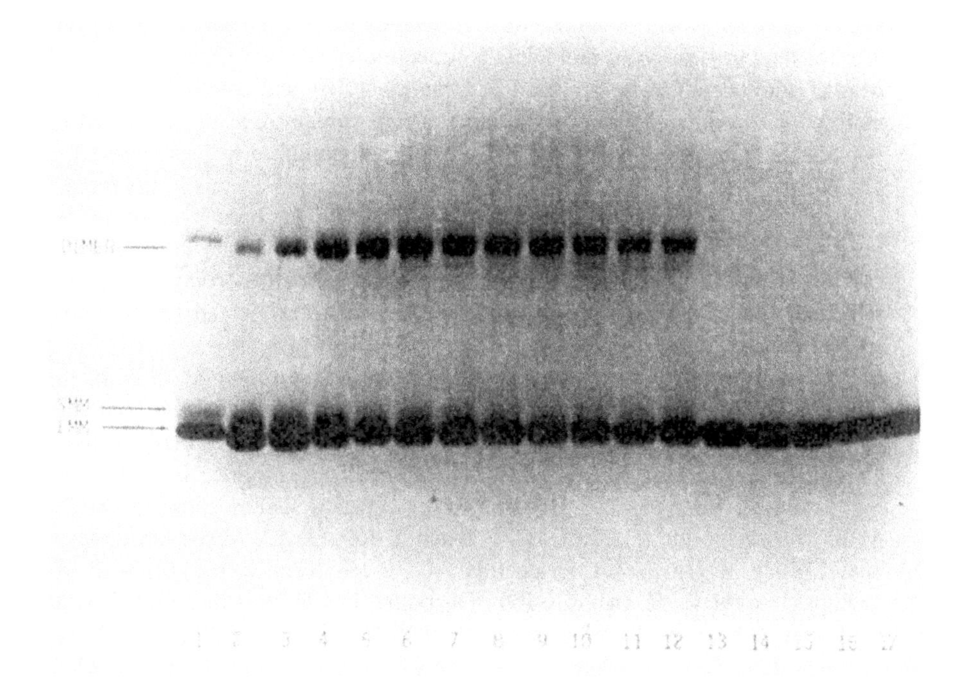

Figure 1 Polyacrylamide gel electrophoresis (PAGE). The gel is run in the presence of sodium dodecyl sulfate and nonreducing conditions are used. Lane 1 contains the pool from the immunoaffinity column. Lanes 2–17 contain various fractions from the copper chelate column (see Fig. 2).

X-100, 0.1 M Tris base (Tris hydroxymethyl aminomethane). After mixing for 1 hour, 5 volumes of distilled water are added. The pH is then adjusted to 7.0, an appropriate amount of an Aerosil (silica used for removing lipids) is added and the slurry is stirred for an additional hour. This process serves to break open the bacteria and solubilize the interferon, much of which is present as insoluble inclusion bodies.

The slurry is then centrifuged to remove the cell debris and unbroken bacteria. The clear supernatant, which contains soluble interferon as well as bacterial proteins, is concentrated about fourfold using ultrafiltration with a 10,000 dalton cutoff membrane. This concentrated material is termed the "crude extract" and is used for further processing.

Immunoaffinity Chromatography

Conventional purification procedures tend to be tedious and time consuming. We have adopted immunoaffinity chromatography as the first purification step. The development of monoclonal antibodies has proved to be a powerful tool for the purification of biologically active molecules (see Chap. 12). Monoclonal antibodies can be prepared in large quantities and coupled to a solid matrix for chromatography. The immunoaffinity column used in this instance is prepared from the LI-8 monoclonal antibody immobilized to Sepharose CL-6B, as first described by Staehelin et al. (3). Later improvements in the large-scale purification of rIFNα-2a by immunoaffinity chromatography were made by Tarnowski et al. (4, 5). The advantages of this technique are its simplicity of use, the high degree of purification (i.e., essentially all *E. coli* proteins are removed), and the robust character of the column (i.e., it can be used for several hundred runs).

The crude extract is pumped onto an immunoaffinity column equilibrated with a buffer composed of 0.3 M guanidine hydrochloride, 0.1 M Tris-HCl (pH 7.0), and 0.3% Triton X-100. The column is washed sequentially with 5 column volumes each of (1) 0.3 M guanidine hydrochloride, 0.1 M Tris-hydrochloride (pH 7.0), 0.3% Triton X-100; (2) 0.5 M sodium chloride, 0.025 M Tris-hydrochloride (pH 7.0), 0.2% Triton X-100; (3) 1.0 M sodium thiocyanate, 0.025 M Tris-hydrochloride (pH 7.0), 0.1% Triton X-100; (4) 0.15 M sodium chloride, 0.1% Triton X-100. The bound interferon is then eluted with a solution containing 0.2 M acetic acid, 0.15 M sodium chloride, and 0.1% Triton X-100. This procedure is based on the strong affinity of the immobilized antibody for interferon at neutral pH. Extraneous proteins that are held onto the column by nonspecific mechanisms (e.g., ionic interactions) are removed using a buffer of high ionic strength and a chaotropic agent. By decreasing the pH, this particular antibody changes its conformation and loses its ability to bind interferon.

According to polyacrylamide gel electrophoresis (PAGE) analysis (Fig. 1, lane 1), three major bands are seen. One represents a dimeric form of rIFNα-2a, in which a disulfide linkage has formed between two molecules of the protein. The fast-moving monomer (FMM) is the desired product, whereas the slow-moving monomer (SMM) may contain incomplete disulfide linkages. Other oligomeric forms (trimer, tetramer, etc.) may be discerned.

When the same sample is electrophoresed under reducing conditions (i.e., boiling the sample in 2-mercaptoethanol prior to application to the gel), only one band is observed. This confirms that all the other bands represent different disulfide forms of rIFNα-2a. It should be recalled (see Chap. 9 by Jones and Stein) that there are four residues of cysteine in this protein. The disulfide linkages form after extraction from the bacteria and upon exposure to atmospheric oxygen, but a portion of the protein molecules do not have the proper configuration (i.e., oligomers and SMM). Although the affinity column does provide essentially pure rIFNα-2a, it is not able to distinguish between these different forms. Hence, further purification is necessary.

Copper Chelate Chromatography

The use of divalent metal ions for the purification of biopolymers was first demonstrated by Porath et al. (6). Transition metals form complexes with the side chains of certain amino acids (e.g., the imidazole group of histidine). In this application, the proteins become immobilized to a solid support containing a chelating moiety charged with copper ions. The complex of metal-protein is dissociated at low pH and the differential affinity of the forms of rIFNα-2a allows the removal of dimer and slow moving monomer (SMM) from the product. Copper chelate gel is prepared in-house from Sepharose CL-6B (Pharmacia, Inc.).

The pH of the antibody pool is adjusted to 7.0 and loaded onto a copper chelate column equilibrated with a solution containing 0.2 M acetic acid, adjusted to pH 6.5 with sodium hydroxide, 0.15 M sodium chloride, and 0.1% Tween-20. The column is washed with a solution containing 0.05 M acetic acid, 0.10 M sodium chloride, and 0.1% Tween-20 until the pH of the effluent is 5.0. The protein is then eluted with a solution containing 0.025 M acetic acid, pH 4.1, 0.10 M sodium chloride, and 0.1% Tween-20. Individual fractions are collected (Fig. 2) and analyzed by PAGE (Fig. 1). The fractions comprising Peak 2 (Fig. 2) that have been freed of dimer and SMM are then processed further.

Carboxymethyl Cellulose Chromatography

The purpose of this ion-exchange chromatography step is to remove the nonionic detergents, which had been used in the preceding chromatography

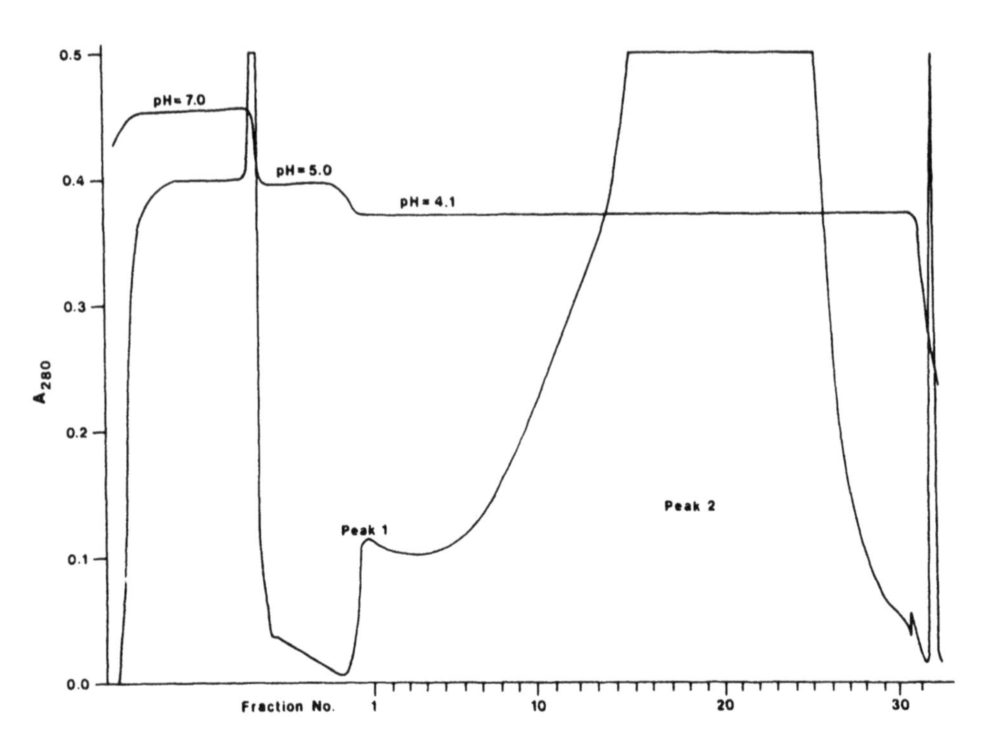

Figure 2 Copper chelate chromatography. Partially purified rIFNα-2a from the immunoaffinity column is treated as described in the text and applied to the copper chelate column. Protein elution is effected by lowering the pH of the mobile phase. Fractions from Peak 2 are selected according to the results of PAGE (see Fig. 1).

steps. A column of the cation-exchange support, CM-52 (Whatman, Inc., Clifton, NJ), is equilibrated with a buffer containing 0.025 M ammonium acetate, pH 4.5 and 0.025 M sodium chloride. The pooled sample from the copper-chelate column is diluted with 5 volumes of water to reduce the ionic strength prior to loading onto the column. The column is then washed with several volumes of equilibration buffer until the absorbance at 280 nm is back to baseline levels. The protein is then eluted with a buffer consisting of 0.12 M sodium chloride in 0.025 M ammonium acetate, pH 5.0 (Fig. 3). The first peak is discarded and the second peak, containing rIFNα-2a, is collected as a pool.

Permeation Chromatography

The purpose of this step is to remove small amounts of contaminating oligomeric rIFNα-2a, as well as high-molecular weight endotoxins. It utilizes a column of Sephadex G-50 equilibrated with a solution containing 0.025 M ammonium acetate, pH 5.0 and 0.12 M sodium chloride. The pool from the CM-52 column is concentrated about fourfold prior to loading onto the column. The major peak, which elutes at about 11 hours, represents the product (Fig. 4). It is preceded by a smaller peak containing oligomeric interferon.

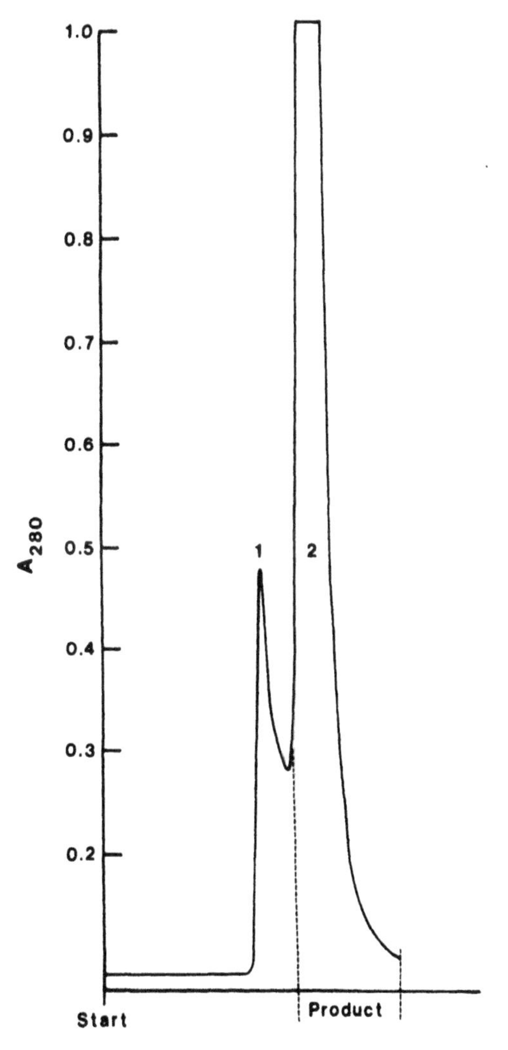

Figure 3 Cation-exchange chromatography. The selected fractions from the copper-chelate column are pooled, reduced ionic-strength and applied to the CM-52 column. Elution is effected by raising the salt concentration and pH of the mobile phase.

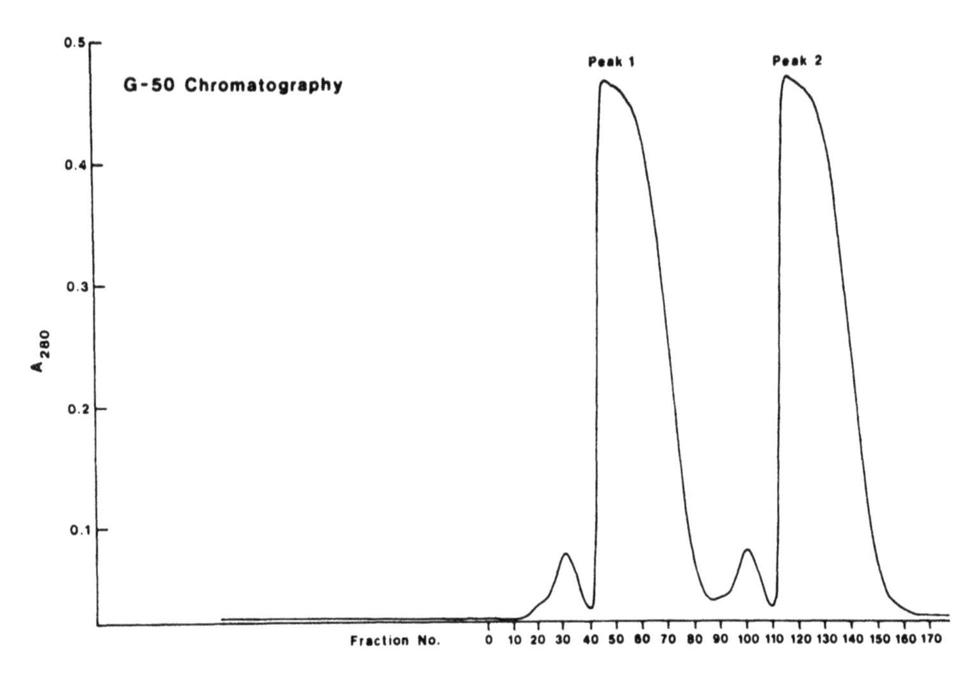

Figure 4 Permeation chromatography. Concentrated sample from the CM-52 column is applied to a Sephadex G-50 column. A second sample run is initiated before the fractions from the first injection have eluted from the column. Fractions containing minor peaks (15–40 and 90–110) are discarded. Fractions containing the product (41–88 and 112–160) are collected and pooled.

To use the column to its maximum capacity, it is possible to do two runs in a day. This is accomplished by loading the second sample 5 hours after loading the first sample (Fig. 4). The permeation mode of chromatography is uniquely suited for this procedure, since a single pass through the column provides the separation.

The Sephadex G-50 pool is concentrated by ultrafiltration using a membrane with a 10,000 dalton cutoff. The concentrate is sterile filtered and the product is stored frozen at –20°C or formulated immediately. The overall yield is about 200 mg/kg of *E. coli* cell paste, which represents a recovery of 30% of the cell content of rIFNα-2a.

ACKNOWLEDGMENTS

We wish to thank Ms. Lori Revello for typing the manuscript and Mr. Leo Bowski for critical review of the paper, as well as for supplying the cells.

REFERENCES

1. Isaacs, A. and Lindemann, J. *J. Proc. R. Soc. London Ser. B* 147, 258–267 (1957).
2. Goeddel, D. V., Yelverton, E., Ullrich, A., Heynsker, H. L., Miozzari, G., Holmes, W., Seeburg, P. H., Dull, T., May, L., Stebbing, N., Crea, R., Maeda, S., McCandliss, R., Sloma, A., Tabor, J. M., Gross, W., Familletti, P., and Pestka, S. *Nature* (London) 287, 411–416 (1980).
3. Staehelin, T., Hobbs, D. S., Kung, H. -F., Lai, C. Y., and Pestka, S. *J. Biol. Chem.* 256, 9750–9754 (1981).
4. Tarnowski, S. J., and Liptak, R. A. In *Advances in Biotechnological Processes*, Vol. 2. Edited by A. Mizyani and A. L. Van Wesel. Alan R. Liss, New York, 1983, p. 271.
5. Tarnowski, S. J., Roy, S. K., Liptak, R. A., Lee, D. K., and Ning, R. Y. In *Methods in Enzymology*, Vol. 119 Academic Press, New York, 1986, p. 153.
6. Porath, J., Carlsson, J., Classon, I., and Belfrage, G. *Nature* (London), 258, 598 (1975).

SUGGESTED READING

Pestka, S. The human interferons—From protein purification and sequence to cloning and expression in bacteria: before, between and beyond. *Arch. Biochem. Biophys.* 221, 1–37 (1983)

Bendig, M. M. The Production of foreign proteins in mammalian cells. In *Genetic Engineering*, Vol. 7. Academic Press, New York, 1988, pp. 91–127.

8

Production of Human Calcitonin by Recombinant DNA Technology

Barry N. Jones

Sterling Research Group
Sterling Drug, Inc.
Malvern, Pennsylvania

INTRODUCTION

Small peptides (below about 40 residues) may be produced on a scale suitable for clinical use by solid-phase synthesis procedures. At the present time, this synthetic approach becomes too difficult and costly to be practical for peptides of higher molecular weight. An alternative is to isolate these peptides from natural sources. For various reasons this too is generally not practical. Recombinant DNA technology promises to provide the most suitable and cost-effective approach for preparing peptides and proteins. There are already several examples of proteins produced by recombinant DNA technology for use as human therapeutics. These include human chorionic gonadotropin (1), growth hormone (2), tissue plasminogen activator (3), Factor VIII (4, 5), interferons (6, 7), and interleukins (8).

There is a major problem, however, associated with the production of small peptides by recombinant DNA technology. The direct expression in *Escherichia coli* of a gene encoding a peptide has not been found to yield significant amounts of the peptide product (9). This failure may be due to either of two factors; poor biosynthesis of the peptide or rapid degradation of the peptide. The latter possibility fits in with the known propensity of *E. coli* to degrade heterologous and abnormal proteins (10).

There are other considerations associated with the production of small peptides, such as those related to posttranslational modifications. For example, many biologically active peptides are amidated at the carboxyl terminus. This can be readily accommodated in solid-phase peptide synthesis, but it presents difficulties in the genetic engineering of amidated peptides. The reason for this and similar problems is that the biosynthetic precursors of small peptides are of molecular weights generally in excess of 20,000. The maturation process then occurs in the context of the precursor protein, as well as the environment of the cell in which biosynthesis occurs. While the *E. coli* host is cost effective for the production of human and other mammalian proteins, special attention must be given for a small peptide, as illustrated with the production of human calcitonin (11) by scientists at Unigene Laboratories.

HUMAN CALCITONIN

Calcitonin was originally identified as a factor having potent hypocalcemic actions (i.e., lowering calcium levels in blood). It is synthesized in the parafollicular cells of the thyroid gland of mammals and the ultimobranchial

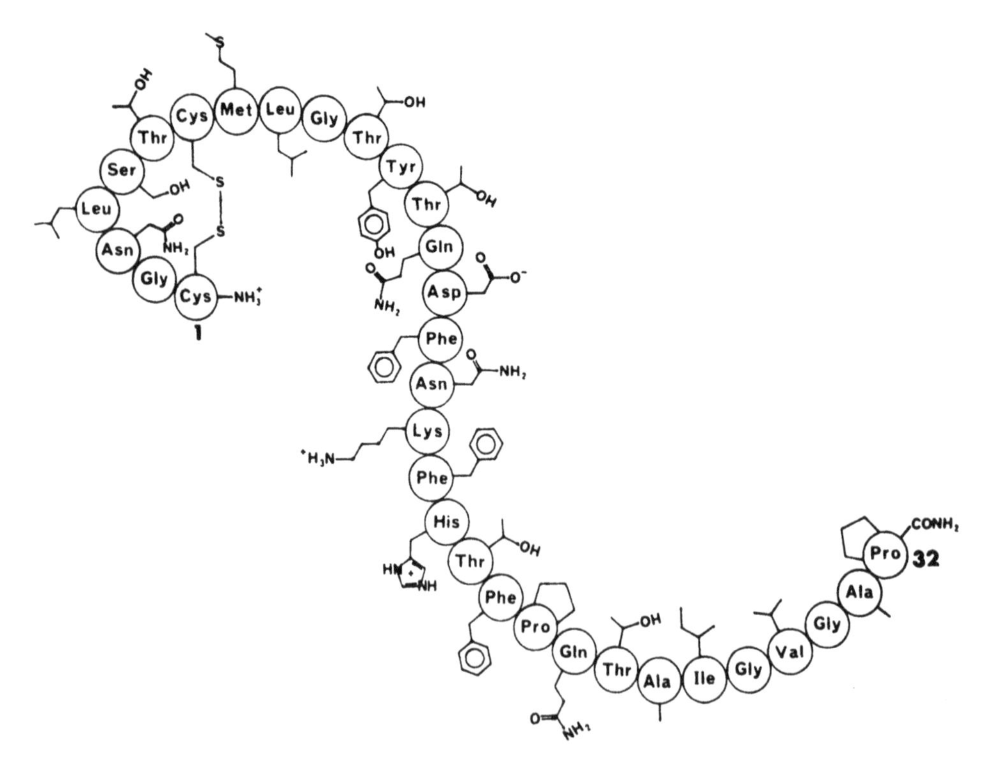

Figure 1 Primary structure of human calcitonin.

glands in lower vertebrates. The main physiological role ascribed to calcitonin is the mobilization of calcium from blood to bone. Calcitonin may also have a role in the central nervous system in the regulation of appetite. Based on its regulation of calcium metabolism, calcitonin is clinically useful in Paget's disease, a disorder characterized by demineralization of bone, leading to an increased incidence of fractures and bone deformity. Another potential therapeutic use is to ameliorate postmenopausal increase in bone porosity. Acceptance of this major clinical application to post menopausal osteoporosis would create a vastly increased demand for the peptide product.

Rat calcitonin was demonstrated to originate as a precursor protein of 136 residues, including a 25-residue leader sequence (12). The mature polypeptide, which is generated by regulated proteolysis, contains 32 amino acid residues (Fig. 1). There is one disulfide bridge between the cysteine residues at positions 1 and 7. As mentioned above, the carboxyl terminal residue, proline, is amidated.

DESIGN OF THE FUSION PROTEIN

To circumvent the poor yield of peptides in recombinant *E. coli*, the approach has been devised of fusing an endogenous *E. coli* protein, such as β-galactosidase, to the desired peptide (10). That is, the structural gene encodes both the protein and the peptide in tandem as a single transcript (see Fig. 2). One drawback of β-galactosidase as the fusion partner is its relatively high molecular weight (MW100,000) and, consequently, the proportion of the product represented by the peptide is low. In order to increase this ratio, it has been possible to use another endogenous protein (proprietary information) of lower molecular weight (ca. 20,000) as the fusion partner. Using this fused structural gene on a recombinant plasmid, it has been possible to achieve levels of expression in which the fusion protein represents greater than 10% of total cellular protein in *E. coli*. The presence and level of the fusion protein in a total cell lysate is readily monitored by polyacrylamide gel electrophoresis in comparison to a lysate of the untransformed host strain; a major band

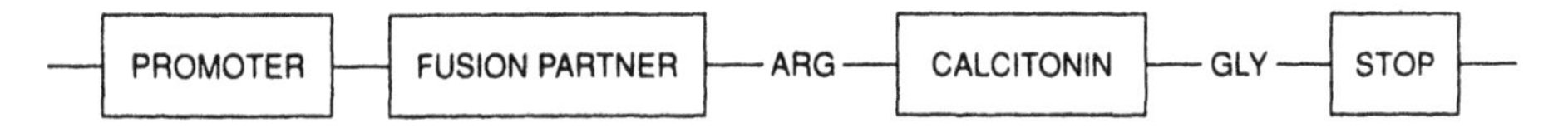

Figure 2 Design of the fusion protein gene. The fusion partner is an endogenous *E. coli* protein having a molecular weight of about 20,000. The structural gene codes for a protein in which the carboxyl terminal residue of the fusion partner is separated by one residue of arginine from the amino terminal residue of calcitonin, which itself is extended by one residue of glycine at its carboxyl terminus. The structural gene is terminated with a stop codon.

should be observed corresponding to the predicted molecular weight of the fusion protein.

The construct was designed in a manner that would allow the most facile method of excision from the fusion protein, as well as achievement of a product identical to native calcitonin (Fig. 1). A suitable promoter provided for high level expression of the fusion protein. The structural gene for calcitonin was ligated so that the amino terminal residue of the mature peptide (cysteine) was separated from the fusion partner by a residue of arginine. This residue would serve as a convenient site for cleavage from the fusion partner using the proteolytic enzyme, trypsin. The fusion gene was engineered to code for an additional residue of glycine beyond the carboxyl terminal residue (proline) of the mature peptide. This glycine residue would serve as the donor of the nitrogen atom found in the proline amide. The addition of a stop codon would result in this glycine residue being at the carboxyl terminus of the expressed protein.

PRODUCTION OF CALCITONIN

The production process is outlined in Table 1. An enzyme-linked immunosorbent assay (ELISA) (i.e., using anticalcitonin antibody) was found to be a convenient way to monitor for the presence and quantity of calcitonin, whether as part of a fusion protein or as the free peptide, throughout the purification process. The first step of purification took advantage of the insoluble nature

Table 1 Process for Production of Human Calcitonin

1. Synthesize an oligonucleotide corresponding to the amino acid sequence of the desired peptide
2. Ligate into a plasmid containing the fusion partner
3. Transform *E. coli*
4. Express fusion protein at high levels
5. Lyse *E. coli*
6. Isolate and solubilize inclusion bodies
7. Sulfonate and citraconylate the proteins
8. Purify the modified fusion protein by strong anion-exchange and size exclusion chromatography
9. Digest with trypsin, remove citraconic blocking group and isolate modified calcitonin by reverse-phase HPLC
10. Amidate the carboxyl terminus and form the disulfide bond
11. Purify the product by reverse-phase HPLC
12. Validate structure and demonstrate biological potency

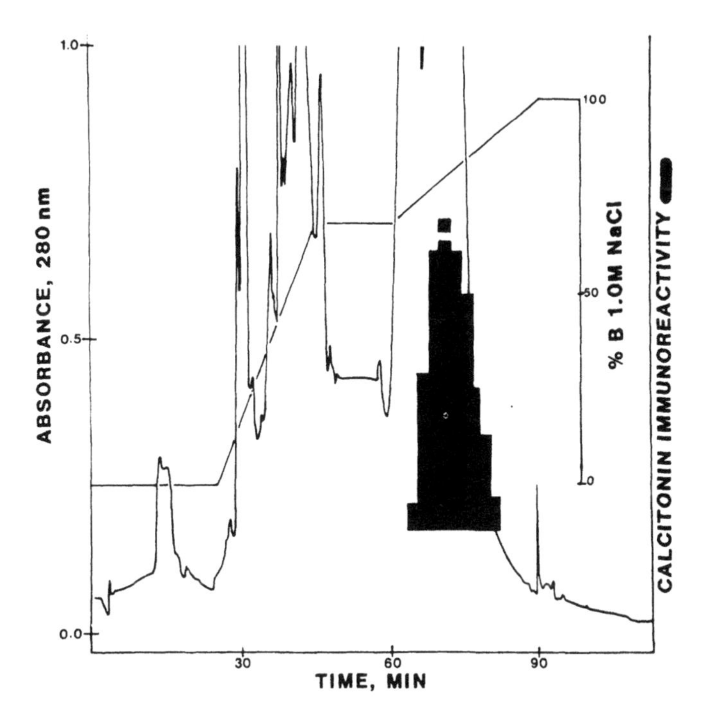

Figure 3 Strong anion-exchange chromatography of sulfonated, citraconylated calcitonin-fusion protein. A Mono Q HR-10 / 10 column was eluted with 20 mM ethanolamine, pH 9.6 using a gradient of sodium chloride, as indicated. Calcitonin immunoreactivity was determined by ELISA using a rabbit polyclonal antiserum.

often observed for recombinant proteins expressed at such high levels (13, 14). After cell lysis, these insoluble inclusion bodies were collected by differential centrifugation. The protein was solubilized in the strong denaturant, 6 M guanidine hydrochloride.

Before proceeding further, it was necessary to address the problem of interprotein disulfide linkages. The cysteine residues of proteins biosynthesized in *E. coli* normally exist in the reduced (sulfhydryl) state. After disruption of the cells and upon exposure to air, these sulfhydryl groups spontaneously oxidize, yielding random disulfide linkages between the desired product and *E. coli* proteins. A convenient procedure for preventing these covalent linkages, which would otherwise complicate further purification, involves sulfolysis of the cysteine residues immediately upon extraction (15).

Another procedure that was done before proceeding with the purification was to block the lysine residues with citraconic anhydride. The rationale for this is as follows. The primary structure of calcitonin (Fig. 1) includes one residue of lysine and no residues of arginine. The strategy utilized in this project was, therefore, to place an arginine residue as the bridge between the fusion partner and the amino terminal residue of the desired peptide product (Fig. 2). Digestion of the fusion protein with trypsin, which cleaves at both lysine and arginine residues, would release the calcitonin portion. Citraconylated lysine residues are not recognized by trypsin, thereby avoiding internal cleavage of the calcitonin portion. Modifications by citraconylation and sulfolysis are readily reversible.

Although the fusion protein represented about 60% of the total protein of the solubilized inclusion bodies, it was decided to further purify the protein prior to enzymatic release of the peptide. This additional purification would help avoid contamination of the final product by fragments of *E. coli* proteins. After sulfolysis and citraconylation (both processes impart negatively charged groups), the fusion protein was found to be soluble at alkaline pH. The sulfonated, citraconylated fusion protein was chromatographed on a strong anion exchange column, using Mono Q on a small scale and Q Sepharose (both from Pharmacia) on a large scale (Fig. 3). Further purification was done by size exclusion chromatography (Fig. 4). The fusion protein was then nearly pure according to polyacrylamide gel electrophoresis.

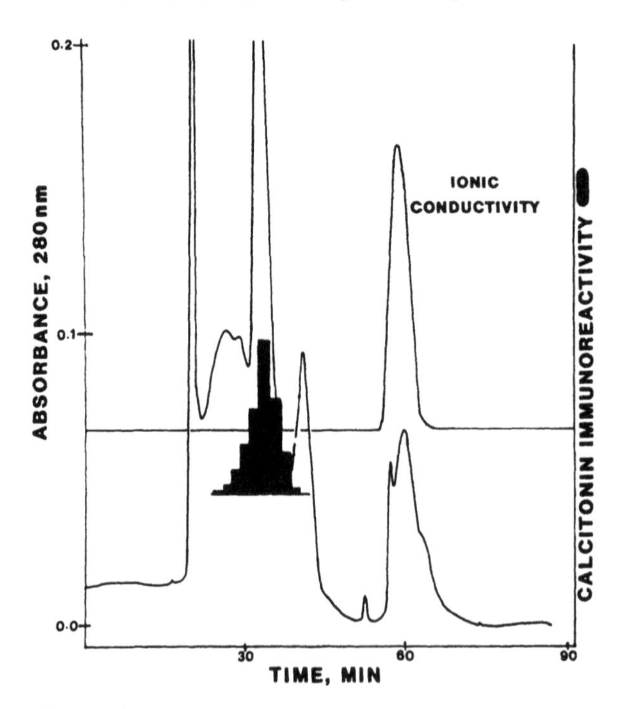

Figure 4 Size-exclusion chromatography of partially purified calcitonin-fusion protein. A Superose column (10 X 300 mm) was eluted with 0.1 M ethanolamine, pH 9.5.

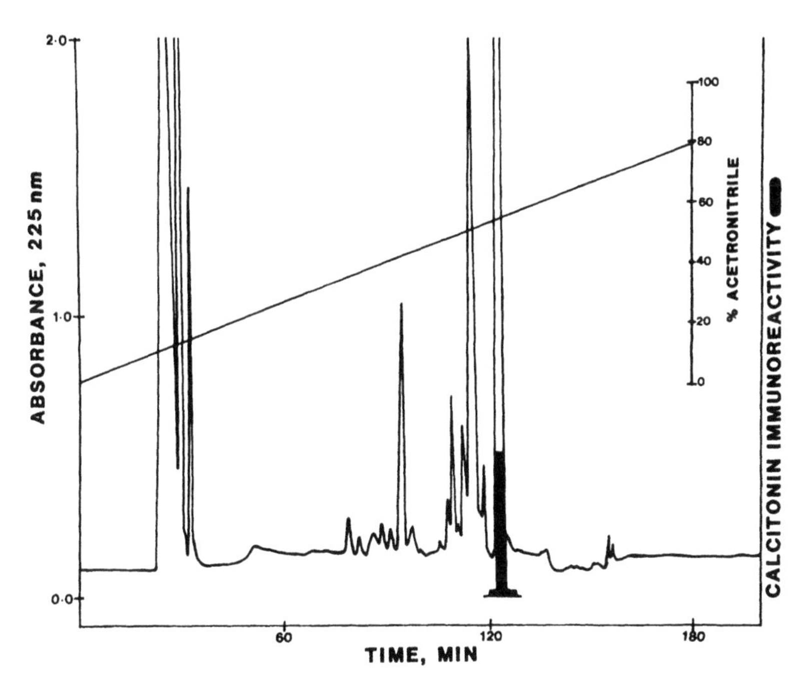

Figure 5 Reverse-phase chromatography of a tryptic digest of purified calcitonin fusion protein. An octadecyl-silica column was eluted with 0.1% trifluoroacetic acid using a gradient of acetonitrile, as indicated.

After digestion of the fusion protein with trypsin, the digest was acidified with acetic acid to remove the citraconic blocking group. The peptide fragments were separated by reverse-phase HPLC (Fig. 5). The peak corresponding to glycine extended-calcitonin was readily detected by the ELISA. The glycine residue was converted into the amide group using an α-amidating enzyme (16, 17). The sulfonate groups were removed and the disulfide bridge was allowed to reform by incubating the α-amidation reaction mixture with 2-mercaptoethanol. The product was shown to be identical to natural human calcitonin by reverse-phase HPLC (Fig. 6). Precursor forms, corresponding to sulfonated or glycine-extended calcitonin, were absent from the product, indicating that the reactions had gone essentially to completion (Fig. 6).

ANALYSIS OF THE PRODUCT

The product, recombinant human calcitonin, was shown to elute from a reverse-phase HPLC column at the same position as authentic natural human-calcitonin (Fig. 6). Both recombinant and natural human calcitonin were di-

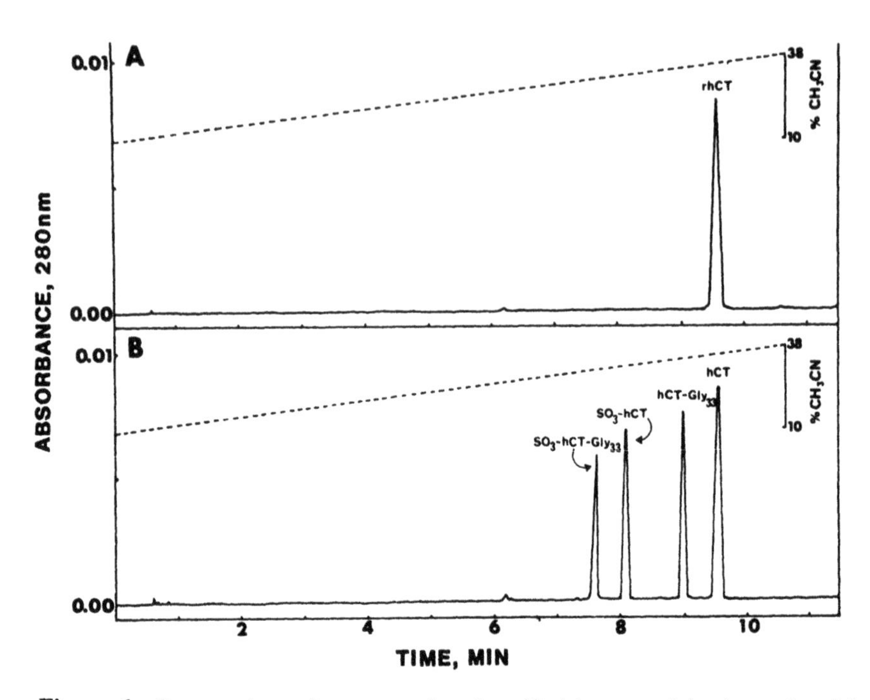

Figure 6 Reverse-phase chromatography of purified human calcitonin produced by recombinant DNA technology. The chromatography positions of the various precursor forms are shown in panel B to demonstrate their absence from the product shown in panel A.

gested with trypsin, and the same two fragments (residues 1–18 and 19–32) were observed by reverse-phase HPLC (not shown). Amino acid compositional analysis and sequencing by automated Edman degradation confirmed the structure of the recombinant peptide. Most importantly, the product had the full biological potency, as measured by lowering of serum calcium levels in rats.

REFERENCES

1. Goeddel, D. V., Heyneker, H. L., Hozumi, T., Arentzen, R., Itakura, K., Yansura, D. G., Ross, M. J., Miozzari, G., Crea, R., and Seeburg, P. H. *Nature* 281, 544–548 (1979).
2. Hsiung, N. M., Mayne, N. G., and Becker, G. W. *Biotechnology* 4, 991–995 (1986).
3. Pennica, D., Holmes, W. E., Kohr, W. J., Harkins, R. N., Vehar, G. A., Ward, C. A., Bennett, W. F., Velverton, E. Y., Seeburg, P. H., Heyneker, H. L., Goeddel, D. V., and Collen, D. *Nature* 301, 214–221.

4. Toole, J. J., Knopf, J. L., Wozney, J. M., Sultzman, L. A., Buecker, J. L., Pittman, D. D., Kaufman, R. J., Brown, E., Shoemaker, C., Orr, E. C., Amphlett, G. W., Foster, W. B., Coe, M. L., Knutson, G. J., Fass, D. N., and Hewick, R. M. *Nature* 312, 342–347 (1984).
5. Wood, W. I., Capon, D. J., Simonsen, C. C., Eaton, D. L., Gitschier, J., Keyt, B., Seeburg, P. H., Smith, D. H., Hollingshead, P., Wion, K. L., Delwart, E., Tuddenham, E. G. D., Vehar, G. A., and Lawn, R. M. *Nature* 312, 330–337 (1984).
6. Goeddel, D. V., Yelverton, E., Ulrich, A. Heyneker, H. L., Miozzari, G., Holmes, W., Seeburg, P. H., Dull, T., May, L., Stebbing, N., Crea, R., Maeda, S., McCandliss, R., Sloma, A., Tabor, J. M., Gross, M., Familletti, P. C., and Pestka, S. *Nature* 287, 411–416 (1980).
7. Goeddel, D. V., Sheppard, H. M., Yelverton, E., Leung, D., Crea, R., Sloma, A., and Pestka, S. *Nucleic Acids Res.* 8, 4054–4057 (1980).
8. Wingfield, P., Payton, M., Tavernier, J., Barnes, M., Shaw, A., Rose, K., Simona, M. G., Demczuk, S., Williamson, K., and Dayer, J. -M. *Eur. J. Biochem.* 160, 491–497 (1986).
9. Itakura, K., Hirose, T., Crea, R., Riggs, A. D., Heyneker, H. L., Bolivar, F., and Boyer, H. W. *Science* 198, 1056–1063 (1977).
10. Goldberg, A. L. and Goff, S. A. In *Maximizing Gene Expression*. Edited by *Reznikoff* and *Gold*, Butterworth Publishers, Stoneham, MD, 1986, pp. 287–311.
11. Gilligan J. P., Warren, T. G., Koehn, J. A., Young, S. D., Bertelsen, A. H., and Jones, B. N. Biochromatogr. 2, 20–27 (1987).
12. Jacobs, J. W., Goodman, R. H., Chin, W. W., Dee, P. C., Habener, J. F., Bell, N. H., and Potts, Jr., J. T. *Science* 213, 457–459 (1981).
13. Schoner, R. G., Ellis, L. F., and Schoner, B. E. *Biotechnology 3,* 151–154 (1985).
14. Williams, D. C., Van Frank, R. M., Muth, W. L., and Burnett, J. P. *Science* 215, 687–689 (1982).
15. Cabily, P., Riggs, A. D., Pande, H., Shively, J. E., Holmes, W. E., Rey, M., Perry, L. J., Wetzel, R., and Heyneker, H. L. *Proc. Natl. Acad. Sci.* (*USA*) 81, 3273–3277 (1984).
16. Eipper, B. A., Mains, R. E., and Glembotski, C. C. *Proc. Natl. Acad. Sci.* (*USA*) 80, 5144–5148 (1983).
17. Gilligan J. P., Lovato, S. J., Young, S. D., Jones, B. N., Koehn, J. A., LeSueur, L. F., Sturmer, A. M., Bertelsen, A. H., Warren, T. G., Birnbaum, R. S., and Ross, B. A. In *Advances in Gene Technology: Molecular Biology of the Endocrine System*. ICSU Press, Cambridge, MA 1986, pp. 38–39.

9

Structural Analysis of Proteins

Barry N. Jones

Sterling Research Group
Sterling Drug, Inc.
Malvern, Pennsylvania

Stanley Stein

Center for Advanced Biotechnology and Medicine
Piscataway, New Jersey

INTRODUCTION

This chapter deals with the elucidation of the primary structure (i.e., the sequence of amino acids) of a protein. There are numerous reasons for determining the sequence of a protein. In the case of a previously uncharacterized protein, the primary structure may be considered to be a key step in the characterization of that protein. Such sequence information would be required in a patent application. If the intent is to produce that protein on a commercial scale for biomedical or other applications, even partial sequence data is a valuable aid in cloning the gene for that protein. The entire sequence of the protein can then be determined from the sequence of the structural gene. For a small peptide (< 40 residues), it may be preferable to chemically synthesize the product and, hence, the complete structure must first be elucidated. Whether the polypeptide is to be produced by recombinant DNA technology or by chemical synthesis, it is desirable to have a complete and accurate determination of the primary structure of the natural material, including any posttranslational modifications.

Structural information on a previously characterized protein may be important for other purposes. For example, it may be of interest to analyze specific functional regions of a protein, such as the active site of an enzyme or the receptor binding domain of a hormone. It is also possible to solve the three-dimensional structure of a protein by x-ray crystallography. Knowledge of the primary structure is a prerequisite for these studies.

Sometimes it is of interest to compare the primary structure of the same protein from several different species. Such a comparison can provide information on the evolutionary or phylogenetic relationships among different species (1). An understanding may be gained of the correspondence between the functional attributes of a protein and slight variations in its primary structure. One illustration of this is the hemoglobin family. Within the human population, the most well known variant is sickle hemoglobin, in which there is a substitution of only one amino acid in the beta-chain; this results in a protein that tends to crystallize in the red blood cells. There are numerous variants of human hemoglobin that result in less marked symptoms.

From the perspective of the biotechnology industry, it is necessary to analyze every batch of a protein produced by recombinant DNA technology or every batch of a chemically synthesized peptide. This quality control application may be extremely involved and utilize sophisticated analytical technology. Even low levels of contamination by peptide or protein molecules with erroneous structures (to be discussed below) may be potentially hazardous.

Methodology

There has been a significant conceptual change in protein structural analysis due to the great advances in recombinant DNA technology. The classical approach was to determine the entire sequence of a protein. Even with modern procedures, this would still be a tedious task, especially for a large protein. The new approach utilizes DNA sequencing technology. The structural gene for the protein is isolated by cloning procedures, sequenced, and then the nucleotide sequence is translated into the amino acid sequence according to the genetic code.

There have been significant advancements in the technology for protein analysis. However, the general overall approach for primary structure elucidation remains essentially unchanged. The volume edited by Shively (2) provides excellent review articles on the techniques discussed below.

Amino Acid Composition

Sample Treatment

A sample of a protein or peptide is hydrolyzed to its constituent amino acids in the presence of strong acid and heat. Most commonly, hydrolysis is done in 5.7 N hydrochloric acid (constant boiling) for 24 h at 110°C. Antioxidants are often added during hydrolysis in order to prevent oxidative destruction of the side chains of some amino acids (e.g., tyrosine, tryptophan and methionine). The amidated amino acids, asparagine and glutamine, are deamidated during hydrolysis to their respective carboxylate analogs, aspartic acid and glutamic acid. In certain applications, peptides are enzymatically digested to their constituent amino acids, and in such cases, deamidation does not occur.

Sometimes, certain amino acids are converted to other derivatives prior either to hydrolysis or to chromatographic analysis. For example, proteins are treated with a reducing agent to convert cystine to cysteine, followed by alkylation of the sulfhydryl group. Performic acid oxidation is used for analysis of cysteine and methionine in their oxidized forms. Specific applications will dictate how the sample is to be treated.

The Chromatographic Analysis

The first automated amino acid analyzer was reported by Spackman et al. in 1958 (3). In their approach, the amino acid mixture is resolved on a strong cation exchange chromatography column; the stationary phase support is sulfonated polystyrene. The mobile phase of citrate buffer is programmed to form a gradient of increasing pH and ionic strength with time. The acidic amino acids elute first, followed by the neutral and then the basic amino acids. The amino acids are detected by postcolumn reaction with ninhydrin, which is continu-

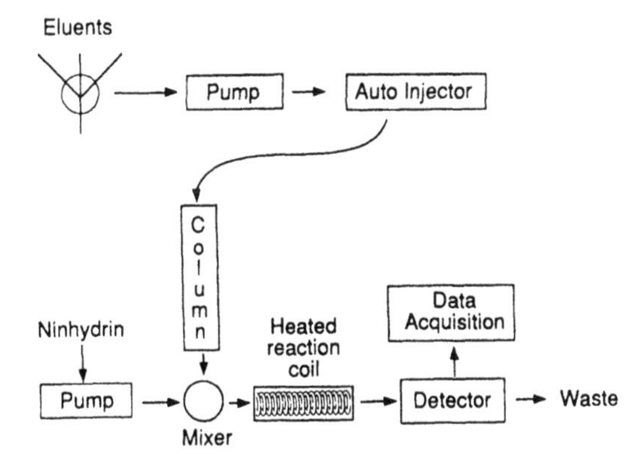

Figure 1 Block diagram of the classical amino acid analyzer. The eluents are sodium citrate buffers of different pH and ionic strength. One is selected at a time using a valve and pumped through the thermostatted cation-exchange column. Sample hydrolysates may be applied to the column manually or with an automatic sample injector. The column effluent is mixed with ninhydrin reagent, which is added with a second pump. The mixture is heated for several minutes and passed through a detector, which measures the absorbance at 570 nm (and 440 nm for proline). The detector signal goes to a chart recorder, integrator or other data acquisition system.

A similar apparatus would be used for amino acid analysis by precolumn derivatization. In this case, the second pump, the mixer and the heated reaction coil would be eliminated. A reverse-phase, rather than an ion-exchange column, as well as the appropriate eluents and detector would be used.

ously added to the column effluent. The mixture is heated and passed through a colorimeter. The resulting blue color is proportional to the amino acid concentration. A block diagram of the instrument is shown in Figure 1 and a chromatogram of an amino acid mixture is shown in Figure 2. To this day, this process remains the one most commonly used for amino acid analysis.

Modifications of and alternatives to this procedure have been developed. In order to achieve a lower limit of detection, the colorimeter can be replaced by a fluorometer in conjunction with the replacement of ninhydrin by either fluorogenic reagent, *o*-phthalaldehyde (OPA) (4) or fluorescamine (5). Precolumn derivatization of the amino acid mixture, as opposed to postcolumn reaction, has gained in popularity. There are several advantages. The amino acid derivatives are readily separated by reverse-phase high-performance liquid chromatography (HPLC). Reverse-phase chromatography is superior to ion-exchange in terms of flexibility, simplicity, and cleanliness of the mobile phases, and resolving power. With respect to instrumentation, it is advantageous not to have a postcolumn pump and reactor which add an extra measure of complication (cf., Fig. 1). Indeed, standard HPLC equipment can be used, whereas the postcolumn system must be dedicated to that particular application.

One of the earliest precolumn derivatization reagents to be applied was OPA (6), which gives fluorescent derivatives. A drawback is that the imino acid, proline, does not react and, therefore is not detected. Phenylisothiocyanate (PITC), which is the reagent used for sequencing (see below), can be used to determine all the amino acids, including proline. Detection of phenylthiocarbamyl derivatives of amino acids by ultraviolet absorbance (approx. 250 nm) does not offer the sensitivity of the fluorescence reagents, but the limit of analysis of about 30 picomoles is sufficient for most applications. In order to utilize PITC at this level, it is necessary to have scrupulously clean reagents and to have a chromatographic system capable of resolving the many interfering substances detected by ultraviolet absorption. To satisfy these needs, Waters Associates, Inc., a leading manufacturer of chromatographic equipment, has developed a line of PicoTag reagents.

The Data

The amino acid composition is characteristic for a given protein. In the case of a small peptide, the composition should agree closely with the true integer numbers for the constituent amino acid residues. With a large protein, there can be large deviations from the true integer numbers for certain amino acids. This inaccuracy is mainly due to two factors. First, the peptide bonds are not all equivalent; the bulky hydrophobic residues are most resistant to hydrolysis. Thus, the amino acids are released from the protein chain at different rates, depending both on the type of amino acid and its location in the chain. Second, some amino acids are partially destroyed during hydrolysis. Threonine

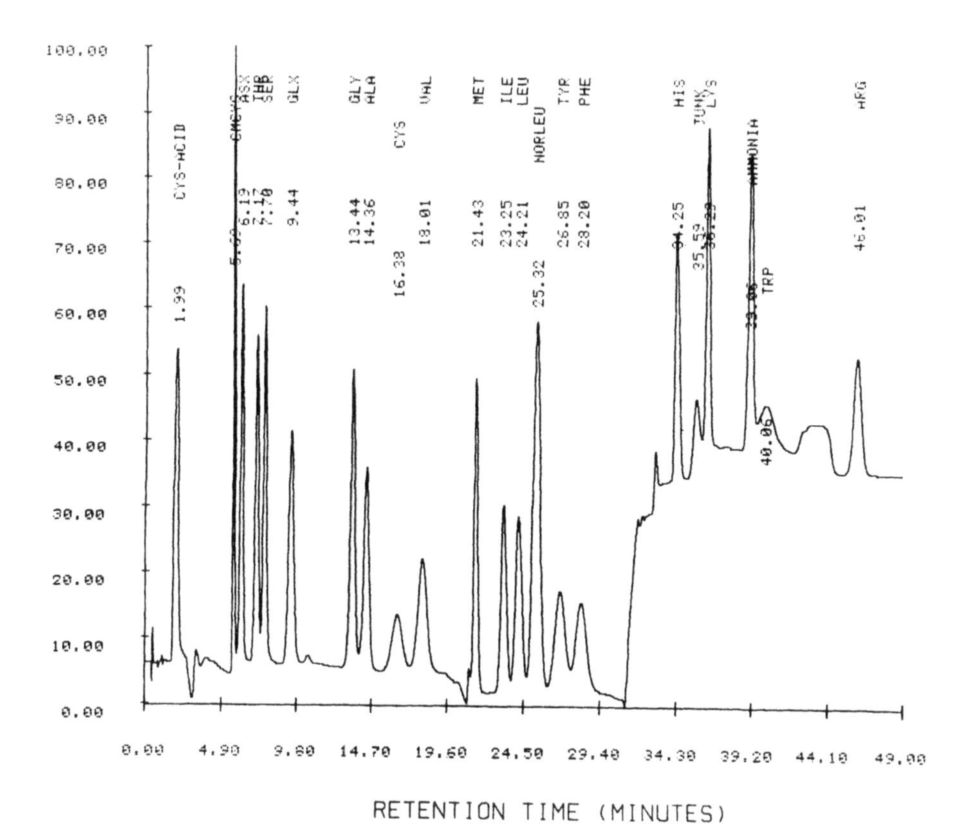

Figure 2 Chromatogram of an amino acid standard mixture. The detector was set at 0.1 AUFS (absorbance units-full scale) at 570 nm and the sample represented 250 picomoles of each amino acid. The amino acids CYS-ACID (1.99 min) refers to cysteic acid, CMCYS (5.69 min) to carboxymethylcysteine, and NORLEU to the internal standard, norleucine. Proline is seen as a tiny peak eluting (10 min) just after GLX. Detection on the 440 nm channel of the dual wavelength detector (not shown in this chromatogram) gives a higher response for proline and this latter channel is used for quantitation of this imino acid. The baseline shift that occurs prior to elution of the basic amino acids (beginning at 32 min) coincides with the increase in ionic strength and pH of the mobile phase. It is due to the presence of ammonia in the citrate buffers.

and serine, which have hydroxyl moieties, are lost by nonoxidative processes. The presence of reducing agents can minimize loss of other amino acids, such as methionine and tryptophan. This instability increases when the free amino acid is released from the protein chain. One way to compensate for these temporally opposite factors is to do a time course of hydrolysis study (e.g., 24, 48, and 72 h). The values for the hydrophobic amino acids are obtained from

Table 1 Amino Acid Composition of Triplicate Hydrolysates of HSA Performed on 4 Separate Days

Amino acid	Theoretical composition[a]	Calculated composition				4-Day average
		Day 1	Day 2	Day 3	Day 4	
Asx	53	57.9 ± 1.8[b]	53.7 ± 3.0	56.3 ± 1.5	59.9 ± 2.8	56.9 ± 2.6
Ser	24	28.8 ± 6.3	32.5 ± 3.5	31.5 ± 2.4	19.9 ± 1.3	28.1 ± 5.7
Gly	12	14.3 ± 0.7	13.1 ± 0.6	13.6 ± 0.7	14.0 ± 0.4	13.8 ± 0.5
Glx	81	87.1 ± 3.5	82.2 ± 5.0	88.1 ± 1.8	87.5 ± 3.5	86.2 ± 2.7
Thr	28	22.4 ± 4.8	29.2 ± 2.0	18.7 ± 0.7	26.0 ± 1.7	24.1 ± 4.6
Ala	62	66.8 ± 2.8	68.3 ± 0.7	57.1 ± 0.9	67.3 ± 2.6	64.9 ± 5.2
Val	41	33.5 ± 1.3	36.8 ± 2.5	28.4 ± 0.6	37.4 ± 1.5	36.5 ± 2.1
Met	6	6.5 ± 0.3	7.1 ± 1.3	5.9 ± 0.2	6.0 ± 1.0	6.4 ± 0.5
Tyr	18	18.4 ± 0.3	17.3 ± 2.3	20.2 ± 1.0	22.9 ± 2.0	19.8 ± 2.3
Ile	8	6.1 ± 0.8	5.9 ± 0.7	6.8 ± 0.2	7.0 ± 0.3	6.4 ± 0.5
Leu	61	60.8 ± 0.9	58.6 ± 1.8	64.2 ± 2.3	64.3 ± 3.8	62.0 ± 2.8
Phe	31	28.6 ± 0.6	28.2 ± 0.6	30.5 ± 1.3	31.7 ± 1.7	29.7 ± 1.7
His	16	11.1 ± 0.7	12.7 ± 0.9	14.0 ± 0.3	16.3 ± 1.2	13.5 ± 2.2
Lys	60	51.0 ± 7.9	64.3 ± 1.6	43.9 ± 1.7	62.2 ± 9.3	55.4 ± 9.6
Trp	1	0.7 ± 0.2	0.9 ± 0.1	0.6 ± 0.1	0.8 ± 0.2	0.8 ± 0.1
Arg	24	25.2 ± 1.7	21.2 ± 2.4	20.0 ± 1.2	19.1 ± 1.2	21.4 ± 2.7
Pro[c]	24	ND	ND	ND	ND	ND

[a]Theoretical composition is predicted from the gene sequence.
[b]Mean ± average or standard deviation.
[c]Proline was not determined.

the longest time point, whereas the values for unstable amino acids are obtained by extrapolating the data to zero time. These inaccuracies are also reflected in a poorer precision, from hydrolysate to hydrolysate, than analytical chemists are accustomed. Typical data for a large protein are given in Table 1.

In many cases, amino acid analysis provides the most accurate means for determining the total protein content of a sample. This is because other methods of measuring protein content can be exceedingly inaccurate. They suffer from variability in response factor among different proteins, as well as from both positive and negative errors from other components in the sample.

Peptide Mapping

This technique refers to the act of cleaving the protein chain and then separating the resultant peptide fragments for both analytical and preparative purposes. The molecular cuts may be done either by enzymatic or chemical means. The theoretical cleavage sites for the protein, human alpha-interferon, with the enzyme trypsin are depicted in Figure 3. This enzyme cleaves at the carboxy side of the basic (positively charged) amino acids, arginine and lysine. Complete digestion of this protein would result in the generation of 16 peptides, ranging in size from 2 amino acid residues (there are 3 such peptides) to 29 residues (positions 84–112 inclusive). The digestion mixture should also contain two residues of Arg, two residues of Lys, and one residue of Glu.

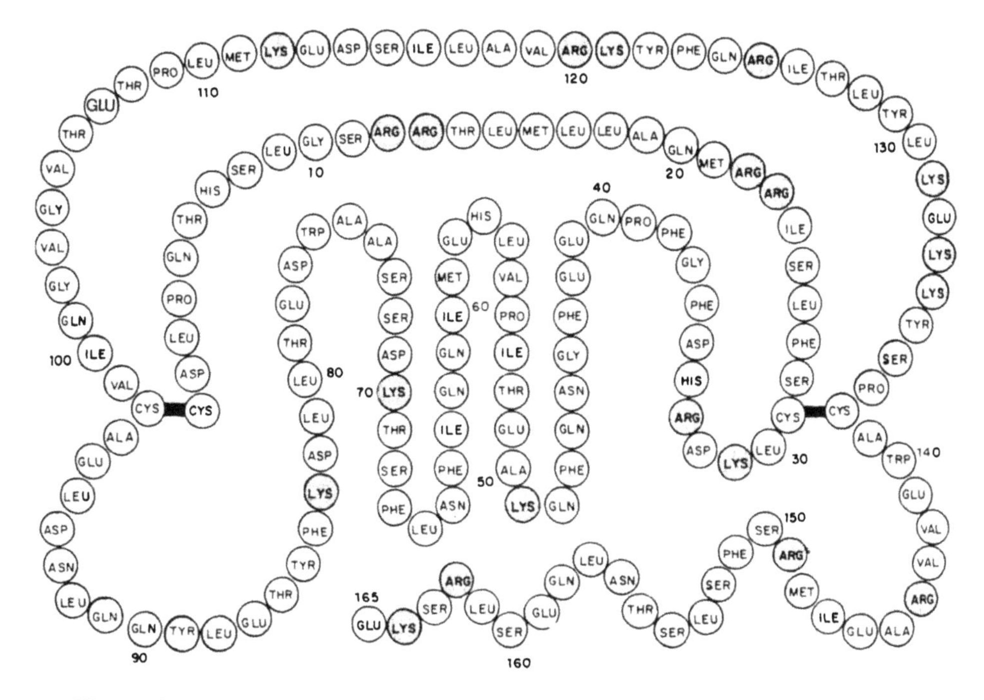

Figure 3 Tryptic fragmentation sites of human alpha$_2$-interferon. The primary structure of this protein is depicted. Cysteine at position 1 represents the amino terminus of the protein. The proteolytic enzyme, trypsin will potentially cleave at all lysine and arginine residues, which are represented as shaded circles. When two or more basic residues are in tandem, trypsin will, generally, cleave either one or the other in a random fashion. Due to tertiary structure constraints, some potential sites may be unavailable for cleavage. Unless the disulfide linkages between the cysteine residues are chemically broken, some of the resultant products of trypsin digestion may contain more than one polypeptide chain. In this illustration, two peptides would be linked by cysteines at positions 1 and 98, while another pair of peptides would be linked through positions 29 and 138. The shortest peptides in the digestion mixture include Asp-Arg (residues 32–33), Glu-Lys (residues 132–133) and Ser-Arg (residues 163–164). The carboxyl terminal residue of Glu (position 165) is released as the free amino acid.

The enzymatic fragmentation often does not go to completion. In such cases, a denaturant such as 4 M urea may be added during the digestion. It is also common to break the disulfide bonds in the protein prior to the digestion. This is accomplished by reductive cleavage with a reagent such as dithiothreitol followed by alkylation with a reagent such as iodoactic acid. The alkylated derivative, carboxymethyl cysteine, is readily quantitated by amino acid analysis (see Fig. 2). Alternatively, the disulfide bonds may be broken after the digestion but prior to chromatographic separation of the fragments.

There are several enzymes that are commonly used for structural analysis of proteins. The enzyme Lys-C endoproteinase, for example, is similar to trypsin but cleaves only at lysine and not at arginine. The enzyme called V8-protease (isolated from *Staphylococcus aureus*), cleaves at the carboxyl side of the negatively charged amino acids, glutamic acid and aspartic acid. Under certain conditions (pH 8 and ammonium buffer) cleavage by V8 protease is specific for Glu.

Chemical fragmentation procedures are also used. The most notable is cyanogen bromide in strong acid (e.g., 70% formic acid), which breaks the amide bond between the carboxyl group of methionine and the amino group of the next amino acid. In the process, methionine is converted to homoserine (also the cyclized form, homoserine lactone). Thus, the progress of the reaction can be followed by amino acid analysis of aliquots of the digest for the disappearance of methionine. Residues of oxidized methionine (i.e., the sulfoxide or sulfone) in the protein, which can form during the isolation procedure, are not attacked by cyanogen bromide. Human alpha$_2$-interferon has five residues of methionine and, accordingly, should yield 6 peptide fragments upon complete digestion. The carboxyl terminal fragment of the protein may be identified as the one lacking a residue of homoserine, as determined by amino acid analysis.

The main purpose of peptide mapping is to generate fragments of the protein for sequencing, as described in the next section. This is necessary, because Edman degradation of a protein generally elucidates only about 20 to 50 residues. Alpha-interferon (Fig. 3) is composed of 165 amino acids and, therefore, the remaining primary structure must be obtained from peptide fragments. In the case of a recombinant protein, the combined sequence analyses of all the isolated peptide fragments confirm the structure of that protein. With a previously uncharacterized protein, however, the sequence information for each of the peptide fragments is not sufficient. What is missing is the knowledge of how the peptides are aligned in the protein. To gain this latter information, aliquots of the protein are digested in two different ways (e.g., trypsin and cyanogen bromide). Sequencing of the two sets of peptide fragments then provides overlapping data for assigning the amino acid positions in the protein. The most common methods for the specific cleavage of proteins are listed in Table 2.

Table 2 Methods for the Specific Cleavage of Peptides and Proteins

Cleavage Reagent	Modification reagent / reaction conditions	Residue modified	Major cleavage site(s)	Reference
Trypsin	None	None	Arg-X; Lys-X; X≠Pro	12
Trypsin	Methyl acetimidate	Lys	Arg-X; X≠Pro	13
Trypsin	Citraconic anhydride	Lys	Arg-X; X≠Pro	14
Trypsin	1,2-Cyclohexanedione	Arg	Lys-X; X≠Pro	15
Trypsin[1]	Ethylenimine	Cys	Arg-X; Lys-X; Cys-X X≠Pro	16
Endopeptidase LysC	None	None	Lys-X	17
Clostripain	None	None	Arg-X	18
Submaxillarius protease	None	None	Arg-X	19
Armillaria mellea	None	None	X-Lys	20
Armillaria Mellea[a]	Ethylenimine	Cys	X-Cys; X-Lys	20
Chymotrypsin[b]	None	None	Trp-X; Phe-X; Tyr-X	12
S. protease V8[c]	Ammonium acetate / pH 4.0	None	Glu-X	21
S. protease V8[c]	Phosphate / pH 7.8	None	Asp-X; Glu-X	21
Postproline enzyme[d]	None	None	Pro-X; X≠Pro	22
Pancreatic elastase	None	None	Ala-X; Gly-X; Ser-X Val-X	23
Thermolysin	None	None	X-Phe; X-Leu; X-Val X-Ile; X-Tyr; X-Met	24
α-Protease	None	None	X-Leu; X-Val; X-Ile	25
Cyanogen bromide[e]	None		Met-X	17, 26
Cyanogen bromide[e]	Heptafluorobutyric acid	None	Met-X; Trp-X	27
Cyanogen bromide[e;f]	Methyl p-nitrobenzene Sulfonate	Cys	Met-X; X-Cys	26
BNPS-Skatole[g]	None	None	Trp-X	28
N-Chlorosuccinimide[g]	Urea	None	Trp-X	29
N-Bromosuccinimide[g]	None	None	Trp-X; Tyr-X	30
2-Nitro-5-thiocyano-benzoic acid	None	None	X-Cys	31
Hydroxylamine[h]	None	None	Asn-Gly	32
Mild acid hydrolysis	Pyridine acetate buffer	None	Asp-Pro	33

[a]Cys residues are reversibly aminoethylated.

[b]Not very specific, cleavage at Met, Leu, His, Asn, Lys, Gln, and Thr can also occur.

[c]Slow cleavage rate when X=Leu, Phe, or Val.

[d]Slow cleavage at Ala-X. Peptide substrates should contain less than 50 residues.

[e]Met is modified to homoserine lactone.

[f]Cys is modified to serine.

[g]Met is oxidized to sulfone.

[h]Cleavage can also occur at Asn-Leu, Asn-Ala, and Asn-Met.

Automated Edman Degradation

The Process

The procedure developed by Edman and Begg (7) is routinely used to determine the sequence of a peptide or protein. The steps in this procedure are presented in Figure 4. In the first step, the Edman reagent, PITC, is covalently coupled to the amino group of the amino terminal residue of the protein. Formation of this phenylthiocarbamyl (PTC) protein is done at an alkaline pH using a base such as trimethylamine. Excess PITC and base are removed from the derivatized protein by evaporation and extraction. In the second step, an anhydrous acid, such as neat trifluoroacetic acid, is added. This causes a cyclization reaction, which results in cleavage of the peptide bond between the first and second amino acid residues. The cyclized anilinothiazolinone (ATZ) derivative is extracted into an organic solvent, such as *n*-butyl chloride, leaving the insoluble remainder of the protein behind. In the third step, the extracted ATZ amino acid is treated with aqueous acid, such as 25% trifluoroacetic acid, which causes rearrangement (conversion) to the more stable phenylthiohydantoin (PTH) derivative. The identity of this derivative, which represents the first amino acid (counting from the amino terminus) in the chain, is then determined by reverse-phase HPLC.

The amino acid representing the second residue in the original protein chain is now positioned at the amino terminus (see Fig. 4, step b). It is possible to repeat the Edman degradation, beginning with the addition of PITC and base, to derivatize, cleave, and identify the second amino acid in the chain. Repetitive cycles of degradation can, typically, be used to determine the sequence of between 20 and 50 residues.

There are several factors which limit the extent to which a protein may be

Figure 4 The Edman degradation. The first three residues of the peptide (or protein) chain are depicted with the R groups representing the side chains of the amino acid residues. In the coupling step, the Edman reagent, PITC, is reacted with the amino group of the first residue (amino terminal) of the peptide, under alkaline conditions, to give the PTC peptide. In the next step, peptide bond cleavage, specifically at the amino terminal residue, is accomplished by treatment with anhydrous acid. The released ATZ amino acid is extracted into an organic solvent, leaving behind the remainder of the peptide, now with residue 2 at the amino terminal position. This peptide is again reacted with PITC to derivatize and release the second residue, thereby placing the third residue at the amino terminal position. While this is occurring, the ATZ amino acid is being treated with aqueous acid for conversion to the PTH derivative in a separate reaction vessel. The degradation is repeated until no further information is gained from the chromatographic analyses.

A. COUPLING

$$Ph{-}N{=}C{=}S + H_2N{-}\underset{R^1}{CH}{-}CO{-}NH{-}\underset{R^2}{CH}{-}CO{-}NH{-}\underset{R^3}{CH}{-}CO{-}\ldots \longrightarrow$$

$$Ph{-}\underset{H}{N}{-}\overset{S}{\overset{\|}{C}}{-}\underset{H}{N}{-}\underset{R^1}{CH}{-}CO{-}NH{-}\underset{R^2}{CH}{-}CO{-}NH{-}\underset{R^3}{CH}{-}CO{-}\ldots$$

phenylthiocarbamyl-peptide

B. CLEAVAGE

$$Ph{-}\overset{H}{N}{-}C({=}S){-}\underset{H}{N}{-}\underset{R^1}{C}H{-}C({=}O){-}\overset{H}{N}{-}\overset{R^2}{CH}{-}\underset{O}{\underset{\|}{C}}{-}NH{-}\overset{R^3}{CH}{-}\underset{O}{\underset{\|}{C}}{-}\ldots \xrightarrow{H^+ \text{ (anhydrous)}}$$

$$\text{Ph-NH-C}\overset{\text{S}}{\frown}\text{C=O (ring: C=N-CH(R}^1\text{))} + H_3^+N{-}\overset{R^2}{CH}{-}\underset{O}{\underset{\|}{C}}{-}NH{-}\overset{R^3}{CH}{-}\underset{O}{\underset{\|}{C}}{-}\ldots$$

anilinothiazolinone

peptide with new amino-terminal residue

C. CONVERSION

$$\text{anilinothiazolinone} + H_2O \xrightarrow{H^+} Ph{-}NH{-}\overset{S}{\overset{\|}{C}}{-}NH{-}\underset{R}{CH}{-}COOH \longrightarrow$$

anilino thiazolinone

$$\text{Ph-N}\langle\overset{S}{\overset{\|}{C}}{-}NH{-}\underset{R}{CH}{-}C({=}O)\rangle + H_2O$$

phenylthiohydantoin

sequenced. One type of limitation is due to the continual loss of sample. A small amount of side reactions occurs that tends to block further cycles of degradation. There is also a small amount of loss of peptide or protein during the wash and extraction steps. The other category of limitation is due to the build-up of background signal. Random internal peptide bond cleavages may occur to some extent and these lead to the generation of new amino termini. There is also a certain amount of carry over (or lag) from previous cycles, since the reactions do not go 100% to completion. The cycle to cycle decrease in the true signal, along with the accumulated increase in background eventually disallow further calls. A comparison of chromatograms from two successive cycles of Edman degradation is shown in Figure 5.

The Instrument

Edman recognized the value of an automated instrument when he developed the spinning cup sequencer (7). This instrument was manufactured by Beckman Instrument Company and remained the hallmark for protein sequencing for two decades. The degradation takes place in a glass cup. The centrifugal force provided a means for mixing, as well as for removal of liquids through a piece of tubing called a scoop at the top lip of the cup. Originally, milligram amounts of protein were required for sequencing. As the development of HPLC analysis allowed for sequencing of microgram amounts of protein, modifications were made to the spinning cup sequencer to accommodate such low amounts of protein. One important concept was to use the cationic polymer, polybrene, as an adhesive to noncovalently attach the protein to the glass surface and, thereby, minimize washout.

A completely redesigned sequencing instrument was developed several years ago (8) and is marketed as the ''gas-phase sequencer'' by Applied Biosystems, Inc. Essentially, the manufacturer miniaturized the plumbing (i.e., the valves and tubing), replaced the glass cup with a fiberglass filter disc assembly and made an overall modernization of the instrument. The latest version of this instrument is called the ''liquid pulse sequencer.'' With but minor modifications to Edman's original chemistry, it is now possible to do microsequencing at the low picomole level.

Other Sequencing Techniques

Enzymatic Sequencing

Exopeptidases are enzymes that degrade peptides from the termini, but do not break any internal peptide bonds. Both aminopeptidases and carboxypeptidases are commercially available. These enzymes tend to be ineffective in their ability to digest proteins, most likely due to steric hindrance resulting from tertiary structure. Exopeptidase sequencing is an extremely useful technique in special applications, such as with peptides having a blocked amino terminus.

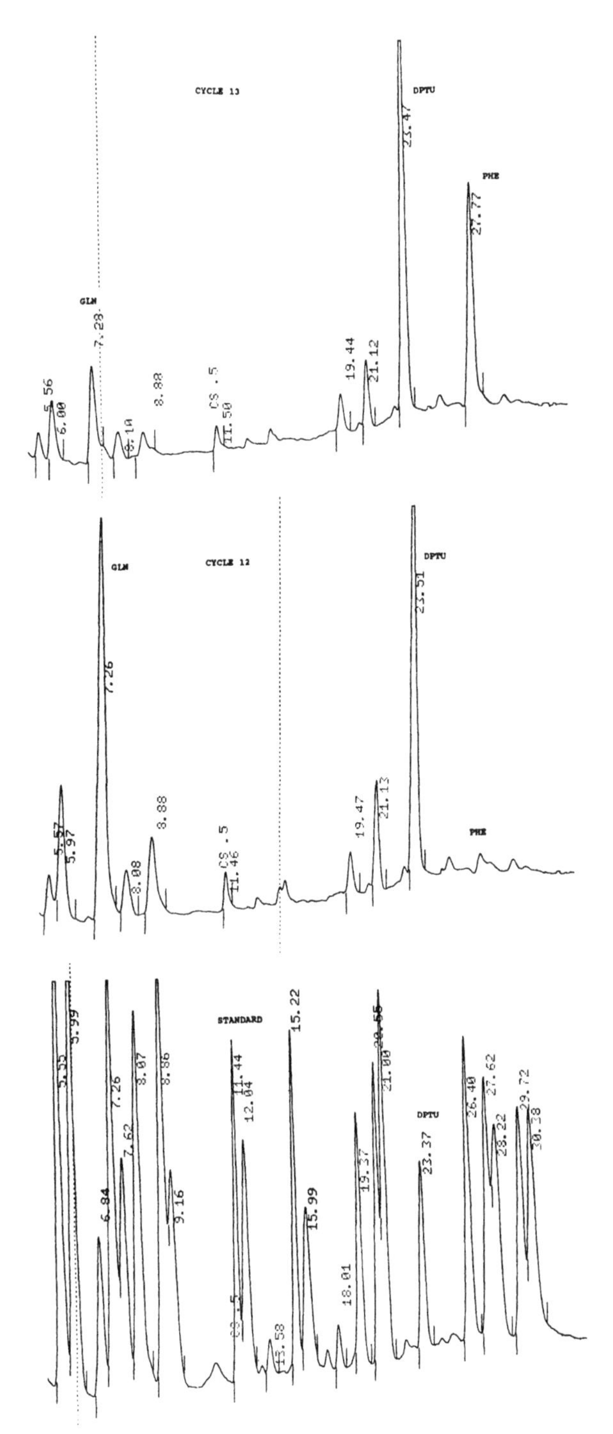

Figure 5 Chromatographic analysis of sequencing cycles. Reverse-phase chromatography of the PTH derivatives of the 20 standard amino acids (40 pmol injected) is depicted in the bottom tracing. Two representative sequencing cycles of a peptide are depicted in the other tracings. The amino acid assignment in cycle 12 (i.e., the twelfth residue counting from the amino terminus) is glutamine and that in cycle 13 is phenylalanine. It should be noted that there is a carryover (~ 20%) of some glutamine from cycle 12 into cycle 13. Also present in these cycles are background amino acid peaks, as well as reagent byproduct peaks. Thus, amino acid assignments are typically made in a semiquantitative approach by observing which amino acids increase and decrease in successive cycles.

In this process, removal of the terminal amino acid exposes the penultimate amino acid, which then is available for enzymatic removal. This process differs from the Edman degradation in that it cannot be controlled in "cycles." Instead, the analyst does a time course of digestion and monitors the increase in concentration of each free amino acid. A typical time course of digestion is demonstrated in Figures 6 and 7 (9). A known quantity of a peptide (~ 1 nmol as determined by amino acid analysis) with a molecular weight of about 5300 was treated with carboxypeptidase Y. Aliquots were removed at various time points and analyzed for released amino acids using the precolumn *o*-phthalaldehyde procedure (Fig. 6). The resulting data are plotted in Figure 7. From this information it is possible to deduce the carboxyl terminal sequence of the polypeptide. As one can see, the amino acid analysis procedure must be precise and, preferably, sensitive.

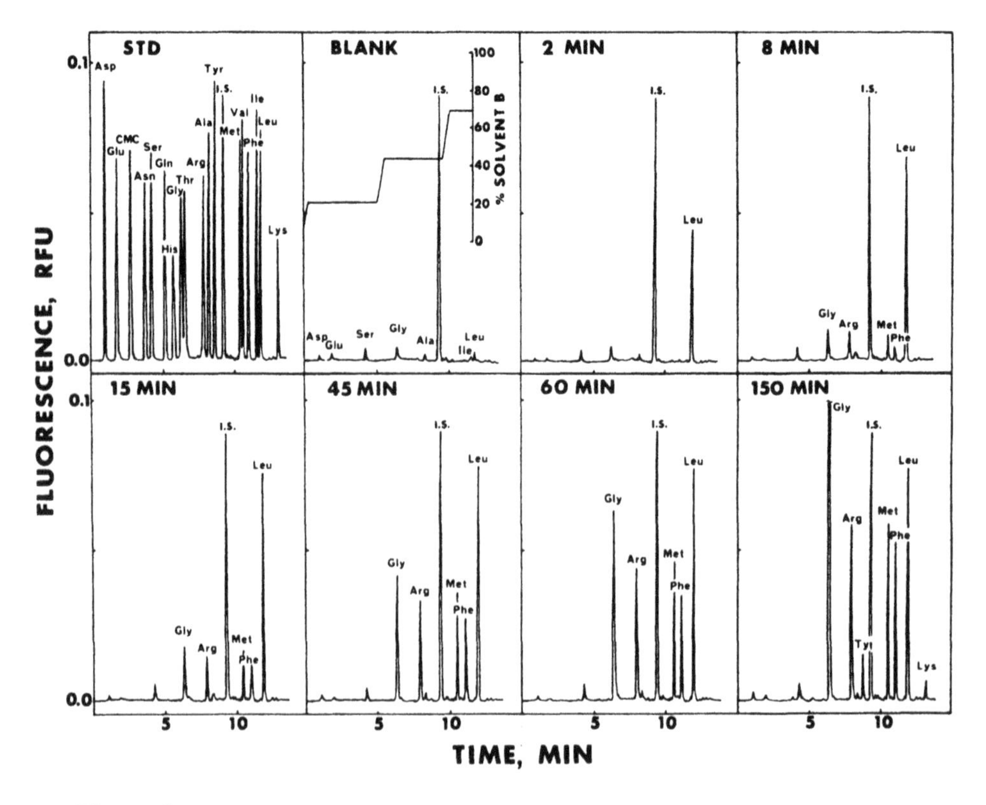

Figure 6 Time course of carboxypeptidase digestion of a polypeptide. Aliquots of the digest were removed and analyzed at different time points using the precolumn *o*-phthalaldehyde amino acid analysis procedure. An internal standard was added to the original digest to compensate for aliquoting and other sources of error.

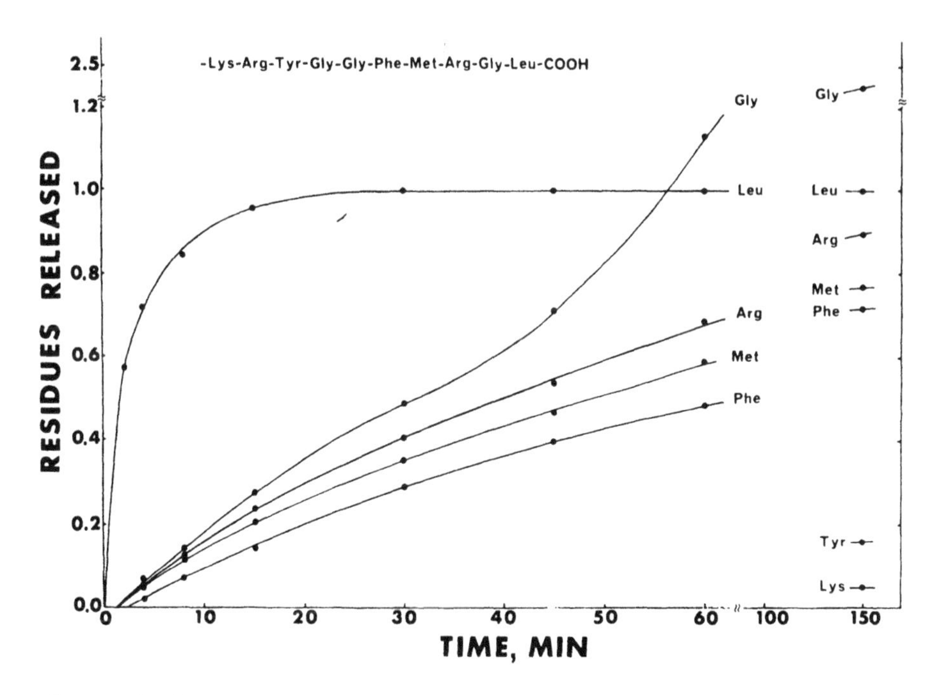

Figure 7 Plot of the time course data from Figure 6. The quantity of each amino acid (in picomoles) was calculated at each time point and normalized to the concentration of the peptide in the digest. Leucine is the carboxyl terminal residue, being released in 1 molar equivalent from the polypeptide. The penultimate residue is glycine. The next several residues are released almost as rapidly, illustrating the importance of precise analytical data. The deduced carboxyl terminal sequence of this polypeptide, which contains about 50 residues, is shown. It is generally not possible to determine more than several residues by this technique.

Mass Spectrometry

This technique had previously not been used extensively for determination of the structure of proteins, due to the complexities and difficulties of the procedures. The traditional approach involves (1) randomly cleaving polypeptides, such as with mild acid, (2) chemically modifying the resultant fragments to make them volatile, (3) separating them by gas chromatography, and (4) analyzing the components by mass spectrometry. An alternative approach involves the generation of dipeptides from the polypeptide to be analyzed using a dipeptidase. These dipeptides are analyzed as above. In the next step, the polypeptide is modified by the addition or removal of one amino acid at either the amino terminus or carboxyl terminus. This "frameshift" polypeptide is then fragmented with the dipeptidase and the new series of dipeptides are analyzed to yield the overlap sequence information.

As an alternative approach "fast atom bombardment" sample desorption for mass spectrometry (FAB-MS) allows for the analysis of polypeptides of molecular weights up to several thousand daltons. Furthermore, with FAB-MS it is not necessary to modify chemically the peptides for increased volatility. Sequential peptide bond cleavages also occur in FAB-MS, thereby generating two series of fragments (from the amino terminus and carboxyl terminus) that can be discerned in the mass spectrum. Application of this technique to the analysis of peptides generated by cyanogen bromide fragmentation of a protein is illustrated in Figure 8 (10). It was possible to deduce the primary structure of a polypeptide from a single mass spectrum.

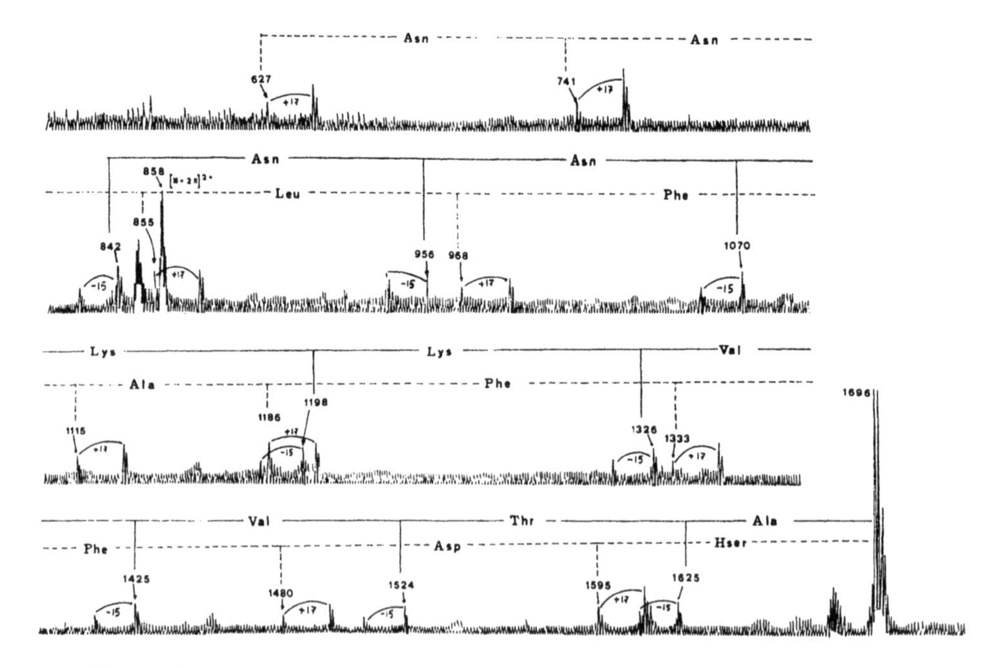

Figure 8 FAB mass spectrum of a peptide isolated from a cyanogen bromide digest. The molecular ion for the peptide is at 1696 daltons. Two series of peptide fragments were generated. In the carboxyl terminal series, one can discern the fragment with the carboxyl terminal homoserine (generated from methionine by cyanogen bromide treatment) deleted at 1595 daltons. The next fragment at 1480 daltons represents the fragment having the last two carboxyl terminal residues removed. The peptides having the last three, four, five, six, seven, or eight residues removed can also be seen and from this series one may deduce the sequence -Asn-Asn-Leu-Phe-Ala-Phe-Asp-Met-COOH. Similarly, from the amino terminal series one may deduce the sequence NH_2-Ala-Thr-Val-Val-Lys-Lys-Asn-Asn-. Thus, the complete sequence of this 14-residue peptide is obvious from a single mass spectrum.

APPLICATION TO ANALYSIS OF rIL-2

Sequence of rIL-2

In this illustration (11), the complete sequence of human interleukin-2 (also known as T-cell growth factor or IL-2) had been deduced by sequence analysis of a cDNA corresponding to the structural gene. The purified protein was submitted to automated Edman degradation, which allowed positive identification of the first 29 residues (Fig. 9). Although the natural protein has alanine at the amino-terminus, the recombinant version, produced in *Escherichia coli*, has an additional residue of methionine at the amino-terminus. This residue is assigned position 0 in Figure 9. In the engineering of the recombinant gene, the initiating codon for methionine must be placed prior to the codon for the first alanine. Different lots of recombinant protein were found to contain this additional methionine residue in about 90% of the product (due to partial post-translational processing). Thus, cycle 1 of the Edman degradation revealed

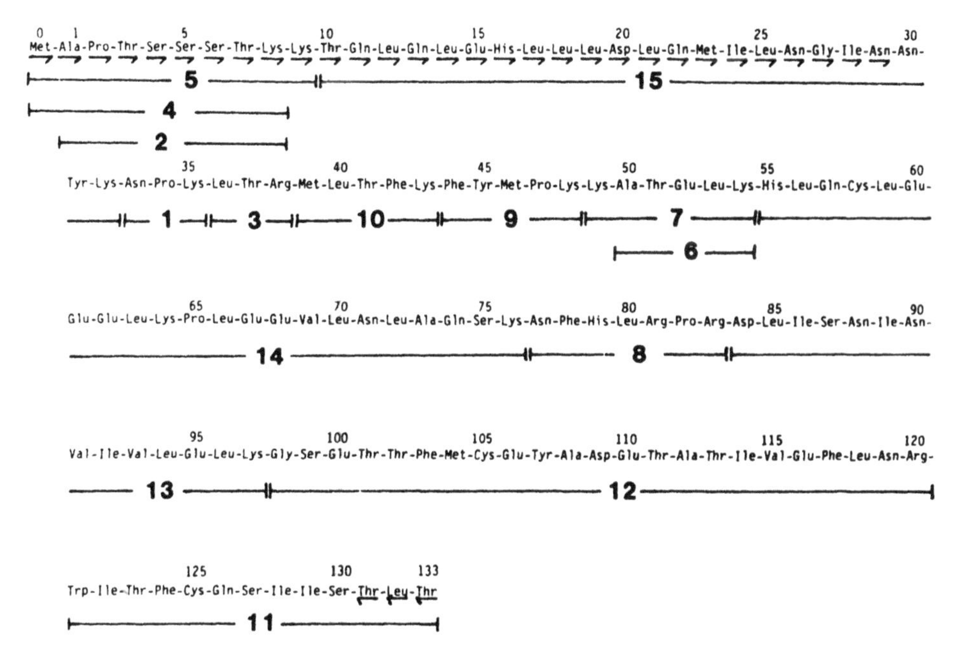

Figure 9 Sequence analysis of recombinant human interleukin-2. The elucidated primary structure is shown in the top panel. Amino-terminal sequence analysis of the intact protein using Edman degradation is indicated by the arrows pointing to the right. Carboxyl terminal sequencing of the intact protein using carboxypeptidase digestion is indicated by the arrows pointing to the left. The numbered regions correspond to the tryptic peptides that were isolated and analyzed (see Fig. 10).

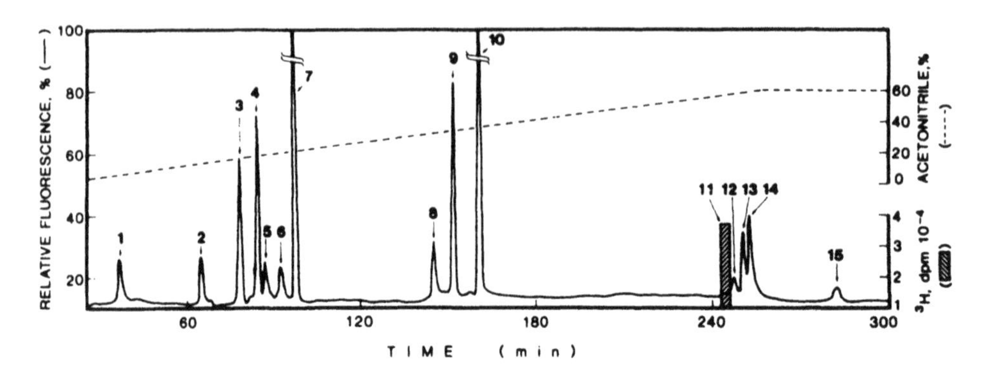

Figure 10 Tryptic peptide map of recombinant human interleukin-2. The protein was alkylated with [^{3}H]iodoacetic acid to label the free cysteine thiol. After treatment with trypsin, the digest was applied to a reverse-phase HPLC column which was eluted using a gradient of increasing concentration of acetonitrile in 0.1% trifluoroacetic acid. The numbering of the tryptic peptides is according to order of elution from the column (*cf.*, Fig. 9).

both PTH-methionine and PTH-alanine in a ratio of about 9:1. The second cycle had PTH-alanine and PTH-proline in about the same ratio. At this time, it is not known whether this modification to the natural protein has any influence on the clinical efficacy of the product, but it certainly is a concern.

The "preview" of a low level of the upcoming amino acid in the predicted sequence, as described above, is informative. It is sometimes found in the analysis of natural proteins that have been partially degraded by a contaminating aminopeptidase. It is also found in chemically synthetic peptides that are contaminated with product having amino acid deletions (due to incomplete coupling) in the polypeptide chain.

The carboxyl terminus of the protein was analyzed using a time course digestion with the exopeptidase, carboxypeptidase Y. As indicated in Figure 9, it was possible to identify the final three amino acids in the protein chain.

To obtain the remaining sequence information, the protein was digested with the endopeptidase, trypsin, which fragments the protein chain at the carboxyl side of the positively charged amino acids (at pH 8), lysine and arginine. The tryptic fragments were separated by reverse-phase HPLC (Fig. 10) and each peptide (numbered according to elution position) was sequenced by automated Edman degradation. The correspondence of each peak (Fig. 10) to a predicted fragment of the protein (Fig. 9) was demonstrated.

It should be noted that three different peptides (numbers 2, 4, and 5) correspond to the amino terminal region of the protein. The peptide present in the greatest amount (peptide 4) includes residues 0–8 of the protein. The peptide spanning residues 1–8, which should be present in about 10% of the product, is number 2. Peptide 5 represents an alternate cleavage by trypsin and includes residues 0–9. A fourth amino terminal peptide, encompassing residues 1–9, was most likely generated, but was present at too low a level to be detected in the tryptic map. This alternative cleavage phenomenon, which occurs when there are two or more sequential, positively charged amino acids, is also observed with peptides 6 and 7.

Disulfide Assignment

Human IL-2 has three residues of cysteine. From analysis of the natural protein, it is known that two of these residues (positions 58 and 105) are in disulfide linkage, while cysteine at position 125 has a free sulfhydryl group. In order to evaluate this matter, the recombinant protein was treated with radioactive [^{3}H]iodoacetic acid (at pH 8.5), which should only derivatize the free sulfhydryl group. The protein was then reduced and alkylated, but this time with ^{14}C-labeled iodoacetic acid. The ^{3}H label was found only in peptide 11 (Fig. 10), which corresponded to Cys^{125}. No significant ^{14}C label was found in peptide 11, whereas peptides 12 and 14 contained this label. This study demonstrated that the same disulfide bonding structure in the recombinant protein is the same as in the natural protein. Any product with a deviant structure, with respect to the three residues of cysteine, would be present at too low a level, if at all, to be detected.

Conclusions on rIL-2

The information gathered in these studies essentially defines the primary structure of rIL-2 produced in *E. coli*. Such information is generally sufficient as a proof of structure when making a submission to a governmental regulatory agency. Batch-to-batch structural analysis may not necessarily have to be so thorough.

The natural protein is known to be *O*-glycosylated at position 4. Were this protein to be produced in eukaryotic cells, it would then be necessary to investigate for carbohydrate attachments. As a general rule, each recombinant protein or synthetic peptide product will require a unique set of analyses for proof or confirmation of primary structure.

SAMPLE PREPARATION

Obviously, a purified protein or peptide must be prepared before any type of analysis can be attempted. Other chapters in this volume (chaps. 4, 6, 7, and 8) deal with this topic. Typically, a protein is purified by column chromatography and the purity is evaluated by polyacrylamide gel electrophoresis. Due to the diverse physical properties of different proteins, various approaches must be investigated in order to obtain the homogeneous protein sample.

The use of gel electrophoresis for preparing samples for further analysis has always been a desired option. Gel electrophoresis is usually the simplest technique for resolving protein mixtures. Even though only small quantities (micrograms) of protein can be run, this can be sufficient for microanalysis. Simply extracting the protein from the gel has generally been found to give poor yields as well as background contamination. Indeed, substances in gel extracts are known to clog gas-phase microsequencing instruments and such samples are not acceptable for analysis. One technique has gained prominence. After electrophoretic separation of the proteins, the gel is placed against a sheet of polyvinylidene difluoride membrane and the proteins are transferred transversely in an electric field. The membrane is stained to visualize the bands, which may be excised for further analysis, such as sequencing (34) or amino acid analysis (35).

Various novel approaches and variations for preparing protein samples appear in the scientific literature and each has certain merits and limitations. For example, disulfide crosslinked polyacrylamide gels can be chemically reduced to more easily release the protein following electrophoresis (36). After gel electrophoresis, it is possible to generate peptide fragments either chemically (37) or enzymatically (38). It is even possible to fragment proteins purified by reverse-phase HPLC in the presence of organic cosolvents (39). The experience of the investigator and the properties of the protein under study will dictate the course of analysis to be taken.

ACKNOWLEDGMENTS

The authors wish to thank Dr. Pietro Pucci for supplying an original photograph for Figure 8. James Sligar kindly supplied the chromatogram for Figure 2.

REFERENCES

1. Dayhoff, M. O. (Ed.). *Atlas of Protein Sequence and Structure*, Vol. 5. National Biomedical Research Foundation, Washington, D.C., 1978.
2. Shiveley, J. E. (ed.). *Methods of Protein Microcharacterization—A Practical Handbook*. Humana Press, Clifton, NJ, 1986.
3. Spackman, D. H., Stein, W. H., and Moore, S. *Anal. Chem.* 42, 1190–1206 (1958).
4. Benson, J. R., and Hare, P. E. *Proc. Natl. Acad. Sci.* (*USA*) 72, 619–622 (1975).
5. Stein, S., Bohlen,P., Stone, J., Dairman, W., and Udenfriend, S. *Arch. Biochem. Biophys.* 155, 203 (1973).
6. Lindroth, P. and Mopper, K. *Anal. Chem.* 51, 1667–1674 (1979).

7. Edman, P. and Begg, G. *Eur. J. Biochem.* 1, 761–768 (1967).
8. Hewick, R. M., Hunkapiller, M. W., Hood, L. E., and Dryer, W. J. *J. Biol. Chem.* 256, 7990–7997 (1981).
9. Jones, B. N., Paabo, S., and Stein, S. *J. Liq. Chromatogr.* 4, 565–586 (1981).
10. Pucci, P., Borgia, R., and Marino, G. *J. Chromatogr.* 331, 425–431 (1985).
11. Lahm, H. -W. and Stein, S. *J. Chromatogr.* 326, 357–361 (1985).
12. Smyth, D. G., In *Methods in Enzymology*, Vol. XI. Edited by C. H. W. Hirs. Academic Press, New York, 1967, pp. 214–231.
13. Hunter, M. J. and Ludwig, M. L. In *Methods in Enzymology*, Vol. XXV. Edited by C. H. W. Hirs and S. N. Timashelf. Academic Press, New York, 1972, pp. 585–596.
14. Atassi, M. Z. and Habeeb, A. F. S. A. In *Methods in Enzymology*, Vol. XXV. Edited by C. H. W. Hirs and S. N. Timashelf. Academic Press, New York, 1972, pp. 546–558.
15. Smith, E. L. In *Methods in Enzymology*, Vol. XLVII. Edited by C. H. W. Hirs and S. N. Timashelf. Academic Press, New York, 1977, pp. 156–161.
16. Cole, R. D. In *Methods in Enzymology*, Vol. XI. Edited by C. H. W. Hirs. Academic Press, New York, 1967, pp. 315–317.
17. Tarr, G. E. In *Methods of Protein Microcharacterization*. Edited by J. E. Shively. Humana Press, Clifton, NJ 1986, pp. 155–194.
18. Mitchell, W. M. In *Methods in Enzymology*, Vol. XLVII. Edited by C. H. W. Hirs and S. N. Timashelf. Academic Press, New York, 1977, pp. 165–174.
19. Shively, J. E. In *Methods of Protein Microcharacterization*. Edited by J. E. Shively. Humana Press, Clifton, NJ, 1986, pp. 155–194.
20. Stone, D., Phillips, A. W., and Burchall, J. J. *Eur. J. Biochem.* 72, 613–618 (1977).
21. Drapeau, G. R. In *Methods in Enzymology*, Vol. XLVII. Edited by C. H. W. Hirs and S. N. Timashelf. Academic Press, New York, 1977, pp. 189–191.
22. Koida, K. M. and Walter, R. *J. Biol. Chem.* 251, 7593–7599 (1976).
23. Ingram, V. M. and Stretton, A. O. W. *Biochem. Biophys. Acta* 60, 20–26 (1962).
24. Heinrikson, R. L. In *Methods in Enzymology*, Vol. XLVII. Edited by C. H. W. Hirs and S. N. Timashelf. Academic Press, New York, 1977, pp. 175–188.
25. Mella, K., Vols, M., and Pfleiderer, G. *Anal. Biochem.* 31, 219–226 (1967).
26. Gross, E. In *Methods in Enzymology*, Vol. XI. Edited by C. H. W. Hirs.. Academic Press, New York, 1967, pp. 238–254.
27. Huang, H. V., Bond, M. W., Hunkapiller, M. W., and Hood, L. F. In *Methods in Enzymology*, Vol. 91. Edited by C. H. W. Hirs and S. N. Timashelf. Academic Press, New York, 1983, pp. 318–323.
28. Fontana, A. In *Methods of Enzymology*, Vol. XXV. Edited by C. H. W. Hirs and S. N. Timashelf. Academic Press, New York, 1972, pp. 419–423.
29. Schecter, Y., Patchornik, A. and Burstein, Y. *Biochemistry* 15, 5071–5076 (1976).
30. Ramachandran, L. K. and Witkop, B. In *Methods in Enzymology*, Vol. XI. Edited by C. H. W. Hirs. Academic Press, New York, 1976, pp. 283–298.
31. Jacobson, G. R., Schaffer, M. H., Stark, G. R., and Vanaman, T. C. *J. Biol. Chem.* 248, 6583–6590 (1973).

32. Hermodson, M. A., Ericsson, L. H., Neurath, H., and Walsh, K. A. *Biochemistry* 12, 3146–3151 (1973).
33. Inglis, A. S. In *Methods in Enzymology*, Vol. 91. Edited by C. H. W. Hirs and S. N. Timashelf. Academic Press, New York, 1983, pp. 324–332.
34. Matsudaira, P. J. *Biol. Chem.* 262, 10035–10038 (1987).
35. Tous, G. I., Fausnaugh, J. L., Akinyosoye, O., Lackland, H., Winter-Cash, P., Vitorica, F. J., and Stein, S. *Anal. Biochem.* 179, 50–55 (1989).
36. Ghaffari, S. H., Dumbar, T. S., Wallace, M. O., and Olson, M. O. *J. Anal. Biochem.* 171, 352–359 (1988).
37. Plaxton, W. C. and Moorhead, G. B. G. *Anal. Chem.* 178, 391–393 (1989).
38. Prussak, C. E., Almazan, M. T., and Tseng, B. Y. *Anal. Chem.* 178, 233–238 (1989).
39. Welinder, K. G. *Anal. Biochem.* 174, 54–64 (1988).

10

Chemical Synthesis of Peptides

Chester A. Meyers

Squibb Institute for Medical Research
Princeton, New Jersey

INTRODUCTION

The central importance of peptides in normal physiologic processes is reflected by their widespread distribution throughout most body tissues and by the expanding number of known peptides with diverse biological actions. Peptides have been discovered in the gastrointestinal tract, central nervous system, immune system, and heart. Their function is twofold; as regulatory elements and as neurotransmitters. The growing awareness of the many important functions of peptides beyond those of the classic endocrine hormones (e.g., insulin, glucagon, vasopressin, gut hormones, etc.) has been accompanied by an increased appreciation of the potential utility of these substances in research and medicine.

Synthetic peptides are of great biomedical interest as tools to study the role of natural peptides in normal and pathophysiological processes. Synthetic peptide analogs can increase the clinical potential of peptides when they exhibit, for example, more selective, potent, long-acting, or antagonistic properties. Antigenic peptide fragments have also been synthesized for use in producing specific antibodies to proteins, and such peptides are under evaluation as synthetic vaccines. There are numerous additional applications for synthetic peptides, all of which depend upon reliable techniques for preparing these compounds.

An understanding of the relationship of a peptide or protein structure to its function is essential for the rational design of molecules with predicted properties that can be exploited for research and in the clinic. The chemical synthesis of peptides has a fundamental role in this area of modern biological research. Systematic variations in structure can be correlated with the properties of the resultant molecules; a direct approach to the identification of functional domains or of specific amino acids having specialized roles in the molecule of interest. These efforts are critically important since there is not yet a full appreciation of how the amino acid sequence of a peptide or protein directs its proper folding into a preferred bioactive conformation or a precise three-dimensional structure, giving rise to its unique properties. Macromolecular interactions, and the mechanisms governing biorecognition, particularly ligand/receptor interactions, must also be understood in order to design structures with predicted properties, and chemically synthesized peptides contribute to this goal as well. These and other applications of peptides have depended in large measure upon the accessibility of purified synthetic peptides ever since the first synthesis, in 1953, of a biologically active peptide hormone by du Vigneaud and his co-workers. The hormone was the cyclic octapeptide, oxytocin. Recent advances in the technology of chemical peptide synthesis, purification, and structural analysis have made such molecules more generally available, and the use of peptides in biomedical research and pharmaceutical development is expanding rapidly.

BACKGROUND

The term polypeptide was proposed in 1906 by Emil Fischer to describe structures consisting of variable numbers of amino acids linked together by amide bonds. The basic structural features of a linear peptide are illustrated in Figure 1. A peptide bond is formed by a condensation reaction between two amino acids; the carboxyl group of one amino acid forms an amide bond with the amino group of a second amino acid, and a molecule of water is eliminated. The amino acids, which are called residues when they occur in peptide linkage, are conventionally numbered consecutively starting at the amino (N) terminus. The repetitive peptide backbone, which is all of the molecule except

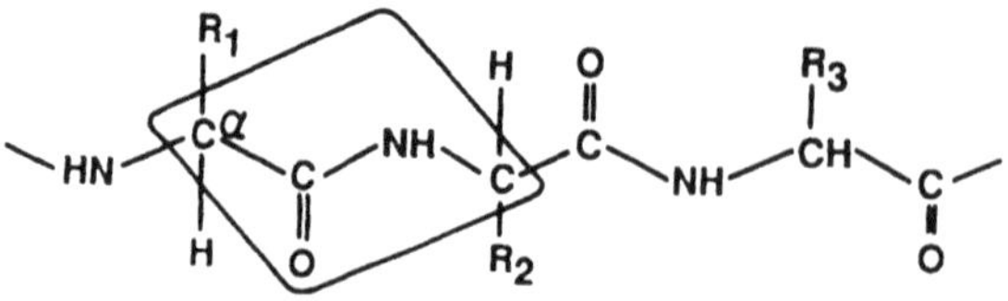

Figure 1 General structure of a peptide. The six atoms involved in the predominantly coplanar peptide bond are enclosed in the solid box. R; amino acid side chains which are numbered consecutively from the amino toward the carboxy-terminus.

for the side-chain (R) groups, is defined by its length and by its direction. Each amino acid residue is distinguished by its side-chain moiety, which is attached to the alpha carbon atom (Fig. 1). The functional side chains encompass a variety of chemical classes. For example, they may be aliphatic or aromatic, polar or nonpolar, acidic, neutral, basic, alcoholic, etc. It is this diversity which gives rise to the unique characteristics of individual peptides, despite the repetitive backbone.

The structural information of a peptide is principally provided by the sequence of its amino acids. Seemingly minor substitutions, deletions, alterations, or rearrangements of amino acids can result in structural variants with profoundly changed properties, especially with regard to biologically active peptides. The number of possibilities for different peptides from the more than 20 amino acids available to biological systems is therefore virtually limitless by these considerations alone, even for only relatively short chains. As chain lengths increase toward proteins, complex secondary and tertiary structures arise through multiple modes of side chain interactions, so that the information content and versatility of these molecules become truly staggering. No definitive number of amino acids or molecular size cutoff has been established to specifically identify the point at which the term peptide should be replaced by protein as a more appropriate description. There is, however, general agreement that chain lengths of up to 80–90 amino acid residues, which correspond to molecular weights up to about 10,000, can be regarded as peptides. Some investigators tend to view the differences as related less to size than to structure. Proteins generally have more restricted, or stabilized three-dimensional structures, and frequently contain distinct domains.

HISTORICAL DEVELOPMENT OF CHEMICAL PEPTIDE SYNTHESIS

The importance of peptides and proteins in biological systems was recognized early, and attempts to synthesize them naturally followed. The development of agents which were sufficiently capable of activating the carboxyl group to promote peptide bond formation with another amino acid met with great initial success. However, if two unaltered amino acids are condensed in this manner, each activated specie can react with the amino group of either amino acid, resulting in an uncontrolled polymerization. The challenge of synthesizing peptides of designed sequence was investigated at the turn of the century by E. Fischer and T. Curtius, who approached the problem from the standpoint of classical organic chemistry. In order to circumvent the problem of random polycondensation, they developed several basic concepts which have persisted as fundamental elements of modern peptide chemistry. Blocking of the amino group (by acylation) of an amino acid prior to activating its carboxyl group resulted in a controlled, directed coupling to an amino acid (or peptide) ester (Fig. 2). The C-terminal ester was necessary as a protecting group in order to restrict carboxyl activation to only the desired (N-blocked) amino acid. This requirement could be avoided by "preactivating" the N-protected amino acids. For example, N-protected amino acid chlorides or azides were prepared, and these could be coupled directly to free amino acids or peptides. Today, the use of stable, often crystalline protected amino acids and peptides that have been preactivated as esters (e.g., hydroxysuccinimide, nitrophenyl, or pentafluorophenyl esters) are available.

An initially frustrating search then followed for an acyl group which could be selectively removed after the peptide bond was formed. Since simple acyl blocking groups themselves form amide bonds, their properties are naturally similar to peptide bonds, so the difficulty of selective cleavage can be appre-

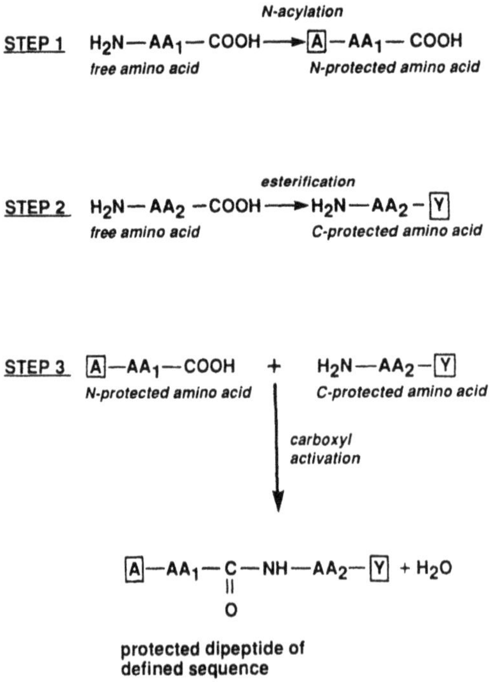

STEP 1 $H_2N—AA_1—COOH \xrightarrow{\text{N-acylation}} \boxed{A}—AA_1—COOH$

free amino acid → *N-protected amino acid*

STEP 2 $H_2N—AA_2—COOH \xrightarrow{\text{esterification}} H_2N—AA_2—\boxed{Y}$

free amino acid → *C-protected amino acid*

STEP 3 $\boxed{A}—AA_1—COOH + H_2N—AA_2—\boxed{Y}$

N-protected amino acid + *C-protected amino acid*

↓ *carboxyl activation*

$\boxed{A}—AA_1—C(=O)—NH—AA_2—\boxed{Y} + H_2O$

protected dipeptide of defined sequence

Figure 2 Principle of controlled coupling of two amino acids. Each amino acid is prepared as an appropriately protected intermediate prior to condensation. Only the reactive species intended to participate in the formation of the peptide bond are unprotected AA; amino acid.

ciated. In 1932, M. Bergmann introduced the carbobenzoxy group (Cbz, Z, benzyloxycarbonyl) which could be removed by a mild catalytic hydrogenolysis reaction without harm to peptide bonds. As a result, the stepwise elaboration of unambiguous peptide chains became possible. A generalized scheme of stepwise peptide synthesis, using a tripeptide as an example, is depicted in Figure 3 to show how selectively removable blocking groups are used in the assembly of large peptide chains.

As in Figure 2, the initial coupling of two suitably blocked amino acids generates a blocked dipeptide. Amino acids that have reactive side chain groups are also blocked (S). For example, lysine, arginine, cysteine, serine, threonine, glutamic, and aspartic acid usually require side chain protection during synthesis. At this stage, the amino protecting group of the H_2N-terminal amino acid is selectively removed. The resultant free amino group is then ready to be coupled with another N-protected amino acid possessing a free carboxyl group. The new fully blocked peptide has been elongated by one amino acid, and the cycle of selective deprotection and coupling can be repeated until the target sequence is completed. A final step to chemically remove all of the protecting groups (global deprotection) yields the synthetic peptide.

Note that the stepwise elongation of the peptide begins with the "last," or C-terminal amino acid and proceeds toward the "first," or N-terminal amino acid. This strategy avoids the need to activate the carboxyl group of peptides, which is known to promote racemization. Instead, at each cycle, a single protected amino acid is activated, and the risk of racemization is greatly reduced. The most commonly used amine protecting groups for peptide synthesis (Table 1) are based on the structure of urethane. These groups substantially suppress racemization of the amino acid to which they are attached, in addition to fulfilling the main requirements of adequate protection and relatively mild, selective removal after the coupling step. For reviews, see Finn and Hofmann (1) and Schroder and Lubke (2).

SOLID-PHASE PEPTIDE SYNTHESIS

A major breakthrough occurred in 1963 when R. B. Merrifield demonstrated that the synthesis of long-chain peptides could be automated by a greatly simplified process that was far more rapid than the classical method. The concept featured attachment of the C-terminal amino acid to an insoluble support (solid phase) which provided a growing point for the peptide chain. This permitted the entire synthesis to be performed in a single vessel where byproducts and unreacted materials could simply be washed away by filtration. The speed and ease of operation offered by the solid-phase method, also referred to as the Merrifield method, has resulted in the preparation and widespread availability

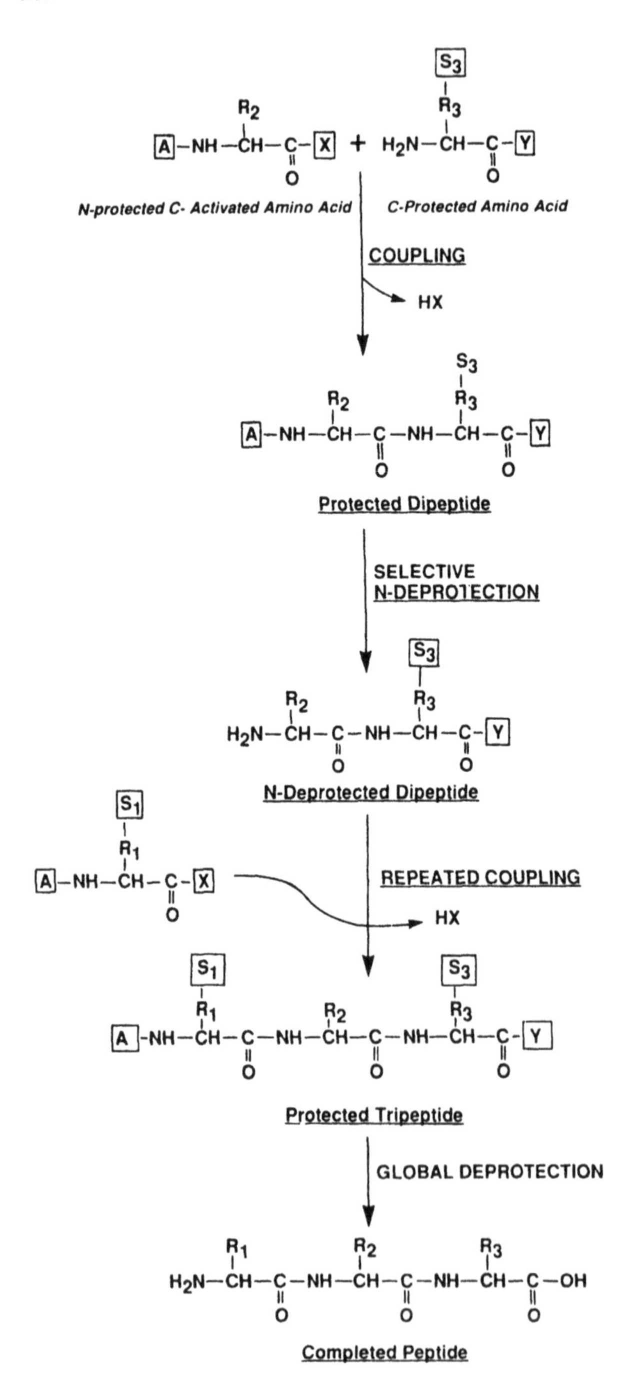

Figure 3 Synthesis of a tripeptide. Appropriately protected amino acids are coupled as in Fig. 2. Selective removal of the alpha amino protecting group permits repetitive stepwise synthesis to proceed. On completion, all protecting groups, including those on reactive side chains (S) are removed. In solid-phase synthesis, the solid support serves as the C-terminal-protecting group (Y), and global deprotection includes cleavage of the peptide from the support. See text.

Table 1 Common Amine Protecting Groups in Peptide Synthesis

Protecting groups	Typical removal conditions
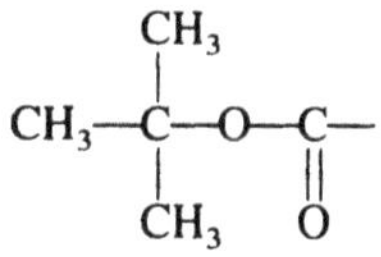 $CH_3{-}C(CH_3)_2{-}O{-}C(=O){-}$ *tert*-Butyloxycarbonyl (t-BOC)	35% TFA / dichloromethane, 30 min
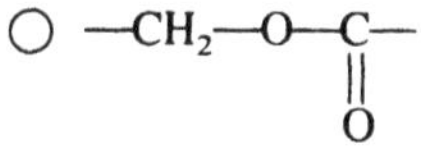 ○ $-CH_2{-}O{-}C(=O){-}$ Carbobenzoxy (Z, Cbz, Benzyloxycarbonyl)	Catalytic hydrogenation
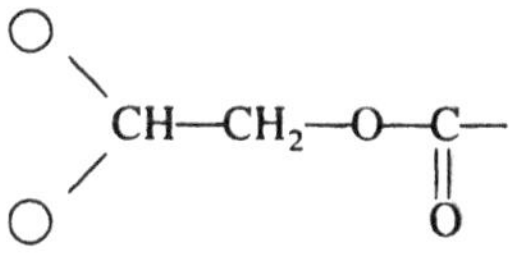 (○)(○) $CH{-}CH_2{-}O{-}C(=O){-}$ 9-Fluorenylmethyloxycarbonyl (Fmoc)	Piperidine
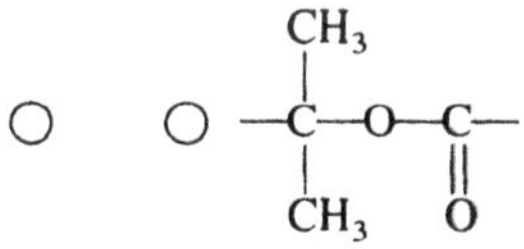 ○ ○ $-C(CH_3)_2{-}O{-}C(=O){-}$ Biphenylisopropyloxycarbonyl (Bpoc)	1% TFA / dichloromethane, 1h

of thousands of peptides for many applications. Merrifield was awarded the Nobel prize in chemistry in 1984 for this revolutionary development. The method is briefly described along with considerations regarding its limitations. For detailed reviews see Barany and Merrifield (3) and Stewart and Young (4).

Synthetic Strategy

The peptide chain assembly begins with the C-terminal residue and is lengthened toward the amino terminus for the same practical reasons that were described for "classical" peptide synthesis in solution. Hence, the carboxyl group of the protected C-terminal residue is covalently attached to one of a variety of available insoluble supports. The supports are most commonly uniform polymeric beads, or resins, that are supplied with a choice of chemically reactive groups. This provides flexibility in designing the synthetic strategy for a particular peptide or intermediate. For example, a single reagent can be used to liberate a completed peptide from its resin either in the form of a C-terminal amide (—$CONH_2$) or as a free carboxylic acid (—COOH), depending upon the initial choice of substituted resin. The linkage must, however, be stable to the conditions of stepwise synthesis and also permit removal of the completed peptide from the solid support with minimal damage to the peptide itself.

Once the protected C-terminal amino acid is suitably attached to the resin, the chemistry of repetitive stepwise solid phase synthesis is fundamentally similar to that described for a solution synthesis (Fig. 3). That is, the amino protecting group [most commonly t-BOC (Table 1)] is removed by treatment with an appropriate reagent, the next N-protected amino acid is suitably activated and coupled to the anchored residue, and so on until the chain is completed. It is the manner in which these steps are performed that gives the solid phase method its tremendous advantages of speed and simplicity over the solution method. The reagents for deprotection, coupling, washing etc. are delivered to a single vessel containing the growing, resin-bound peptide chain. The vessel is fitted with a frit for removal of reagents, solvents, and byproducts by filtration. This replaces the laborious and time-consuming purification steps that are necessary at each cycle of solution synthesis, and also avoids the accompanying losses in product. The process is readily adapted to automation, and several instruments are commercially available, many with computer programs that capitalize on the experience gained by many investigators in the preparation of numerous peptide sequences. Thus, conditions for the incorporation of individual residues can often be optimized.

Following incorporation of the final N-terminal residue, the protected peptide must be liberated from the resin and all of the blocking groups must be removed (Table 2). For many peptides, the side-chain blocking groups may be chosen such that they are labile to the conditions of peptide cleavage from the resin, so that both deblocking and removal of the finished product from the support can be achieved in a single step. Alternatively, there are strategies to permit the removal of protected peptides which may be needed, for example, as a fragment to be coupled with another peptide (fragment condensation method). It is also not uncommon for the desired final product to retain one or more protecting groups. In general, they can enhance overall lipophilicity which can be desirable, for example, in enhancing the delivery of peptides into cells across cell membranes. Amino and carboxy-terminally blocked peptides may also better approximate internal amino acid residues than do the corresponding free peptides. As an example, the well-known trypsin substrates, such as tosyl arginine methyl ester (TAME), are very poor substrates when the Arg amino terminus is unblocked.

An inherent feature of stepwise synthesis is that partially completed peptides can be removed to serve as intermediates for the synthesis of analogs. In conventional solid-phase synthesis, a portion of the resin is simply removed from the reaction vessel at any step and then separately continued to incorporate the desired changes (e.g., replacements, additions, omissions). Since the peptides are elaborated from the C → N terminus, it is far easier to make changes near the N-terminus of a growing peptide. A change near the C-terminal starting point means that the rest of the synthesis needs to be repeated, so that in effect, each C-terminal analog involves a complete, or nearly complete, separate synthesis. Several recent innovative concepts have helped to increase the efficiency in preparing large numbers of peptides, which can make desired analogs accessible regardless of the position in the peptide at which the changes are needed. However, the fundamental chemistry used in newer synthetic strategies has not substantially changed. It must be emphasized that the speed and automation which has resulted in a major proliferation of synthetic peptides must not be construed as having rendered the synthesis of even fairly short sequences as completely routine. Each sequence is unique and subject to its own potential problems in its preparation. Thorough characterization of purified peptides, as described in a later section, is in each case crucial for the proper interpretation of subsequent experiments, and it is typically the purification rather than the synthesis that today is rate limiting in preparing peptides.

HF Cleavage

The most routinely used method for simultaneous removal of amino acid side-chain protecting groups and cleavage of the bound peptide from its solid support is treatment with anhydrous liquid HF at low temperature in the presence of scavengers. The method requires an apparatus made of materials that are resistant to the strongly acidic HF. The scavengers serve to minimize side reactions that can occur via the formation of reactive components during the HF treatment. Typically, scavengers have been chosen to structurally resemble reactive amino acid side chains, thus affording their protection through direct competition. For example, excesses (often 10% by volume of HF) of anisole, dimethylanisole, or phenol are used to protect phenolic tyrosine side chains. Similarly, sulfides (e.g., dimethyl sulfide) and thiols (e.g., dithioethane) are added to protect Met and Cys residues, and indole type scavengers have been used for peptides containing tryptophan. An insightful two-step HF cleavage method was developed by Tam and Merrifield to exploit separate reaction mechanisms for deprotection and cleavage from the resin. The aim was to remove a majority of blocking groups in the first step under conditions of lowered HF concentration that would be less likely to generate harmful reactive ionic species. Following an important organic extraction step, a more typical concentrated HF reaction liberates the peptide from the resin.

After removal of the volatile HF, free peptides are usually precipitated and washed with an organic solvent (ether, ethyl acetate). Then they are dissolved in a suitable solvent and conveniently separated from the solid resin beads by filtration. The peptide solutions may be lyophilized or stored frozen until further purification.

Fmoc Chemistry

The Boc-benzyl-based chemistry (Tables 1 and 2) has been firmly established as an integral part of solid-phase peptide synthesis. When newly devised synthetic chemistries are introduced, a sound conceptual basis and a few convincing examples of its use in preparing peptides may not be sufficient to generate widespread acceptance for several reasons. The new N-protected amino acids, many with other side chain blocking groups as well, must be stable, well characterized and readily available at competitive cost. They may also need to be supplied in a preactivated form. Cost alone is likely to be relatively high at first, as a commercial demand has not yet been established. The individual behavior of each amino acid during coupling, and the stability and selective removability of the proposed blocking groups must all be assessed under varied conditions of synthesis and in many types of sequences. Many years may be required to approach the level of understanding and experience that has been gained with the established chemistry that is often routinely successful. The new method must offer major practical advantages over the existing chemistry to be widely adopted. The most recent alternative strategy to have apparently gained such acceptance is the base-labile Fmoc system. In 1970, Carpino and Han introduced the 9-fluorenylmethoxycarbonyl (Fmoc) group (Table 1) as an amine protecting group having the advantages of other urethan-type groups, but with the added property of being removable in mild base while remaining stable to acid treatment. This opened an opportunity to prepare peptides under unusually mild conditions by avoiding extremes of pH at each cycle. For solid-phase synthesis, it also offered removal of the peptide from the resin and simultaneous removal of the side-chain-protecting groups using trifluoroacetic acid instead of the harsh and more dangerous HF treatment needed for benzyl-based chemistry. The acid-labile *t*-butyl group, introduced in 1960, has provided an important and well-known alternative to the more harshly removed methyl and ethyl esters often used for carboxyl protection. It is an ideal side-chain protecting group for Fmoc-based solid-phase synthesis. The Fmoc amino-protecting group was recently reviewed by Atherton and Sheppard (6). Its use for solid-phase peptide synthesis is rapidly expanding, and most modern instruments are now compatible with, or even dependent upon this chemistry. To be sure, there are shortcomings in need of further development (e.g., better side-chain-blocking groups are needed for some amino acids, particularly Arg and Cys; peptides containing these residues are still problematic for Fmoc-based synthesis), but this synthetic strategy promises to improve and simplify the preparation of ever larger peptides.

LIMITATIONS

The classical and solid-phase methods of peptide synthesis share a number of fundamental constraints, and each also has certain unique shortcomings which collectively limit the scope and applicability of modern peptide synthesis. The difficulties in the chemical synthesis of peptides ultimately derive from the variety of reactive functional groups and numerous chiral centers in the amino acids themselves. The challenge of reversibly and selectively blocking these groups at varied stages of the synthesis, preparing and coupling optically pure amino acid and peptide intermediates, and purifying final products under conditions that preserve full biological activity are indeed formidable.

Historically, a modest number of selectively removable blocking groups have gradually been developed for most of the reactive moieties in the common amino acids (Tables 1 and 2). The most successful and widely used alpha amine blocking groups (Cbz, *t*-BOC, Fmoc) have, as mentioned, been based upon the urethane structure which suppresses racemization during carboxyl activation. The methods for removal of these groups at each cycle of stepwise synthesis are considered to be relatively mild, although the growing peptide is typically exposed to strong acids in organic solvents. The conditions for repetitive removal of the amine protecting group can dictate the severity required for subsequent removal of side-chain-blocking groups, since the latter are usually required to be stable to the conditions of amine deprotection throughout the synthesis. This is true, of course, when both types of blocking groups share a common mechanism (e.g., acidolysis) for their removal. The degree of severity between the selective deblocking procedures spans several orders of magnitude since the NH_2 deprotection is repetitive. The problem may be avoided by employing groups which are removable by different mechanisms, but the availability of such options is extremely limited. As mentioned, most peptides are still prepared by solid-phase methods using *t*-BOC for amine protection and benzyl-based side-chain protection, both of which are removed by treatment with strong acids. In this regard, the expanding role of Fmoc in modern peptide synthesis is most welcome.

The major deleterious effects of strong acids include: (a) breakage of peptide bonds; (b) loss of peptide from the resin support in solid-phase synthesis;

Table 2 Common Side-Chain Protecting Groups

Reactive amino acid side chain	Protecting groups
$—CH_2—OH$ (Serine); $—CH(CH_3)—OH$ (Threonine)	Benzyl ethers
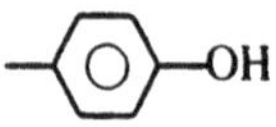 $—C_6H_4—OH$ (Tyrosine)	Substituted benzyl ether derivatives
$—(CH_2)_3NH—C(=NH)—NH_2$ (Arginine)	Tosyl, nitro
$—(CH_2)_4—NH_2$ (Lysine)	Carbobenzoxy derivatives, *t*-BOC
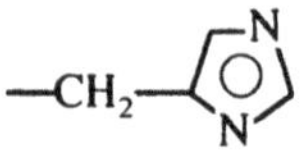 $—CH_2—$imidazole (Histidine)	Tosyl, dinitrophenyl, benzylmethoxy
$—CH_2COOH$ (Aspartic acid); $—CH_2CH_2COOH$ (Glutamic Acid)	Benzyl esters
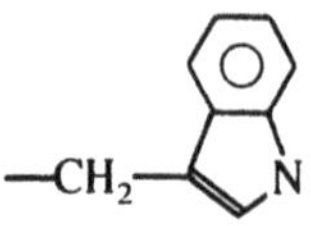 $—CH_2—$indole (Tryptophan)	Formyl
$—CH_2—SH$ (Cysteine)	Methylbenzyl, acetamidomethyl

(c) side reactions causing peptide modifications during removal of blocking groups (the groups themselves can become alkylating agents and modify susceptible amino acids such as Trp, Tyr, Met, and Cys). Further losses in final products also occur from incomplete deprotection, and this risk is increased when milder acidic conditions and shortened times of exposure are used to minimize the problems associated with strong acids.

In stepwise solid-phase synthesis, the completeness of both coupling and deprotection at each cycle is far more critical to the successful elaboration of the desired pure product than in solution synthesis. In the latter method, each intermediate is purified, whereas in stepwise solid-phase methodology undesired peptides that result from incomplete coupling, deprotection, or other causes, accumulate throughout the synthesis. The finished peptide is then recovered as a so-called crude product that must be purified from a mixture of peptides, some of which may be very closely related to the desired product and can therefore be exceedingly difficult to separate. A good example is in the case of deletion peptides. When incomplete coupling occurs at any given cycle in the synthesis, but with subsequent coupling steps proceeding normally, a peptide byproduct missing a single amino acid residue will be generated. This can, of course, take place at any elongation step in the synthesis, potentially creating an assortment of peptides each differing from the intended peptide by only one amino acid. To minimize this problem, there are a number of preventive and remedial strategies. The first is to use large molar excesses (typically 3-5-fold) of activated amino acids over peptide to force the coupling at each cycle to go as close to 100% completion as possible. Careful monitoring (e.g., by the quantitative ninhydrin method) can help to identify incomplete coupling so that the investigator has the choice of recoupling or implementing other corrective measures. Recoupling may be attempted using combinations of different solvents, reneutralization of the free amine on the peptide, alternate coupling mechanisms, etc. If the coupling cannot be driven to completion, the further growth of a peptide chain can be deliberately terminated at a given cycle by irreversibly blocking "leftover" amino groups (e.g., by acetylation). The shorter, blocked peptide contaminants that would result should be less closely related to, and therefore easier to separate from the full-length product. One last preventive measure is that it is important to use the purest reagents and solvents obtainable.

These procedures, along with a general requirement for fairly large volumes of solvents for routine resin washes, has occasionally raised some concern that the solid-phase method is wasteful, and hence relatively costly. In practice, this is not the case, particularly when the enormous savings in time and manual labor are considered. Also, for large-scale synthesis, recovery of materials for reuse is possible. Actual cost and yield comparisons with solution synthesis argue that solid-phase synthesis is, in fact, efficient and cost effective.

Other Developments

In 1985, R. A. Houghton published a method for rapidly preparing large numbers of peptides using the standard solid-phase procedures for coupling and deprotection (7). The standard resin was sealed into separate packets which resembled small tea bags made of polypropylene, so that solvents and reagents could freely flow in and out. Thus, many individual packets could be placed in the reaction vessel of an automated instrument, or manual methods could be chosen, to simultaneously perform the many common, repetitive steps at each cycle of solid-phase peptide synthesis. For the coupling step, the packets were removed and individually added to solutions containing appropriate protected, preactivated amino acids. The packets were then returned to the reaction vessel for further routine steps. On completion, peptides were liberated from the resin by standard treatment with anhydrous liquid HF/anisole, except that the apparatus had 20 reaction chambers for simultaneous cleavage of as many peptide resin packets. The modified HF apparatus and the packets, which can be filled with a variety of resin types (e.g., with spacer arms to physically distance the peptide chain from the polymer, or with linkages that yield either carboxylic acids or amides at the C-terminus of the peptide following HF cleavage), are now commercially available.

Another multiple peptide synthesis system, developed at the Du Pont company and sold under the name RaMPS, utilizes a series of reaction vessels fitted with manually operated valves and syringes. The system uses Fmoc chemistry and up to 25 peptides can be made simultaneously, generating about 150 mg of each.

One other simultaneous synthetic system that deserves mention was devised by Geysen et al. to prepare tiny amounts of short peptides (e.g., six amino acids) on the end of plastic pins that align with 96-well microtiter plates (8). Standard solid-phase synthesis was used to elaborate the peptides on the pins. The completed sequences were deprotected, but not cleaved from the solid support (pins), and the N-termini were acetylated. The procedure clearly is not intended to prepare amounts of peptides that can be purified or characterized. Rather, the pins are used directly for enzyme-linked immunosorbent-assays (ELISA), where the peptides each represent a short region of a protein sequence to which one or more antibodies have been raised. In this way, the epitopes on the protein recognized by a particular antibody can be mapped. Since there is no direct determination as to the success of the synthesis on individual pins, the results of the procedure normally require confirmation by another method. Nevertheless, the technique can rapidly provide a rationale for the preparative synthesis of pure peptides from specifically identified regions of the protein of interest, which can be used for confirmatory testing and further studies.

Other modern automated instruments are capable of fast syntheses using standard, but optimized chemistries, and by employing multiple reaction vessels or flow-through columns.

Biological Peptide Synthesis

Methods for the biological synthesis of peptides and proteins are also developing rapidly and can offer some important advantages, particularly for the preparation of fairly large sequences. There are several biosynthetic approaches, but all rely on ribosomal assembly systems. Isolation of peptides and proteins from natural sources (e.g., blood and other body tissues) has been important in supplying products such as insulin and growth hormone. Antimicrobial peptides, such as gramicidin and bacitracin, are produced by industrial fermentation processes. Recombinant DNA technology has become a sophisticated tool for altering cells to produce large quantities of important peptide and protein products. Bacterial fermentation is routinely used when feasible, and very large and/or postranslationally modified proteins that are generally not otherwise accessible, have been prepared in mammalian cells and certain expression systems. Human insulin, growth hormone, interferon, and tissue plasminogen activator are good examples of how these methods have been successfully exploited in the pharmaceutical industry. Hybridomas have primarily been used for production of monoclonal antibodies, but immortalized cells secreting peptide products could become an important production technique. A fourth aptly named method, cell-free translation, utilizes isolated ribosomes and the essential components for the assembly of peptides. While this technique requires much development, it offers the intriguing possibility that unnatural peptides, such as analogs containing D-amino acids, may eventually be generated by this approach.

Modern biological and chemical methods of peptide synthesis together provide unprecedented speed and flexibility in elaborating products for multiple applications. The chemical approach is currently essential for developing analogs since peptides containing D-amino acids or unnatural amino acids (e.g., chemically altered side chains, derivitized backbones) can be prepared. This method is best suited for the preparation of short peptides up to about 40 residues, since longer syntheses, though achievable, are accompanied by lower product yields, longer and more difficult purification procedures, and higher costs. Conversely, the biological approach is, in many cases, the best way to prepare very large peptides and proteins. High development costs, together with the present constraint of producing structures comprised only of the naturally occurring L-amino acids, restrict the use of biological synthesis in analog development programs. Nevertheless, an established method to produce a particular peptide can be more efficient and cost effective than chemical methods, especially for industrial-scale production.

Two other synthetic approaches, semisynthesis and enzymatic synthesis, combine elements of both the biological and chemical methods. In semisynthesis, a biologically produced peptide or protein provides the raw material for modifying its structure. A section of the chain is removed, often by using specific enzymes, and is then replaced with an altered section prepared, for example, by chemical synthesis. Porcine insulin was converted to human insulin this way, illustrating the potential value of the method. Enzymatic synthesis exploits the reversible nature of protease catalysis by shifting the reaction in favor of peptide bond formation. This approach harnesses the extraordinary stereoselectivity of enzymes and minimizes or eliminates the need for side-chain-protecting groups. It has been successfully used to prepare enkephalins, substance P, epidermal growth factor fragments, and the dipeptide sweetener aspartame. The method is primarily limited by the substrate specificity and protein character of the enzymes.

PURIFICATION AND CHARACTERIZATION

Many linear peptides up to 15–20 residues can be prepared by routine protocols to yield "crude" products that are often between 70 and 85% pure. The high resolving power of reverse-phase high-performance liquid chromatography (RP-HPLC) is now frequently used to rapidly assess whether or not a particular synthesis predominantly yielded the target peptide. In general, a clearly major single component at 214-nm among any number of minor components is sufficient to warrant further purification and characterization of the main product. For most chemical and biological studies, highly purified peptides are required to assure that the presence of contaminants do not affect the experiments. The same high resolution attainable with analytical reverse-phase HPLC can very often be exploited for preparative peptide purification. The technique is both rapid and scalable. In addition, it is one of few methods where a generalized procedure is suitable for purifying peptides of varied sequence, and therefore with different properties. In cases where further purification is necessary, altering the solvent conditions used for RP-HPLC is sometimes sufficient. If RP-HPLC is inadequate or inappropriate, purification can usually be achieved by a combination of size exclusion chromatography, ion exchange chromatography, isoelectric focusing, and affinity chromatography.

Isolation of the main product from a completed synthesis does not address whether that product is indeed the expected peptide. As discussed earlier, a variety of chemical side reactions are ever-present risks which could generate modifications in the desired molecule. Even the proper sequence of amino acids is not totally safe to assume, despite the stepwise nature of the chemical synthesis. This could result from errors that range from mislabeling of amino acids to instrument malfunction. Optical purity should also be established. For the more complex folded peptides, verification of correct folding should be ascertained. Biologically active peptides have to be tested in relevant assays for specific activity.

Improvements and refinements in peptide synthesis have largely minimized many risks. The use of highly purified reagents, optimal reaction conditions established for individual residues, and careful monitoring for completeness of deprotection and coupling throughout the synthesis have had a major impact on the overall success of modern peptide synthesis. While a greater proportion of the large numbers of attempted syntheses are successful as a result of many such developments, it remains crucial to establish the molecular structure of the final product. This can prove to be a difficult challenge for some peptides, and a variety of techniques are needed to address different structural aspects.

A first approximation of peptide structure is routinely provided by quantitative amino acid composition analysis. The peptide is hydrolyzed to its component amino acids, which must be separated chromatographically either before or after they are derivatized for detection. The common amino acids released in acid hydrolysates can be separated in as little as 10 minutes by modern HPLC methods. The most widely used systems for separation are cation exchange chromatography and reverse phase HPLC. The former system typically employs postcolumn derivatization of the separated free amino acids for either colorimetric, or for increased sensitivity, fluorescent detection. RP-HPLC usually relies upon precolumn derivatization with a fluorescent reagent. These methods have been highly automated for efficient, accurate quantitation of both amino acid composition and peptide/protein concentration. The actual amino acid ratios, however, provide only limited structural information. In practice, overall composition serves best to uncover relatively serious problems in the peptide, such as a missing or incorrect amino acid, or substantial product heterogeneity. Some side chain modifications escape detection altogether, since they can be reversed during the hydrolysis. Thus, after adjusting for losses during hydrolysis, the purified peptide should have, at minimum, the expected amino acid composition.

More detailed structural information is provided by sequencing the peptide. Quantitative Edman degradation chemistry on increasingly sophisticated automated instruments is currently the most widely used method for sequence analysis, and HPLC is the primary tool for identifying the PTH amino acids generated at each cycle. Not only will the expected sequence of a synthetic peptide be confirmed, but deletions and certain other undesirable components arising from the synthesis or purification may also be identified. The conditions of Edman chemistry are, however, also severe and chemically modified amino acids can again go undetected. Additional analytical methods are required to definitively establish the molecular structure.

The most powerful technique yet developed to address this problem is undoubtedly mass spectrometry (MS). The fast atom bombardment (FAB) method of ionization in particular has revolutionized the use of MS for analyzing peptides. In general, FAB allows molecules to be ionized in the solid or liquid state and desorbed in the vapor state for detection. The FAB mass spectrum of a peptide usually provides three types of information: (1) The molecular ion provides a precise molecular weight and sometimes the composition of the peptide; (2) ions at low masses are characteristic of individual amino acids; (3) ions at intermediate masses provide amino acid sequence data. The advantages offered are considerable. Molecular weights determined to within a few tenths of a mass unit can verify a predicted structure down to a single atom. Modifications to the structure are not only discovered, but often the difference in mass from that predicted can disclose the chemical nature of the modification. Mass spectrometry is well-suited for analyzing N-terminally blocked peptides, and for establishing the identity of the blocking group, whereas Edman chemistry depends upon the presence of a free amino terminus. In fact, FAB-MS offers an alternative mechanism to N- (or C-) terminal sequencing when the predicted sequence is available, with the added advantage of having equal probability of obtaining sequence information from anywhere in the molecule. For larger peptides and proteins, *unseparated* mixtures of peptide fragments can be used to generate useful information. Ambiguities can often be resolved by generating a second mass spectrum following one cycle of Edman degradation of the peptide mixture, which is one example of how the two methods can complement one another. Finally, the technique is also suitable for mapping disulfide bridges in peptides that contain them. FAB-MS is not yet routinely available in many laboratories, primarily due to high cost. An informative review was recently published by Carr et al. (9).

As alluded to earlier, ambitious syntheses of larger and more complex peptides require more extensive and difficult characterization. It is not uncommon for automated Edman-type sequencers to provide reliable information for the first 20–40 residues of a peptide, but longer peptides and proteins generally need to be fragmented to obtain further information on internal structure. Fragmentation typically is accomplished by any of a variety of specific enzymes or chemical reagents. The peptides generated can be partially or completely separated by chromatography, most often using reverse phase HPLC. The elution profile can serve as a "fingerprint" for comparison to a profile produced after identical treatment of a known peptide of like structure. Identical patterns of fragmentation provide convincing evidence that the presumed structure is correct. For such a case, there would be no need to actually collect and identify each of the fragments. On the other hand, if one peak, for example, had been shifted, the difference from the expected structure of just that peak could be determined. Alternatively, many or all of the peaks could be collected for complete structural characterization, particularly where there is no reference compound available. This technique is referred to as "peptide mapping." Since the predicted sequence is known, the collected peptides can frequently be identified by amino acid analysis, which is faster than sequence analysis. However, in this instance the actual correct order of amino acids could only be assumed. The fingerprinting and mapping methods can also reveal the disulfide attachments in folded peptides. As a final note, fingerprinting of peptides without fragmentation can be successful to some extent by nuclear magnetic resonance (NMR) spectroscopy. While assignment of signals produced by large peptides is too complex for general application, reproducing the spectrum of a known structure again strongly supports the structure of the test sample.

During the coupling step in either solution or solid phase peptide synthesis, the potential for racemization cannot be ignored, regardless of the methodology chosen, since none is completely free from risk. Proper selection of protecting groups and coupling conditions, along with the previously mentioned careful attention to the use of highly purified solvents and reagents, can greatly minimize this and many other side reactions.

There are several methods for assessing the optical purity of synthetic peptides, most of which have been thoroughly reviewed (10,11). It has long been possible to detect the presence of enantiomers in the starting free amino acids to be used in the synthesis; a widely used method for this is gas-liquid chromatography of derivatized amino acids on a chiral stationary phase (often glass capillaries coated with a valine-containing compound). The method is sensitive to 0.01%, as is another technique introduced by J. Manning and S. Moore in which the free amino acids are reacted with activated L-leucine to form diastereomeric dipeptides that can be separated on an amino acid analyzer or by RP-HPLC. These methods are also suitable for analysing free amino acids liberated from hydrolyzed peptides, but there is the added complication that the extent to which racemization occurs during the hydrolysis must be determined. Subjecting amino acid mixtures to the conditions of hydrolysis as a control does not address the known influence of neighboring residues in a peptide sequence on the rate at which some amino acids racemize; that is, the propensity for certain amino acids to racemize is sequence-dependent. One clever approach to this problem is to conduct the peptide hydrolysis in deuterated HCl. Any amino acid which racemizes during hydrolysis becomes isotopically labeled. The resultant free amino acids are derivatized, separated by gas chromatography, and detected by mass spectrometry to quantitate D/L ratios with accuracy to 0.1%.

Indirect methods that depend upon aminopeptidases to digest only L-amino acid residues (1) have generally been limited by substrate specificity and by the need to sensitively identify incompletely digested products after treatment with proteolytic enzymes. However, the remarkable ability of RP-HPLC to separate closely related peptides is being applied to many types of racemization studies, including those that take advantage of enzyme stereoselectivity. HPLC has been used to resolve fairly large peptides with seemingly subtle differences. For example, the diastereomeric 14-residue somatostatin (SS) peptides, D-5F-Trp^8-SS and L-5F-Trp^8-SS, were synthesized simultaneously using a racemic mixture of D, L-5F-Trp, and then separated by HPLC on a C_{18} reverse-phase column. Also, the 36-residue human pancreatic polypeptide (HPP) synthesized in 1980 (12) was separated by RP-HPLC from bovine PP (BPP), which differs from human by only two amino acids, one of which involves a conservative change from Asp^{23} to Glu^{23} (the other is a Val^6 to Glu^6).

The detection of racemization in a synthetic peptide must sometimes be established with the assistance of synthetic diastereomeric peptide standards on a case by case basis. A region where, for example, a D-amino acid has been incorporated may need to be compared to an L-amino acid-containing counterpart to demonstrate that its detection in the sample is possible, and to what limits. The sensitivity varies with the degree of separation, since samples may be overloaded until resolution becomes obscured. Sequences known to be racemization prone may also be specifically examined using specially prepared synthetic standards. The configurations of the 5F-Trp^8-SS analogs mentioned above were in fact assigned by comparing their HPLC retention times to a separately synthesized authentic "all-L" analog standard.

For most applications, the advances in stepwise peptide synthesis have greatly minimized the occurrence of racemization. In contrast, activation of peptide fragments (in fragment condensation approaches to the preparation of large polypeptides) invites greater risk which demands increased attention to analysis of the products, and also to the synthetic strategy. For instance, the C-terminal residue of the fragment to be activated can often be deliberately selected as glycine; the only one of the 20 amino acids that lacks a chiral center and therefore cannot racemize.

SYNTHETIC ANALOGS

Modern methods of chemical peptide synthesis are fast and reliable so that numerous peptide sequences are readily accessible, and systematic changes to natural peptides which produce analogs are therefore feasible. Many types of changes in the peptide are possible, some of which may occur in nature, and others which are unnatural and are currently available only through chemical synthesis. A primary motivation for preparing peptide analogs is that certain properties of the natural parent compound are often suboptimal for use as pharmaceuticals or other applications (e.g., synthetic enzyme substrates for developing routine assays). Peptides have evolved with sufficient potency, selectivity, and stability to function in highly regulated physiological systems. They have certainly not been optimized for use as drugs, so that there are considerable opportunities for maximizing the properties of natural peptides through the use of peptide analogs. Nevertheless, it is important to note that many unmodified peptides have therapeutic applications. Table 3 contains a list of unmodified peptide drugs that are commercially marketed for human use. For a review of peptides as pharmaceuticals, see Samanen (13).

Examples abound where peptides have been altered to dramatically enhance activity, selectivity, solubility, and stability, or to convert them into antagonists. At present, relatively few peptide analogs have reached the status of an approved drug, in part because the ability to generate large numbers of analogs is a fairly recent development. However, in the past decade over 170 clinical trials with peptides and peptide analogs have been performed and many are encouraging. Two posterior pituitary peptide hormones, oxytocin and vasopressin, are examples of synthetic peptides and analogs that have been successful pharmaceuticals (Table 4). Oxytocin, the first biologically active polypeptide to be synthesized, causes uterine contractions and is used as a drug for the medical induction of labor. The synthetic preparation avoids possible contamination by the antidiuretic hormone vasopressin, which in effect provides a measure of increased selectivity. While the same argument holds for synthetic vasopressin, which is also commercially available, a synthetic analog of this molecule has been developed whereby the potency, selectivity, and duration of antidiuretic action are substantially increased in comparison with the unmodified peptide. Like the parent hormone, the vasopressin analog is also a marketed drug.

Table 3 Therapeutic Applications of Unmodified Peptides Approved by the U.S. FDA for Humans

Peptide	Applications
Adrenocorticotropic hormone (ACTH)	Inflammatory conditions; arthritis, allergies, respiratory diseases, autoimmune disorders
Bacitracin	Topical treatment of bacterial infections
Calcitonin	Paget's disease, control of hypercalcemia
Gramicidin	Topical treatment of bacterial infections
Growth hormone (GH)	Replacement therapy in GH-deficient children
Insulin	Diabetes mellitus
Oxytocin	Induce labor, control postpartum bleeding, milk letdown
Vasopressin	Antidiuresis in diabetes insipidus

Table 4 Potential Uses of Some Peptides or Their Analogs

Peptide	Structure	Potential applications
Adrenocorticotropic hormone (ACTH)	Ser-Tyr-Ser-Met-Glu-His-Phe-Arg-Trp-Gly-Lys-Pro-Val-Gly-Lys-Lys-Arg-Arg-Pro-Val-Lys-Val-Tyr-Pro-Asn-Gly-Ala-Glu-Asp-Glu-Ser-Ala-Glu-Ala-Phe-Pro-Leu-Glu-Phe	Severe inflammation Epileptic seizures Amnesia Depression Senile dementia (see Table 3)
Angiotensin II (AII)	Asp-Arg-Val-Tyr-Ile-His-Pro-Phe	Hypertension Cardiac failure Thirst Enhance cancer chemotherapy
Aspartame (NutraSweet)	Asp-Phe-OCH_3	Food and beverage sweetener
Bombesin	pyroGlu-Gln-Arg-Leu-Gly-Asn-Gln-Trp-Ala-Val-Gly-His-Leu-$MetNH_2$	Hyperthermia Peptic ulcer Tumor suppression
Calcitonin	Cys-Gly-Asn-Leu-Ser-Thr-Cys-Met-Leu-Gly-Thr-Tyr-Thr-Gln-Asp-Phe-Asn-Lys-Phe-His-Thr-Phe-Pro-Gln-Thr-Ala-Ile-Gly-Val-Gly-Ala-$ProNH_2$ (disulfide bridge Cys¹–Cys⁷)	Hypercalcemia Gastric secretion Analgesia Osteoporosis
Cholecystokinin (CCK) octapeptide	Asp-Tyr(SO_3H)-Met-Gly-Trp-Met-Asp-$PheNH_2$	Appetite control Pancreatitis Ulcers Diarrhea
β-endorphin [βh-LPH(61–91)]	Tyr-Gly-Gly-Phe-Met-Thr-Ser-Glu-Lys-Ser-Gln-Thr-Pro-Leu-Val-Thr-Leu-Phe-Lys-Asn-Ala-Ile-Ile-Lys-Asn-Ala-Tyr-Lys-Lys-Gly-Glu	Pain (cancer, childbirth) Narcotic abstinence Depression Shock
Leu-enkephalin	Tyr-Gly-Gly-Phe-Leu	Analgesia Diarrhea
Met-enkaphalin	Tyr-Gly-Gly-Phe-Met	Analgesia Diarrhea
Leutinizing hormone releasing hormone (LHRH)	PyroGlu-His-Trp-Ser-Tyr-Gly-Leu-Arg-Pro-$GlyNH_2$	Block or induce ovulation, spermatogensis Prostatic cancer
α-Melanocyte stimulating hormone (α-MSH)	Ser-Tyr-Ser-Met-Glu-His-Phe-Arg-Trp-Gly-Lys-Pro-Val	Fever

Neurotensin	pyroGlu-Leu-Tyr-Glu-Asn-Lys-Pro-Arg-Arg-Pro-Tyr-Ile-Leu	Hypertension Hyperthermia Pain Gastric secretion
Oxytocin	Cys-Tyr-Ile-Gln-Asn-Cys-Pro-Leu-$GlyNH_2$ (disulfide bridge Cys–Cys)	Induce labor Postpartum bleeding Lactation
Parathyroid hormone (PTH)	Ala-Val-Ser-Glu-Ile-Gln-Phe-Met-His-Asn-Leu-Gly-Lys-His-Leu-Ser-Ser-Met-Glu-Arg-Val-Glu-Trp-Leu-Arg-Lys-Lys-Leu-Gln-Asp-Val-His-Asn-Phe-Val-Ala-Leu-Gly-Ala-Ser-Ile-Ala-Tyr-Arg-Asp-Gly-Ser-Ser-Gln-Arg-Pro-Arg-Lys-Lys-Glu-Asp-Asn-Val-Leu-Val-Glu-Ser-His-Gln-Lys-Ser-Leu-Gly-Glu-Ala-Asp-Lys-Ala-Asp-Val-Asp-Val-Leu-Ile-Lys-Ala-Lys-Pro-Gln	Osteoporosis Hypertension
Somatostatin (SS)	Ala-Gly-Cys-Lys-Asn-Phe-Phe-Trp-Lys-Thr-Phe-Thr-Ser-Cys (disulfide bridge Cys–Cys)	Diabetes mellitus Bleeding ulcers Acromegaly Pancreatitis Hemorrhage
Substance P	Arg-Pro-Lys-Pro-Gln-Gln-Phe-Phe-Gly-Leu-$MetNH_2$	Pain Inflammation Bronchoconstriction
Vasoactive intestinal polypeptide (VIP)	His-Ser-Asp-Ala-Val-Phe-Thr-Asp-Asn-Tyr-Thr-Arg-Leu-Arg-Lys-Gln-Met-Ala-Val-Lys-Lys-Tyr-Leu-Asn-Ser-Ile-Leu-$AsnNH_2$	Diarrhea Hypertension Pancreatitis CNS function
Arg^8-Vasopressin	Cys-Tyr-Phe-Gln-Asn-Cys-Pro-Arg-$GlyNH_2$ (disulfide bridge Cys–Cys)	Diabetes insipidus Hypertension Edema Senile dementia Learning and memory

Some selected analogs will serve to illustrate the major types of modifications that can be introduced into peptide sequences in order to develop potent, long-acting, conveniently administered therapeutic drugs with minimal side effects.

Modifications

A variety of modifications can be introduced into peptides to influence their properties. Alterations in the backbone, side chains, and termini have been widely used alone or in combination to produce biologically active peptide analogs with substantially different characteristics from the parent (native) peptide. For example, [D-Ser(tBu)^6, des Gly^{10}]-LHRH ethylamide has over 50 times the potency of leutinizing hormone-releasing hormone (LHRH) in stimulating gonadotropin release. A cyclic α melanocyte-stimulating hormone (α-MSH) analog, [Cys^4, Cys^{10}]-αMSH, is over 10,000 times as active as α-MSH in the frog skin bioassay. Multiply substituted enkephalin analogs (analgesic pentapeptides) have exceeded 14,000-fold activities over the natural peptide. Similar examples exist for analogs that are highly selective for one of the multiple actions displayed by the parent substance, and for analogs that have short plasma half-lives or that are long-acting for other reasons.

Typically, the most successful analogs, in terms of their having desirable therapeutic properties, are the culmination of a logical progression of analogs designed to uncover the relationships between the structure and the biological recognition and activation events that are observed. As an example, there have been extensive studies on numerous synthetic analogs of LHRH in several laboratories aimed at producing both agonists and antagonists. This 10-residue hypothalamic hormone is named for its ability to release leutinizing hormone (LH) and follicle-stimulating hormone (FSH) from the pituitary. Selective, potent agonists of LHRH are under development to improve the ability of the native hormone to induce ovulation, thereby helping women with temporary infertility. Agonists also promote testicular descent in boys and spermatogenesis in men. In addition, a clinical potential for controlling prostatic carcinoma is currently being evaluated. This apparent paradoxical use of an agonist can be explained by the observation that chronic administration of large doses of LHRH agonists have opposite gonadal effects to those observed with acute doses of the peptide. Antagonists of LHRH may be useful for contraception.

The changes to the native decapeptide in the LHRH analog cited above [D-Ser(tBu)^6, des Gly^{10}]-LHRH ethylamide, are actually not surprising since they resulted from systematic studies which showed that the Gly in positions 6 and 10 were principally involved in maintaining the preferred conformation of the peptide for binding to pituitary receptors. In addition, the His^2 and Leu^3 residues were identified as an "active center." Following these key discoveries, all analogs with substantially greater agonist activities have been changed in positions 6, 10, or both. Further, the inhibitory analogs which have been reported were modified in positions 2 and 3 to abolish gonadotropin release while maintaining receptor binding. The bulky *tert*-butyl side-chain in the [D-Ser(tBu)^6, des Gly^{10}]-LHRH ethylamide resulted from the observation that the activities of analogs with D-amino acids in position 6 increased with the size of the side chain. [D-Ala^6]-LHRH was fivefold more active than LHRH, and the corresponding D-Leu, D-Phe, D-Trp, and D-$(CH_3)_5$Phe analogs had increased activity up to 13 times that of LHRH in vivo. The bulky side chains of position 6 apparently stabilize the binding conformation of LHRH.

Peptide Fragments

Peptide hormones and neurotransmitters are recognized by specific cell surface receptors from the information residing in their individual amino acid sequences. Receptor binding and receptor activation leads to biological actions, and in some peptides separate regions of the molecule are responsible for these events. By deleting or altering just the section which signals receptor activation, an analog can be developed that retains receptor binding but cannot trigger a biological response. Such an analog can act as an antagonist by competing with endogenous ligand for receptor binding. Thus, removal of the side chain from the last amino acid in the octapeptide angiotensin II (AII) converts it into an AII antagonist.

Determination of the minimal amino acid sequence of a peptide that is necessary for biological activity can be an important first step in designing useful agonists or antagonists. For example, the search for an antagonist of parathyroid hormone (PTH), a single-chain polypeptide of 84 amino acid residues (Table 4), began with a systematic search for the shortest biologically active fragment. All of the biological activity (elevated levels of calcium) was shown to reside in the fragment 1–34. Amino acids were then deleted stepwise starting from the C-terminal Phe^{34}, with a corresponding gradual decrease in potency, until measurable receptor binding and biological activity were lost in the fragment 1–25. In contrast to this gradual decline in function, deletion of only the first two amino acids from the N-terminus abruptly abolished bioactivity, yet the 3–34 segment retained the ability to bind weakly to parathyroid hormone receptors. This fragment satisfies the definition of a receptor binding antagonist. Having identified that separate domains were responsible for receptor binding and receptor activation, the feasibility of developing more potent antagonists was established. It is worth noting that years of systematic structure/function studies on the 3–34 fragment resulted in a synthetic analog with full receptor binding properties and complete inhibition of PTH-stimulated bioactivity in vitro. Surprisingly, when tested in vivo, the analog was found to have retained partial agonist activities which rendered it ineffective as an antagonist. This has turned out to be a commonly occurring problem that has been encountered during the development of antagonists to LHRH, vasopressin, glucagon, and angiotensin II.

Further truncation from the amino terminus of the 3–34 segment ensued in an effort to identify the principle binding domain of PTH. Residues 25–34 were identified as the essential sequence required for receptor binding. The combined studies led to a PTH analog spanning residues 7–34 that displayed effective in vivo antagonism of the renal response to PTH, with no observable PTH agonist activity.

There are a variety of ways that the messages for biological signaling are distributed within the amino acid sequence of a peptide, or among different peptides. In some cases, different sequences appear to produce the same or similar biological responses (e.g., Leu- and Met-enkephalin, discussed below), whereas a single sequence can produce multiple responses in different tissues. For example, the list of peptides found in both gut and brain includes gastrin, cholecystokinin (CCK), bombesin, substance P, somatostatin, vasoactive intestinal polypeptide (VIP), neurotensin, and enkephalin. Some (substance P, VIP) are known to have dual roles as endocrine hormones and neurotransmitters. Somatostatin, a cyclic 14-amino acid peptide, has many types of actions including inhibition of growth hormone (GH), insulin, glucagon, gastrin, and gastric acid release.

Just as the receptor binding and activation sequences were separated in PTH, the signals for differing biological actions may be found along the length of a peptide, and the expression of multiple functions may depend upon in vivo processing of fairly large sequences in certain tissues. In the opioid pentapeptide Leu-enkephalin (Tyr-Gly-Gly-Phe-Leu), the intact sequence appears necessary to signal an analgesic response as evidenced by the lack of activity in synthetic fragments truncated at either the N- or C-terminus. Met-enkephalin (Tyr-Gly-Gly-Phe-Met), which also causes analgesia, comprises the first five amino acids of another analgesic peptide, β-endorphin. The 31-residue β-endorphin, in turn, occurs (as residues 61–91) at the extreme C-terminus of β-lipotropin (β-LPH), a 91-residue peptide originally isolated from the pituitary. β-LPH, along with the important steroidogenic adrenocorticotropic hormone (ACTH), in which the sequence for α-melanocyte stimulating hormone (α-MSH) resides, exist in tandem on a much larger precursor molecule (molecular weight 30,000) appropriately named pro-opiomelanocortin (POMC). Physiologically, as larger sequences are processed to smaller ones, it is not always clear whether all of the released fragments have biological roles other than to serve as precursor forms of the active agents. Precisely how precursors are processed, and which active agents are indeed liberated in various tissues must be experimentally determined, as this information cannot reliably be inferred solely by inspection of a sequence. A case in point was the obvious temptation to initially suspect that β-endorphin may serve as an endogenous precursor to Met-enkaphalin. Instead, extensive studies have indicated that: (1) the primary pathway in the processing of pituitary β-endorphin includes its inactivation by acetylation at its N-terminus; (2) only small amounts of Met-enkephalin are released from β-endorphin upon digestion with brain homogenates; and (3) other enkephalin-containing peptides, unrelated to

POMC, have been established as precursors to the enkephalins. For reviews, see Douglass et al. (14) and Stern et al. (15).

Peptide fragments can also be useful in enhancing the selectivity of a biologically active peptide when messages that signal different activities are separated along the sequence. The neuropeptide substance P, which is comprised of only 11 amino acids, appears to contain localized active regions. The N-terminal fragment 1–7 has actions in the central nervous system (pain modulation), as does the intact peptide, but does not display peripheral actions on smooth muscle or affect blood pressure. These peripheral activities are observed with a C-terminal substance P fragment from residues 7–11. Similarly, a section from residues 4–10 of the 39-amino acid peptide ACTH has selectively enhanced behavioral effects over the adrenal stimulation properties of the full-length hormone.

Structural Changes

The messages contained within the amino acid sequences of a peptide are not always, or even usually, separated neatly into domains for receptor binding and for the triggering of various biological actions. Rather, amino acids from different regions of the molecule may be critical to the maintainance of a preferred conformation for receptor binding or activation, as was suggested for the LHRH example described earlier. The requirement for preferred conformations is most obvious in cyclic peptides such as somatostatin (14 residues) or vasopressin (9 residues) in which the ring structure is formed by a disulfide bond, or "bridge" between the two cysteine residues in each peptide. Analogs which prevent formation of the ring generally are not active. Noncyclic (linear) peptides, such as the LHRH decapeptide, may also adopt preferred conformations that are stabilized by hydrogen bonding and hydrophobic interactions. Alternate strategies must be used for such peptides in order to develop analogs with useful properties.

The role of individual residues in short peptides may be examined by testing (in appropriate bioassays) many analogs, each with a change (i.e., replacement, deletion) at one position in the natural peptide. As already described, modern methods of chemical peptide synthesis make this approach ever more accessible and attractive. Thus, early structure/ function studies on somatostatin (SS) began with systematic replacement of each amino acid by L-Alanine. Each analog was compared with SS for its ability to inhibit the release of pituitary growth hormone (GH) and pancreatic insulin and glucagon. These actions can potentially be exploited for the treatment of such diseases as diabetes mellitus and acromegaly. Somatostatin also inhibits gastric acid secretion and may therefore be developed to treat ulcers (Table 4). Selective analogs are needed for clinical effectiveness in these diseases, since unmodified SS has so many actions that could translate into multiple side-effects. In the above Ala-replacement study, only Ala^2- and Ala^5-SS retained full biological potency. Most of the other alanyl analogs had negligible activity, although the peptides changed at positions 4, 10, 12, and 13 exhibited moderate activity. The study provided information about which of the side chain groups were most important to the proper functioning of the native hormone. In a similar systematic study, each L-amino acid residue in somatostatin was successively replaced by its D-isomer. These changes uniformly compromised the potency of these peptides with the striking exception of D-Trp^8-SS, which was about six times more active than somatostatin in inhibiting release of GH, insulin, and glucagon. Interestingly, none of the D-amino acid-containing analogs had a longer duration of action than somatostatin in vivo despite the blocking of predicted peptidase cleavage sites. Guided by the results of such studies, further changes at specific positions in the peptide followed. Position 8 was among the most obvious places to focus attention, and many subsequent analogs also included the D-Trp^8 modification to enhance their properties.

A more subtle interpretation of results led to some remarkable position 4 somatostatin analogs with enhanced potency, selectivity, and stability (16, 17). First, the analog with L-Ala substituted for the naturally occurring Lys in position 4 was observed to retain moderate, but significant activity (20%) in the pituitary and pancreas (Table 5). This indicated the lack of an absolute requirement for a basic or polar side chain at this position in maintaining the biolog-

Table 5 Hormone Release-Inhibiting Activities (%) of Position 4 Somatostatin (SS) Analogs in Rats

			Growth hormone	
Positon 4 residues	Insulin in vivo	Glucagon in vivo	in vitro	in vivo
Lys (SS)	100	100	100	100
D-Lys	1	1	22	
des-Lys	1.5	—	2.7	
Arg	60	100	100	
Ala	20	20	20	
Thr	1.6	<5	< 1	
Glu	1.1	—	4.9	
Phe	2.8	15.5	203	342
F_5-Phe	285	263		299
Phe(D-Trp^8)	156	193		360
p-NH_2-Phe	38	43		405
p-NH_2-Phe(D-Trp^8)	139	142		1524

Source: From Refs. 16 and 17.

ical activity of somatostatin. In addition, the D-amino acid series of substituted analogs had produced D-Lys^4-SS which showed a tendency for selective action on the pituitary (about 20% inhibition of GH release versus about 1% of insulin and glucagon release inhibition). The possibility of exploiting the selective action and enhancing the observed potency prompted further systematic changes in position 4 of somatostatin. The Lys^4 was replaced by an aromatic (Phe), alcoholic (Thr), and acidic (Glu) residue, and it was also deleted (des Lys^4-SS). The single change of Lys^4 to Phe^4 produced a somatostatin peptide that was over three times more active than unmodified somatostatin on the inhibition of GH release in vivo, yet which maintained the degree of selectivity over in vivo pancreatic activity that had been suggested by the results of the D-Lys^4-SS analog. Also, the Phe^4-SS blocked GH release with apparently greater potency in vivo than in vitro, which would be consistent with its having long-acting properties. When tested for duration of action on GH in vivo, Phe^4-SS was effective for 2 hours, even after intravenous administration, compared with only 15 minutes for somatostatin. Peptides that are resistant to degradation could be present in greater local concentrations at the receptor, and they may be slowly released from a storage depot in vivo. In the case of Phe^4-SS, a trypsin-sensitive Lys residue was replaced by a chymotrypsin-sensitive Phe residue, so that resistance to enzymatic degradation would not ordinarily be expected. However, conformational changes, which are almost certainly operative here, can also alter degradative processing and may also increase receptor binding affinity.

This study led to the logical preparation of other position 4 analogs in combination with the D-Trp^8 modification, and ultimately resulted in [p-NH_2-Phe^4, D-Trp^8]-SS which restored some of the side-chain character of the naturally occurring Lys^4 residue. This analog had a striking 15-fold increase in activity over somatostatin in the pituitary, exhibited the same degree of selectivity toward GH inhibition as Phe^4-SS, and was also active for about 2 h in vivo.

Peptide Backbone Analogs

Conformationally restricted peptide analogs can be made to stabilize the bioactive conformation, thus potentially enhancing potency or selectivity. D-amino acid analogs like the D-Trp^8-SS described above can serve this function. Other synthetic options include; (1) cyclization through cystine disulfide bridges (or related analogs with nonreducible bridges), N- to C-terminal amide bond formation, or through other side chains (e.g., Asp or Glu bonded to Lys); (2) incorporation of individually restricted amino acids including N- and α-methyl amino acids, β-disubstituted residues, and cyclic amino acids such as proline; (3) individually restricted side chain analogs such as those with dehydroamino acids.

The empirical nature of analog design described earlier for PTH, LHRH, and somatostatin can initially be applied to conformationally restricted analogs as well, in order to identify the structures that maintain bioactivity. Once the range of possible bioactive conformations is limited, computer models of potentially relevant structures may be generated. NMR spectroscopy on small restricted analogs may also help to determine bioactive conformations, allowing truly rational design of analogs based upon refined three-dimensional models to proceed.

CONCLUDING REMARKS

The chemical synthesis of peptides has developed into a rapid, reliable, automated procedure available to an increasingly broad spectrum of investigators. An unprecedented access to large numbers of synthetic peptides has revolutionized their applications in biological research and clinical medicine. New developments in the synthesis, purification, and characterization of peptides promise to make ever more challenging sequences more readily available. As these tools contribute to a better understanding of peptide and protein structure, as well as their roles in normal physiology and disease, the number of therapeutically useful peptide analogs for a wide variety of ailments will dramatically increase. The rational design of pharmaceuticals will further include the development of nonpeptide mimetics based upon the active structures of peptides. The commercial applications are, of course, not limited to medicine, as exemplified by the artificial dipeptide sweetener, aspartame. Even that is only one of an emerging category of so-called tastants (peptides that taste bitter, sweet, and "delicious" have been described). The remarkable advances in peptide synthesis just over the past decade are a reminder of how recently the methods that sparked such widespread interest have developed, and therefore how many exciting new developments can be expected in the next decade.

REFERENCES

1. Finn, F. M. and Hofmann, K. In *The Proteins*, vol II. Edited by H. Neurath, R. L. Hill, and C. Boeder. Academic Press, New York, 1976, pp. 105–253.
2. Schroder, E. and Lubke, K. *The Peptides*, vol. 1. Academic Press, New York, 1965.
3. Barany, G. and Merrifield, R. B. In *The Peptides*, vol. II. Edited by E. Gross and J. Meienhofer. Academic Press, New York, 1980, pp. 1–284.
4. Stewart, J. M. and Young, J. D. *Solid Phase Peptide Synthesis*. Pierce Chemical Company, Rockford, IL, 1984.
5. Tam, J. P. and Merrifield, R. B. In *The Peptides*, vol. 9. Edited by S. Udenfriend and J. Meienhofer. Academic Press, New York, 1987, pp. 185–248.

6. Atherton, E. and Sheppard, R. C. In *The Peptides*, vol. 9. Edited by S. Udenfriend and J. Meienhofer. Academic Press, New York, 1987, pp. 1–38.
7. Houghton, R. A., *Proc. Nat'l. Acad. Sci. (USA)* 82; 5131–5135 (1985).
8. Geysen, H. M., Meloen, R. H., and Barteling, S. J. *Proc. Nat'l Acad. Sci. (USA)* 81; 3998–4002 (1984).
9. Carr, S. A., Hemling, M. E., and Roberts, G. D. *In Macromolecular Sequencing and Synthesis: Selected Methods and Applications*. Edited by D. H. Schlessinger. Alan R. Liss, New York, 1988, pp. 83–99.
10. Kemp, D. S. In *The Peptides*, vol. 1. Edited by E. Gross and J. Meienhofer. Academic Press, New York, 1979, pp. 315–383.
11. Benoiton, N. L. In *The Peptides*, vol. 5. Edited by E. Gross and J. Meienhofer. Academic Press, New York, 1983, pp. 217–284.
12. Meyers, C. A. and Coy, D. H. *Int. J. Peptide Protein Res* 16, 248–253 (1980).
13. Samanen, J. M. In *Polymeric Materials in Medication*. Edited by C. G. Gebelein and C. E. Carraher. Plenum Press, New York, 1985, pp. 227–247.
14. Douglass, J., Civelli, O., and Herbert, E., *Ann. Rev. Biochem.* 53; 665–715 (1984).
15. Stern, A. S., Lewis, R. V., Kimura, S., Kilpatrick, D. L., Jones, B. N., Kojima, K., Stein, S., Undenfriend, S., and Shively, J. E. In *Biosynthesis, Modification, and Processing of Cellular and Viral Polyproteins*. Edited by G. Koch and D. Richter. Academic Press, New York, 1980, pp. 99–110.
16. Meyers, C. A., Coy, D. H., Murphy, W. A., Redding, T. W., Arimura, A., and Schally, A. V., *Proc. Nat'l. Acad. Sci. (USA)* 77; 577–579 (1980).
17. Murphy, W. A., Meyers, C. A., and Coy, D. H. *Endocrinology* 109; 491–495 (1981).

11

Production and Analysis of Proteins by Recombinant DNA Technology

Sidney Pestka

Robert Wood Johnson Medical School
Piscataway, New Jersey

INTRODUCTION

Recombinant DNA technology is a procedure by which a fragment of DNA from one organism can be cloned into another, or even into the same organism. This permits large amounts of DNA to be obtained for study, and, under proper conditions, large quantities of specific proteins can be synthesized. These objectives could not be achieved for most genes from higher organisms before the advent of recombinant DNA technology. The availability of relatively large fragments of DNA for analysis as well as for the preparation of specific probes opened up many new avenues to the molecular biologist. The primary structure and organization of genes became accessible; the control of specific genes could be evaluated; and the effect of specific genomic or cDNA fragments on physiology and development could be studied by inserting these in proper vectors and subsequently into cells or even into intact animals through the use of transgenic mice or embryocarcinoma cells. Specific probes for DNA elements can be used to study the linkage of genomic DNA to various diseases by analyses of restriction fragment length polymorphisms. Furthermore, as noted above, proper constructions will permit large quantities of specific proteins to be generated. Often these proteins are either unavailable or available in minute amounts through other means. For example, the interferons, the interleukins, and many other polypeptides and proteins have now been produced which were previously scientific curiosities whose physical properties could not be studied because of the limited quantities available. In many cases, recombinant DNA technology has revolutionized this so that many of these proteins are now available in large amounts for therapeutic use. The pharmaceutical industry is of course enthusiastic about proteins with therapeutic properties. However, the chemical and other industries are no less enthusiastic about the potential of recombinant DNA technology in assisting large-scale production of chemical feedstocks, enzymes, and other components. This chapter will provide an outline of the general methods of recombinant DNA technology and its application. Because this field is moving extremely rapidly, effective technology today is not in the vanguard for long. Nevertheless, the general outline should provide the reader with an understanding of genetic engineering and its potential applications. Furthermore, a later section of this chapter is devoted to new techniques which may produce additional rapid advances.

GENETIC ENGINEERING

General Principles

Vectors

Genetic engineering is a method by which genes can be manipulated. Frequently it is a technique used to isolate genes from one organism and insert them into another. However, it is not uncommon to isolate genes from one organism then to reinsert them into the same organism for study of their regulation and/or physiology. Figure 1 outlines the essence of genetic engineering. Recombinant DNA technology requires two components: a vector into which a fragment of DNA can be inserted and a host in which the vector containing foreign DNA can propagate. A large number of vectors have been identified for effective use in recombinant DNA technology. Table 1 summarizes these vectors. These include plasmids, which are circular, small pieces of DNA that independently replicate in bacteria (Fig. 1.) These small circular pieces of DNA replicate independently of the DNA of the host *Escherichia coli* bacterial cell. Usually, these pieces of DNA replicate faster and produce more copies than the host chromosome. There can be 10 to 100 copies of these plasmids per cell. It should be noted that some plasmids do not replicate to a high copy number and so are represented in the host cell by one to five copies. When designing recombinant DNA experiments for specific purposes, it is important to choose a plasmid with an appropriate copy number for the specific goals. For production of large amounts of a protein it is desirable to employ a plasmid with an extremely high copy number. In contrast, however, if, indeed, the final protein product were toxic to the cell, a low copy number plasmid

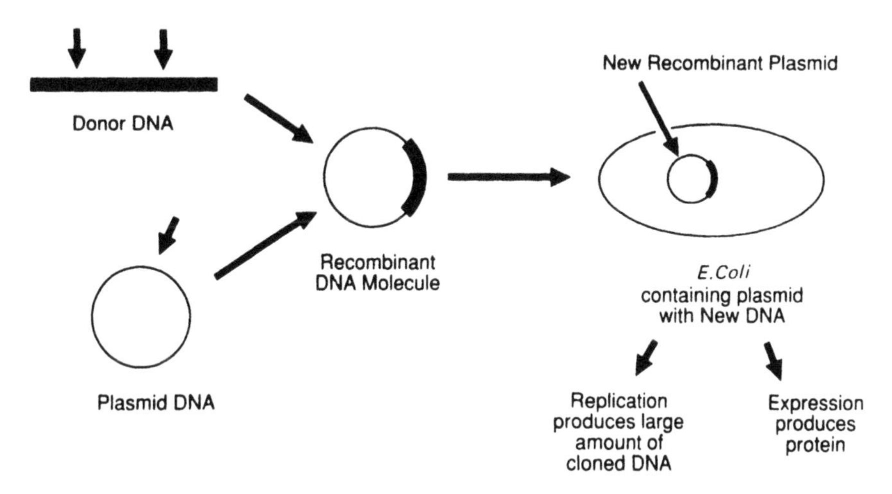

Figure 1 Recombinant DNA. The general strategy of recombinant DNA techniques is outlined here. The plasmid vector represents an independently replicating small DNA molecule when introduced into the bacterium *E. coli*. The arrows represent sites of restriction enzyme cleavage. Restriction enzymes (see Fig. 3 and text) are used to open the circular plasmid vector so that it can accept a foreign piece of DNA (donor DNA). The foreign piece of DNA is then inserted and ligated into the vector to generate the recombinant DNA molecule. This new plasmid is then inserted into the bacterium *E. coli* where it can replicate to produce large amounts of the DNA for study or after proper construction (see text) express the proteins of interest.

might be considered. Other techniques for expressing toxic proteins are also discussed below. Plasmids represent the most common vector for recombinant DNA technology for they are easy to prepare, and they replicate readily. Plasmids with insertions of foreign DNA can be separated from plasmids without insertions in a simple fashion.

Figure 2 illustrates the plasmid pBR322. The plasmid contains two antibiotic genes: the gene for tetracycline resistance (TET) and the gene for resistance to ampicillin (AMP). When these plasmids are inserted into a bacterium

Table 1 Host Vector Systems for Recombinant DNA

Vectors	Host
Plasmid (pBR322)	*Escherichia coli*
Phage (λ Charon 4A)	*Escherichia coli*
SV40 virus	Monkey cells
Bovine papilloma virus	Mouse cells
Retroviruses	Animal cells

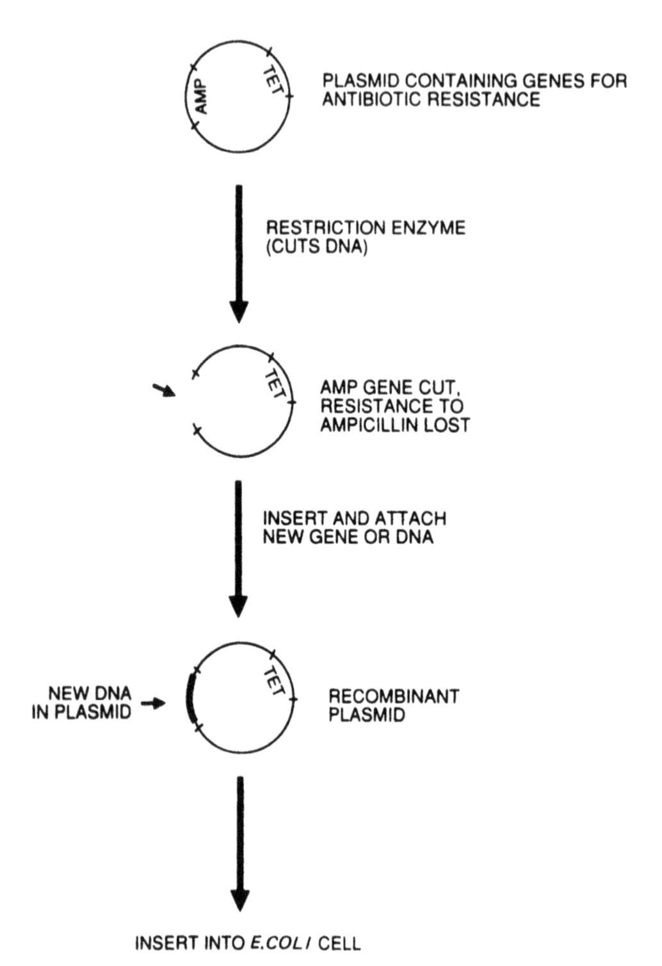

Figure 2 Insertion of a gene into a typical plasmid vector. The typical plasmid, pBR322, is shown here. It contains two genes that code for resistance to the antibiotics tetracycline (Tet) and ampicillin (Amp). For insertion of a foreign DNA fragment the circular plasmid is cut with a restriction enzyme to open it. A DNA fragment cut with the same restriction enzyme so that it contains the complementary sticky ends (see Fig. 3 and text) can then be inserted into this site. Upon cutting the plasmid within the Amp gene, resistance to the antibiotic ampicillin is lost. This provides a method for differentiating between plasmids with and without inserts. The newly generated recombinant plasmid is resistant to tetracycline, but no longer resistant to ampicillin. The recombinant plasmid is then inserted into *E. coli* so that it can be propagated.

that is sensitive to tetracycline and to ampicillin (i.e., killed by these antibiotics), the plasmid renders the host bacterium resistant to these two antibiotics. Accordingly, cells containing the plasmid can be easily selected from the rest of the *E. coli* in the culture by simply selecting cells which have become tetracycline and/or ampicillin resistant. For preparation of a recombinant vector, it is convenient to insert the foreign fragment of interest into either the gene for tetracycline or for ampicillin resistance. If the insert is in the gene for ampicillin resistance, the bacterium with such a plasmid is sensitive to ampicillin, but resistant to tetracycline. Thus, it is a simple matter to select (differentiate) those cells containing an *E. coli* plasmid with an insert from all cells containing the plasmid without an insert. If the insert is in the ampicillin resistance gene, those cells containing the plasmid with the insert will be resistant to tetracycline but sensitive to ampicillin. Those cells containing the parental plasmid without the insert will be resistant to both tetracycline and ampicillin. Cells containing no plasmid will be sensitive to both tetracycline and ampicillin. Therefore, by growing cells in tetracycline, all cells containing plasmids with or without an insert will be seen as colonies on agar in a Petri dish. By replica plating the cells that have grown in agar containing tetracycline onto agar in a Petri dish containing ampicillin, colonies which do not grow on ampicillin would be those that contain an insert in the ampicillin resistance gene.

Restriction Enzymes

Recombinant DNA technology would not have progressed without the discovery and availability of restriction endonucleases. These restriction endonucleases are enzymes that recognize specific sequences within the DNA strands. In general, the sites recognized by these restriction endonucleases have a palindromic symmetry as shown in Figure 3. One of the best studied restriction endonucleases, *Eco*RI, has the recognition site GAATTC. It is the same sequence on each strand of the double-stranded DNA. The cut is made between the first G and the A on both strands. Thus the DNA is terminated with a "sticky end" after being cut with a restriction endonuclease. The sticky end of one strand overlaps the sticky end of the other strand in the complementary sequences which allow DNA bases to pair with one another. The sticky ends produced by one enzyme are usually specific for that enzyme and do not bind or stick to those produced by most other enzymes. If the sticky ends produced with one enzyme will base pair with the sticky ends produced by another enzyme, then the fragments produced by them may be ligated together. During the early stages of recombinant DNA technology in the mid- to late 1970s, restriction endonucleases provided the primary method for ligating strands together through their sticky ends. However, because of the rapid progress and

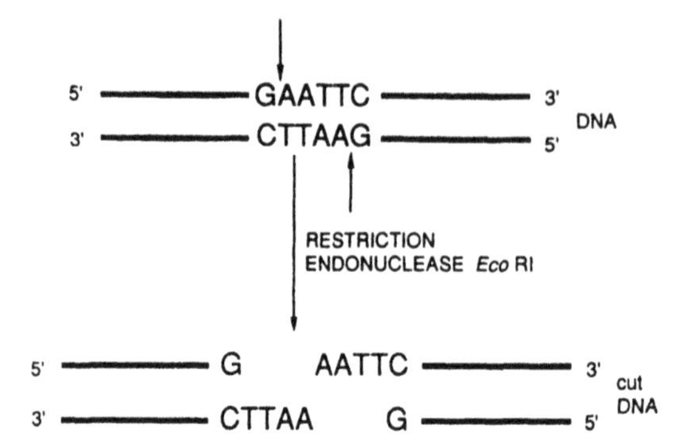

Figure 3 Restriction enzyme *Eco*RI. The restriction enzyme *Eco*RI cuts the double-stranded DNA at the specific site shown (5′ GAATTC 3′) producing overlapping sticky ends that can anneal with each other specifically because the bases are complementary. The remainder of the DNA molecule remains double stranded. Fragments of DNA produced from other sources of DNA by this same restriction enzyme (restriction endonuclease *Eco*RI) can be ligated to other DNA fragments cut with this same enzyme, but not to fragments produced by other enzymes that yield different sticky ends not complementary to these. Because restriction endonuclease *Eco*RI recognizes specifically only this sequence in DNA, but not any permutation of the sequence, it would cut large fragments of DNA in average size fragments of 4096 nucleotides (4^6). Other restriction enzymes recognize sequences of 4, 5, 6, or 8 bases, thus resulting in more frequent or less frequent cuts than produced by restriction enzyme *Eco*RI.

advances in the synthesis of synthetic DNAs, synthetic oligomers of DNA or linkers can now be easily made that can be used as a bridge to ligate strands with different sticky ends to each other.

Transformation of Cells: Inserting the Vector into the Intact Cell

A vector with foreign DNA (recombinant DNA) can be introduced into cells by a variety of procedures. In animal cells, it can be accomplished by injecting the vector directly into cells with a microneedle or by exposing cells to aggregates of DNA produced by calcium phosphate precipitation; under these conditions a small amount of the DNA is taken up by cells. If a virus is the vector, the cell can simply be infected with the recombinant virus. In the case of plasmids, however, it is necessary to weaken the cell wall of the bacterium by treatment with calcium chloride or rubidium chloride. This procedure weakens the cell wall of the bacterium so that a small percentage of the cells take up the vector DNA. The cells are then plated on Petri dishes where the appropriate cells containing the vector can be selected with antibiotics as discussed above.

Isolation and Identification of Genomic and cDNA Clones

A wide variety of procedures can be used to isolate and to identify genomic and cDNA clones. In general, these procedures can be divided into two categories: functional and probe techniques (Table 2). The functional techniques depend on measuring a function that can be elicited by the incorporation of the proper cDNA clone or by the alteration of a function when that particular clone is present. The probe method depends on sequence information about the protein or RNA of interest. This information is generally obtained from the sequence of the protein itself.

Cloning is the process by which a single homogeneous copy of an organism or recombinant DNA vector is obtained. When fragments of DNA are inserted into a vector a population of recombinants is obtained each of which contains a single DNA fragment. Plasmid recombinants are grown in bacteria such as *Escherichia coli*. When the bacteria are grown on agar plates each single colony represents an individual DNA fragment inserted into the vector, in other words, a single clone.

General Procedures

A few descriptive terms and procedures relative to the isolation of clones are in order. Annealing of a single-stranded DNA to another single-stranded DNA (ssDNA) is a highly specific process based on the base-pairing of the nucleotides (the building blocks of DNA). Because A in one strand always matches T in the other, and a G in one a C in the other, this annealing process is specific. Since there are four bases, a fragment of 20 nucleotides will have a probability of one in 1×10^{12} (4^{20}) of being identical to another random sequence of 20 nucleotides. The annealing process can be carried out with RNA

Table 2 Procedures for Identifying Clones

Procedures
Probe techniques
Oligonucleotide probes from protein sequence information
Small: 18–30 nucleotides, multiple
Large: 30–60 nucleotides, single
Functional assays
Hybrid-arrested translation (cell-free, oocytes)
Hybrid selection
Direct translation of synthetic RNA
Direct transfer of functional attributes to cells
Plus/minus hybridization
Antisense RNA

to ssDNA as well in which case a hybrid RNA-DNA is formed and the process is termed hybridization. If a long piece of ssDNA or RNA (>100 nucleotides) is annealed to ssDNA, it is very unlikely a sequence will match by chance alone. The measurement of hybridization is carried out by using radioactive (usually ^{32}P) DNA or RNA as the probe to anneal to ssDNA. The annealing process can then be measured by determining the radioactivity bound to ss-DNA by a number of methods. When the ssDNA is bound to a filter, the quantity of radioactivity that becomes bound is then directly measured. Matches can be exact or inexact where there may be mismatches. The annealing conditions can be adjusted to permit many, few, or no mismatches by carrying out the annealing in high, medium, or low sodium chloride concentrations, respectively. High sodium chloride concentration would be 1.5 M NaCl; low 0.015 M NaCl. Stringency of annealing or hybridization refers to the degree of matching: high stringency refers to a high match; low stringency refers to annealing with significant mismatches permissible.

The DNA or RNA used for annealing that is labeled with the radioisotope is called the probe. These probes can be obtained from natural sources of RNA or DNA, or they may be synthesized chemically or enzymatically. DNA probes can be easily synthesized automatically by DNA synthesizers, many of which are commercially available. A radioactive label (^{32}P as the phosphate) can be attached to these probes by a number of methods. One method commonly employed is the use of the enzyme, polynucleotide kinase, which transfers a terminal phosphate from adenosine triphosphate (ATP) to the 5′-end of the DNA oligonucleotide or of the RNA.

Generating Probes from Protein Sequence Information

For abundant proteins or proteins that can be purified in microgram to milligram amounts, sequence information is readily obtained. Amino acid sequencing, first accomplished by Sanger for insulin, has been extensively modified so that it can be carried out virtually automatically from the amino to the carboxy terminal end by the Edman method. From the sequence of the protein, it is possible to design nucleic acid oligonucleotide probes which correspond to the codons (or their complement) that represent the amino acids of the sequence. Small oligonucleotide probes of 18 to 30 in size have been most commonly used. When designing a probe, two procedures can be used. The codon utilization frequency can be used to design probes which represent the most common codon usage for the organism involved. Such tables are available from a number of sources. Alternatively, it is possible to make every possible oligonucleotide when small probes are prepared. This latter procedure (mixed probes) is particularly useful when amino acids represented by one or at most two codons predominate in the sequence from which the probe is generated.

The presence of methionine and tryptophan in a sequence is a particularly propitious occurrence from which to generate probes since each of these amino acids is represented by a single codon. One can end the oligonucleotide at a codon representing an amino acid containing four synonym codons (e.g., GGU, GGC, GGA, GGG for proline) by simply utilizing the first two positions (GG) which are identical for all the codons (see Chapter 3, Table 1).

A more recent procedure that has been found successful is to generate a long oligonucleotide probe, which is either a single or a mixture of at most a very few probes. In this case the oligonucleotide synthesized is approximately 50 bases in length with the most judicious base chosen for the third position from knowledge of the codon usage. In the case of amino acids represented by six codons (arginine and serine) two possibilities can be synthesized in these positions. In cases where two codon frequencies are equally common, both can be used in generating the long probe. With such a sufficiently long probe, relatively high stringency hybridization conditions can be utilized even with a significant number of mismatches within the sequence. The probe is highly specific despite a significant number of mismatches because of its large size. As a general rule, if one were in fact wrong in the third position in most codons, the hybridization probe would contain about 66% of the positions identical to the gene of interest. If the codon usage provided approximately a 50% accuracy in the choice for the third position, then the homology would be 83%. Probes in the range of 66 to 83% are highly effective for providing specific and sensitive detection of genomic or cDNA sequences.

In practice, the DNA recombinants in individual bacterial colonies containing the recombinant plasmids grown on agar are transferred to cellulose by pressing a circular filter of cellulose or nitrocellulose onto the colonies to make an imprint on the filter. The filter is then treated to release the DNA (plasmids) and bind the DNA in single-stranded form to the filter. The probe is then annealed to the filter. After washing to avoid binding of radioactivity of the probe nonspecifically, the radioactivity remaining is visualized by autoradiography of the filter. Colonies that correspond to areas that "light" (are radioactive) represent positive clones. About 100–200 colonies can be screened on a filter 10 cm in diameter.

Identification of Clones by Functional Assays

As noted in Table 2, a number of functional assays can be used to identify specific recombinants. These include hybrid arrested translation, hybrid selection, and direct conference of functional attributes.

Hybrid-Arrested Translation

Hybrid-arrested translation is based on the observation that mRNA molecules must be single stranded to enable translation. Duplex formation inhibits translation. Thus, if one has a purified mRNA molecule, or if one knows the protein of interest, a population of recombinant molecules can be hybridized to the mRNA. When the proper recombinant is present, the specific protein synthesized will be absent or reduced in the translation mixture (Fig. 4). Hybrid-arrested translation requires that the translational product be easily identifiable in one way or another. The protein product can be measured as an autoradiographic band on a polyacrylamide gel electrophoretic analysis or by its biological activity (see Chapter 4, Figure 15).

Hybrid Selection

Hybrid selection is based on the ability of recombinant plasmids to select the specific mRNA of interest from a population of unfractionated mRNA molecules. The recombinants, usually in pools of ten to several hundred, are bound covalently to a solid matrix such as cellulose. The mRNA is then added to the solid matrix containing the plasmid recombinant mixtures which may contain the inserts of interest. When a matrix contains a recombinant that will hybridize to the specific message desired, the specific message will then be bound (Fig. 5). The specific message can be eluted, translated, and then measured by either functional or radioisotopic techniques. It was in this manner that the recombinants for the human leukocyte interferons were originally identified. Hybrid selection unfortunately is not an efficient method as currently performed. The procedure tends to inactivate approximately 90% or more of the message, and so high levels of message are required.

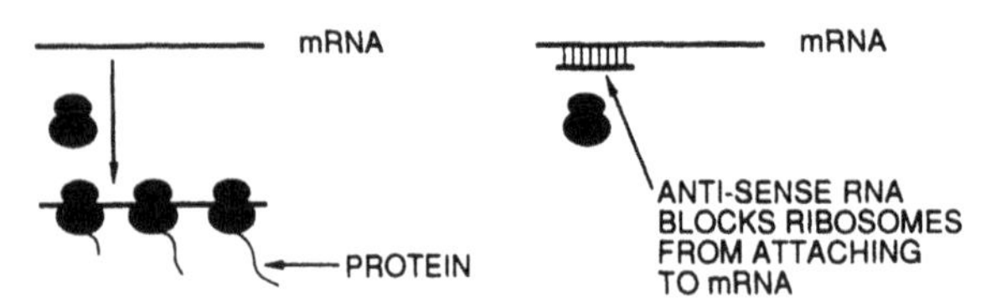

Figure 4 Hybrid-arrested translation. Messenger RNA (mRNA) is the template used by ribosomes for translation into proteins. Binding of the ribosomes to the mRNA is essential for translation to proceed. If a complementary piece of RNA or DNA is annealed to the mRNA as shown on the right of the figure, binding of the ribosome can be prevented so that translation is blocked. Annealing of a complementary oligonucleotide downstream of the binding site will also reduce translation in vitro, but less effectively in the intact cell. This procedure is used to determine if a complementary piece of DNA from a clone is the correct one. If the DNA from the clone inhibits translation of the mRNA specifically, then the protein product of this translation will not be formed, but other proteins encoded by the total mRNA from the cell will be produced in normal amounts. Thus, the specific inhibition of a single protein is diagnostic of a fragment annealing specifically to the mRNA encoding that protein. Similar experiments can be performed in cells by generating "antisense" (i.e., complementary) RNA within a cell.

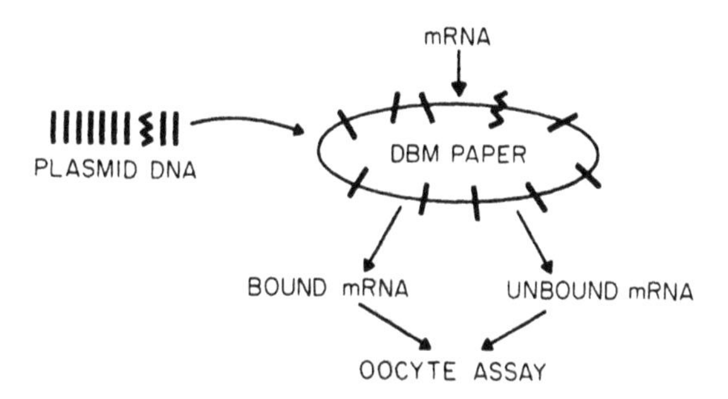

Figure 5 Schematic illustration of hybrid selection. Plasmid DNA is isolated from a pool of colonies (e.g., 10 colonies are shown in the figure) and bound covalently to diazobenzyloxymethyl (DBM) paper. To detect a recombinant containing an interferon sequence (shown as a zig-zag line in the illustration), mRNA prepared from cells synthesizing interferon was hybridized to the bound DNA. After unhybridized RNA was removed by washing, the specifically hybridized mRNA is eluted and translated in a cell-free extract or as shown here in frog oocytes. By this procedure, those pools of clones in which at least one recombinant contained the sequence of interest were identified. Thereafter, individual clones in each pool are separately screened by the same procedure. As a control, unbound mRNA is also microinjected into the frog oocytes to ascertain that the mRNA was not destroyed by the manipulations.

Direct Translation of mRNA Produced by Synthetic cDNA Clones

If cDNA libraries can be prepared in a vector that permits the synthesis of RNA in the test tube, it is possible to translate the RNA directly in cell-free extracts or in oocytes. As above, initially pools of clones can be screened, then individual colonies.

Direct Transfer of Functional Attributes

This procedure involves the transformation of cells with recombinant plasmids or other vectors. The cells are then screened or selected for the proper functional property. For example, bacterial auxotrophs which require arginine for growth can be selected on arginine-free medium after transformation with a library of plasmids containing the genes of *E. coli*. Those cells containing the gene for arginine biosynthesis absent in the auxotroph will grow on arginine-deficient medium. Therefore, bacteria containing this gene can be selected directly. This procedure can be applied to animal cells as well. Furthermore, if a gene is introduced in animal cells that confers a specific antigenic property such as a new receptor on its surface which can be identified either by functional assays or by reaction with labeled antibodies, such cells can be selected by fluorescence-activated cell sorting or other procedures.

Plus/Minus Hybridization Procedures

This procedure depends on the availability of cells that are known to have the mRNA of interest and a closely-related cell known to be devoid or greatly deficient in this specific mRNA molecule. This approach may be conveniently used in cells which can be induced to produce a particular mRNA and its corresponding protein. A plus/minus pair might be uninduced fibroblasts and a culture of fibroblasts induced to produce beta-interferon by infection with a virus. Hybridization can be performed with either RNA or with cDNA prepared from the RNA. In essence, clones hybridizing to the mRNA from cells that are positive for it (+) and not hybridizing to the mRNA or cDNA from the cells in which this message is not present (−) represent the clones of interest. This procedure is illustrated in Figure 6. In this figure a variation of the procedure is shown that allows a single hybridization to be performed rather than double hybridizations with plus and minus mRNAs or cDNA preparations. Since this procedure generally does not select a specific clone definitively, it is used to reduce the total number of clones to be analyzed about

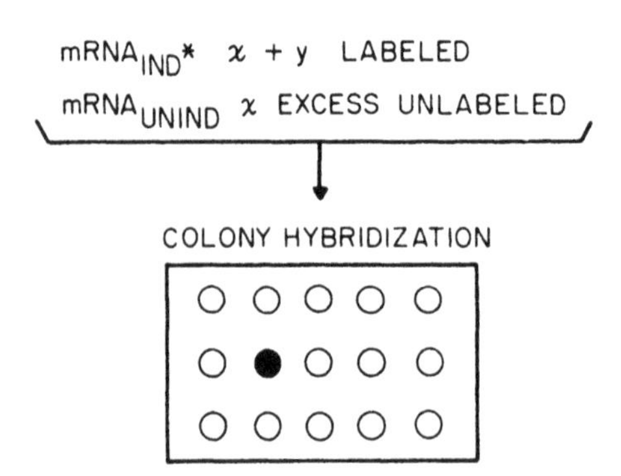

Figure 6 Plus/minus hybridization. The $mRNA_{unind}$ represents messenger RNA from cells which are not producing the protein or the mRNA of interest. The $mRNA_{ind}$ represents mRNA from the same cell type which was induced to produce the protein and mRNA of interest. *E. coli* cells are transformed with the recombinant plasmids generated by procedures shown in Figs. 1 and 2. Colonies derived from individual transformed bacterial cells were transferred to filter paper, fixed, and probed with a ^{32}P-labeled mRNA preparation from cells producing the protein of interest (containing $mRNA_{ind}$) and therefore enriched for the specific mRNA ($x + y$). The hybridization is performed in the presence of excess unlabeled mRNA from uninduced cells (x). This procedure identifies bacterial colonies containing plasmid DNA coding for the protein specifically formed in the induced cells. The induced-specific mRNA sequences (y) include the sequences for the protein of interest as well as others that are induced concomitantly with this protein.

80–90%. Thus, the work load with the other procedures can be reduced greatly if 90% of the clones can be excluded from consideration.

Future Methods with Antisense RNA

Antisense RNA (formally analogous to hybrid-arrested translation above) represents a powerful new procedure to study the physiology as well as the function of many proteins. It also offers an opportunity to select specific recombinants without knowledge of the protein structure or the genes themselves. What is necessary is that a functional assay be available for the protein or RNA of interest. Unfortunately, to date these procedures have not been effectively advanced to make them generally useful. However, it appears that eventually this may provide a powerful and unique opportunity to clone recombinants rapidly and directly.

PRODUCTION OF RECOMBINANT PROTEINS

The production of recombinant proteins can be performed in a variety of organisms. The specific organisms chosen depend on the requirements for the final product. When the product in *Escherichia coli* is identical to the desired product, this organism is usually the one of choice. It can be grown readily and large amounts of protein can be produced. Not all proteins seem to be produced to a high level in *E. coli* so that other organisms may need to be evaluated in specific cases. If secretion of the protein is important, then expression vectors which function in yeast or *Bacillus subtilis* hosts can be explored. The expressed protein can be designed to be secreted efficiently from these organisms. Since posttranslational modifications such as glycosylation, amidation, or other alterations may be important, this is a consideration for choice of the host. The yeast expression systems permit some of these posttranslational modifications. Nevertheless, if a protein virtually identical to the human protein is required, then expression in human or animal cells is optimal. Expression in human cells would provide a product virtually identical to the product isolated from human cells themselves. Other animal cells may produce very similar products. Because glycosylation varies from animal cell to animal cell, even within the same species, it may be necessary to evaluate a number of cell types and the conditions of synthesis to obtain the specific final product in modified form.

Animal cells by and large do not provide sufficiently high densities and high yields of recombinant products compared with bacterial expression systems. Accordingly, if yield is a significant consideration then bacterial or yeast vectors are usually the organisms of choice. Nevertheless, because of the wide variety of host-vector systems which are currently available, it is possible to choose one that meets the requirements for the final product. Future improvements in culturing techniques may overcome this drawback.

CONSTRUCTION OF EXPRESSION VECTORS FOR HIGH-LEVEL EXPRESSION

Expression vectors for bacteria and animal cells require a promoter region where the RNA polymerase which initiates transcription can bind to the recombinant. They also require the coding region for the protein of interest. With these two essential features as shown in Figure 7, any protein may be expressed.

The expression levels of proteins vary immensely. Although there has been great success in producing a large variety of proteins through genetic engineering, it is surprising that no specific rules exist for consistently obtaining high level expression. In general, empirical techniques are utilized to generate high expression together with fermentation enhancement procedures. Government granting agencies have been reluctant to support such studies in developing the rules for high-level expression, feeling that indeed commercial interests in the

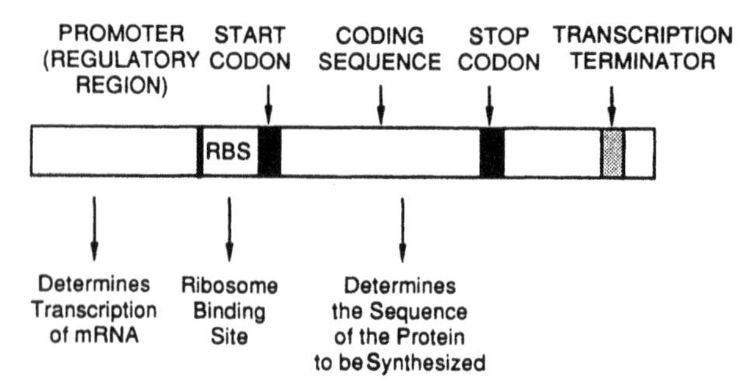

Figure 7 Expression vector of *E. coli*. A general expression vector for producing proteins in *E. coli* is shown here. The vector must contain the following components for expression of proteins. A promoter region is the sequence of DNA that is recognized by the enzyme RNA polymerase that transcribes the DNA into mRNA that can then be translated into protein. The DNA must have a sequence corresponding to a ribosome binding site so that the mRNA formed contains this site to which ribosomes bind just prior to the start of translation. The initiation site or start codon for translation is an AUG which is usually about 7 to 13 nucleotides downstream from the ribosome binding site. Following the AUG is the coding sequence for the mature protein to be produced: this sequence determines the protein to be synthesized. The stop codon is represented by one of three codons (UAA, UAG, or UGA) that signal termination of translation, completion of the protein, and its release from the ribosome. For most efficient utilization of the plasmid for expression, it is useful, but not essential that a transcription terminator be present. This is a sequence that controls the size of the mRNA produced. Without this sequence, the mRNA would be excessively large having a great deal of extraneous unnecessary sequences beyond the coding sequence. This segment is inserted into a plasmid as shown in Fig. 2 to form the expression vector.

private sector would provide the impetus for these efforts. However, this has not occurred, and it appears to be more effective for an organization to try a number of empirical avenues than to try to understand the general rules involved. This is a time-consuming and expensive effort.

Some rules for effective expression are clear. It is necessary to have a strong promoter prior to the coding sequence of the protein to be expressed. An effective ribosomal binding site is essential. A consensus ribosomal binding site has proved useful for this purpose for vectors used in *E. coli* although a large variety of ribosomal binding sites in fact have proved effective. It is useful to have a transcription terminator following the termination codon of the coding sequence in bacteria. This provides for the synthesis of smaller mRNA molecules so that the cellular transcriptional machinery (enzymes for making mRNA) is most efficiently used with the plasmids. It is conceivable, however, that in some cases premature transcription termination may provide for increased degradation of the mRNA molecule. More information is needed about these events. One of the initial expectations was that the frequency of codon usage in a mRNA molecule would correspond to the codon frequency of the host wherever possible. Although this seems logical and may be most efficient in the long run, it has turned out not to be essential. High expression of proteins has been accomplished with codons representing rare tRNA species in *E. coli*.

Another important consideration is the secondary structure of the mRNA molecule. It appears necessary to have the ribosomal binding site in a nonbase-paired region. That is, hairpin loops or similar internally hybridized structures may interfere with the ribosomal binding step. It has also been reported that reducing the number of potential GC base pairs in the early coding region is useful for increasing expression. Despite these suggestions, there remains no clear, decisive understanding of the underlying principles involved.

Fermentation

Fermentation of *E. coli* can be carried out in virtually any volume required. Because of the high expression levels of recombinant proteins in *E. coli*, however, usually fermentors of 500 L are necessary for large-scale production of recombinant proteins. The importance of fermentation to the final expression levels of recombinant proteins in *E. coli* is generally not appreciated. Testing the expression of a vector in various strains of *E. coli* under a variety of conditions provides an empirical judgment as to the best host-vector and medium for expression. The results of these empirical manipulations have been highly successful even in the absence of a fundamental understanding of the principles involved.

Fermentation in *E. coli* can occur under conditions of forced oxygenation, stirring, and feeding, so that extremely high densities of bacteria can be produced. Since food sources as well as pH, minerals, and other factors play an important role, these must all be evaluated on the bacterial yield and the yield of final product, factors which do not necessarily correspond. Since the host-vector systems usually have an antibiotic requirement for growth in *E. coli*, one antibiotic is usually present in the culture medium, and maintains selective pressure for retention of the recombinant. This may have a deleterious effect on the final product, particularly if it is difficult to remove completely. Trace amounts of antibiotics must be removed to avoid problems in individuals allergic to these components. Methods to avoid the use of selected antibiotics can be used, such as high-copy number plasmids, plasmids supplementing auxotrophic requirements of the organisms, and other means.

Isolation of Cells

Escherichia coli or other bacteria or microorganisms grown in fermentors are simply harvested by centrifugation. Animal cells in suspension can be harvested in the same way or by a combination of filtration and/or centrifugation. A number of new avenues have been developed to produce animal cells at high concentrations such as the use of coacervates in which the cells are grown or the use of solid substrates for anchorage-dependent cells. In these cases, the cells can simply be removed from the medium by permitting settling of the solid supports.

PURIFICATION OF RECOMBINANT PROTEINS

The purification of recombinant proteins follows the general principles for all proteins. The use of classical procedures is often highly effective, particularly with bacteria that are producing a high level (sometimes even 20 to 40%) of the recombinant protein. In cases where extraordinarily high level expression occurs, purification is almost trivial, for homogeneity can be obtained with less than a tenfold purification. This contrasts with the 100,000-fold purification o'ten necessary to purify the same products from their natural sources such as the interferons.

Monoclonal antibodies provide an extraordinarily efficient method for the purification of proteins because they can be obtained in large amounts and because the monoclonal antibody affinity columns can be reutilized almost indefinitely. High-performance liquid chromatography (HPLC) is also a useful procedure for refining the final product. For details of the purification of recombinant proteins, see Chapter 7 of this volume.

Before large amounts of recombinant proteins were generated in *E. coli*, it was not appreciated that disulfide bond formation would be a problem. In many cases, the disulfides that are formed in bacterial recombinant host-vector systems are not the proper linkages. The internal milieu of *E. coli* is by and large a reducing environment so that the oxidation of the cysteine residues does not occur internally, but only after breakage of the cells. Because of this phenomenon, proteins which appear homogeneous on analytical reducing gel electrophoresis or other procedures are often mixtures of a primary translation product (the correct full amino acid sequence), comprising many disulfide-bonded secondary variants. These can often be separated by high performance liquid chromatography.

ANALYSIS OF AND SPECIFIC CONSIDERATIONS RELEVANT TO RECOMBINANT PROTEINS

A number of specific considerations apply to recombinant proteins which are not relevant to the natural proteins. Nevertheless, in general, analysis of recombinant proteins follows procedures applied to other proteins. The focus on analysis of recombinant proteins in the last few years has simply illustrated that our analytical techniques for proteins have lacked detail and specific criteria. It is because of recombinant DNA technology that proteins are finding many more uses in the pharmaceutical, chemical, and other industries. Prior to the advent of genetic engineering, relatively few proteins were used in the pharmaceutical industry. When used, often these were in the crude or impure forms although some were purified preparations. Examples include insulin, growth hormone, and the interferons. With the advent of recombinant DNA technology it became important to develop analytical criteria to evaluate proteins and, in particular, to compare them with the natural products. It became evident that the stringent control of the natural products had never been a focus of concern, probably because they were natural products unavailable as homogeneous materials. Because of an underlying concern that the recombinant products might be different in a wide variety of ways, particularly in unexpected properties, it was important to develop an understanding of purity, secondary structure, activity, and primary structure. It is not yet possible to demand tertiary structure information about recombinant proteins since it would be impossible to achieve this in the near term with a significant number. As a long-term goal, this would be an ideal characteristic to define. Indeed, it is the tertiary structure which defines the function, activity, and other major properties of the molecule.

Criteria for Purity

A major consideration for all proteins is their purity. It should be emphasized that there is no such thing as a pure protein. A pure protein is analagous to a point in geometry. In physical reality a point is an idealism; it does not exist. It exists only in the mind of the mathematician. Similarly, a pure protein is an ideal that can be achieved only with pencil and paper, not in reality. In reality, a protein can be purified to near homogeneity, but one can never reach 100% purity. There will always be some contamination. The contamination detectable is dependent on the analytical methods available. As analytical methods have become more sensitive, contaminants have been easier to detect and thus have provided a paradoxical situation where formerly "pure" proteins have been found to be impure. In any case, with this in mind, it is therefore important to set specific criteria for purity. These criteria should include the total percent by weight of the preparation which is represented by the protein; the limits of additional contaminating materials that may be or are in the preparations, such as DNA, RNA, antibiotics, preservatives, buffer salts, detergents, and the like. If the protein is from a natural source, it is important to exclude active components such as viruses or other infectious agents. If the source is from a genetically engineered organism, it is important to exclude the components of the microorganism as well as any part of the host-vector system.

Posttranslational Modification: Problems and Perils

Many animal proteins, particularly those secreted from cells or found on the cell surface, are modified in a number of ways. They may be cross-linked, contain modified amino acids such as hydroxyproline; they may be glycosylated or have numerous other modifications of the primary amino acid sequence. Often the primary sequence suffices to provide biological activity of the protein, although the biological activity of a bare protein lacking the posttranslational modifications of the natural product may not have activity identical to its natural counterpart. This can only be determined by comparison of the recombinant product without modification with the comparable natural product. Because the natural products are frequently unavailable in large amounts for analysis and evaluation, such careful comparisons are often neglected, particularly with the rapid developments in recombinant DNA technology. Furthermore, because recombinant proteins have been developed rapidly even before total analysis of the natural products has been accomplished, the posttranslational modifications of the natural products have often not been defined well. As a result, in many cases, there have been guesses as to the proper amino terminus of the natural product and whether, in fact, the natural product is glycosylated or not. For some recombinant proteins these guesses have later proved to be wrong, so that recombinant proteins that differ significantly from the natural products have been produced, purified, and brought into clinical trial. Such a situation should be avoided. Thus, I believe it is essential before any recombinant product is brought to the point of clinical trial that its natural counterpart be analyzed carefully from the point of view of its primary structure, secondary structure, and major posttranslational modifications. This should be an ideal goal. In some cases, it may be necessary to make a judicious guess in order to make available new therapeutic agents. Nevertheless, it should be clearly understood that, even if these therapeutic avenues are pursued, as soon as new information is obtained about the natural products, the recombinant products should be modified with this new knowledge in mind.

Once a path for a particular protein is pursued in the pharmaceutical industry and other commercial enterprises, it is evident that a major investment has been made into that particular avenue. The approval process for the development of a variant of the initial recombinant product is an expensive and time-consuming procedure which is not undertaken lightly and has an inherent cost in monetary and human effort almost approaching the total cost of the initial development. Thus, it is not without knowledge of this major commitment that I make these suggestions. However, if the best products are to be continually available for the treatment of human diseases and for other uses, these considerations should weigh utmost with the regulatory agencies as well as with the commercial industries involved. For the pharmaceutical industry, this is a more critical problem than it would be for the chemical industry where enzymes are utilized to produce feedstocks; or for the detergent industry where the enzymes are used in laundry and cleaning preparations. Nevertheless, these considerations are real and will weigh heavily on the industries utilizing genetic engineering in the future.

The posttranslational modifications in the case of the interferons have not yet been completely defined for the α-interferons or any other interferon. Although we first reported after the purification of the human α-interferons that five analyzed were not glycosylated, we later reported, after analyzing several additional α-interferons, *O*-glycosylation in several human α-interferons.

As noted above, if glycosylation is important for the recombinant protein, then expression in a eukaryotic system is warranted. To obtain the ideal recombinant protein most comparable to the natural protein, it would be most suitable to express the protein in an animal cell identical or close to the animal cell from which the natural protein is isolated. This should apply to any posttranslational modification.

Biological Assays

It is essential, when developing a recombinant product, that proper assays be available. It is also important for proper standards to be available so that a direct comparison can be made from one laboratory to another. International standards, unfortunately, are not available for most recombinant proteins. It is not likely, in fact, that international standards will be available for many recombinant proteins in the future. Even for the interferons, where a large number of international standards are available, it is unlikely that a significant number of new standards will be produced for the large number of new human and animal interferons.

In light of this, it is necessary that scientists make available to their colleagues recombinant products that they have produced in their laboratory. These will be suitable for research standards for comparison of the data from one laboratory to another. It also will expedite scientific work in general. Journals should require that when recombinant proteins are described in publication, samples of these proteins should be provided to qualified scientific colleagues on request.

Special Problems of Recombinant Products

Mature Proteins

Secreted proteins frequently are synthesized with a signal peptide which defines its destiny to be exported from the cell and which is removed on secretion. Thus, the natural protein isolated and purified from natural sources does not usually contain the signal peptide. To prepare a product similar to the natural one, it is therefore necessary to generate a recombinant product devoid of the signal peptide. If other primary sequence changes occur, such as removal of an internal peptide in the synthesis of insulin, then these must also be taken into account when making the mature product. In the race for priority, very often individuals have synthesized proteins without due consideration to these factors and have produced proteins that are not identical with the mature natural product in significant ways, either because of haste or a lack of knowledge of the nature of the natural product. In at least one case, because the amino terminal sequence of the natural product was not previously identified, a recombinant protein, human interferon-γ (Hu-IFN-γ), was produced with the wrong amino terminus. Not only has this provided confusion in the literature and has confused regulatory authorities, but these proteins are now in clinical trial in many countries. As noted above, because of the immense expense involved in changing an avenue once it has been taken, there is reluctance to make major alterations once new observations have been made. The earlier an error is admitted and acknowledged, the easier it is to correct it. As these errors are embedded in the literature and in the therapeutic arsenal, they will be harder to correct.

Lack of Posttranslational Modification in *E. coli* Products

Escherichia coli products do not undergo posttranslational modifications such as glycosylation, cross-linking, or phosphorylation. Thus, if modified proteins are desired, it is best to express these in the animal cell closest to the cell from which the natural product is obtained as noted above. If, in fact, another microorganism is the origin of the natural protein, such as a yeast or fungus, then the expression vector should be used in a microorganism comparable to that one if the posttranslational modification is necessary for activity. Frequently, posttranslational modifications are not necessary for the activity of the product. The interferons represent a clear case in point where the β- and γ-interferons (IFN-β and IFN-γ) are glycosylated. Nevertheless, the protein without glycosylation made in *E. coli* has activity comparable but not identical to the natural product. Since the *E. coli* products can be made in large amounts, to make these proteins available for general therapeutic use, it is reasonable to evaluate the *E. coli* product. It appears, however, that the *E. coli* IFN-β is less stable than its natural counterparts. To produce a more stable product, one group has genetically engineered the Hu-IFN-β to modify it so that one amino acid, cysteine #17, is replaced by another, serine #17. This has produced what appears to be a more stable product comparable to the natural protein. The change, however, has introduced possibilities for antigenic variation so that this protein may be more immunogenic. As this protein is tested clinically, this should be evaluated with care.

Lack of Formyl-methionyl Processing

Proteins in *E. coli* and in most other prokaryotes are initiated with the modified amino acid formyl-methionine. Animal proteins are initiated with methionine. Thus all proteins are synthesized with an additional formyl-methionine or methionine in the first position. The natural mature protein, however, may be devoid of the terminal methionine. If this is the case, it is important to remove the methionine residue from the terminus of these proteins. In general, *E. coli* spontaneously removes the formyl group, but often the initiating methionine is incompletely removed. The final protein product, having an additional methionine at its amino terminus, obviously differs from the natural protein. Indeed, in many cases, the protein with the additional methionine has been produced and is in clinical trials. If the proteins are produced in animal cells, this is not a problem, for the recombinant product, if appropriately engineered, will be identical to the natural protein.

Disulfide Bonds are Sometimes Scrambled

New proteins and, in fact, most recombinant DNA products are predominantly expressed in *E. coli* host-vector systems. The internal milieu of *E. coli* is generally a reducing environment (as noted above). Accordingly, most products are not oxidized while in *E. coli*. Proteins which contain multiple disulfide bonds are often in the completely reduced state until the *E. coli* cells are broken and the protein then oxidized by dissolved oxygen from the air after release from the cell. The pathway for disulfide bond formation is only beginning to be understood for a few proteins, and the total pathway is still a puzzle for most. In the case of the human interferons, it appears that the disulfide bonds of human interferon-α (Hu-IFN-α) produced in *E. coli* can be the correct ones or scrambled ones; interchain disulfide bonds can also form to yield unnatural polymeric products. Accordingly, even when a single primary product is produced, a large heterogeneous mixture of different secondary configurations may be present in the complex mixture. Accordingly, when analyzing the final product, the disulfide bond configuration of the product should be compared with that of the natural protein. However, the disulfide bond configuration of no natural α- or β-interferon or interleukin is known. In any case, whatever the best configuration, it is important to note that scrambling may occur in cases where multiple disulfide bonds exist and that interchain disulfide linkages can form polymers when these proteins are expressed to high levels, as frequently occurs in *E. coli* host-vector expression systems.

How to Get Correct Structures

In order to generate the correct structure, it is necessary to determine the structure of the natural product. Because the natural product is difficult to obtain, this expensive procedure is often avoided. It is, however, necessary for both commercial interests and funding agencies to recognize this and face it directly. I would hope that more funds would be made available by all sources to accomplish the isolation and characterization of the natural products so that the recombinant products may be the proper ones in structure and in modification. The use of animal cell expression systems can help alleviate this problem, for the proteins produced in animal cells would closely resemble the natural proteins.

ISOLATION OF RECOMBINANTS FOR HUMAN LEUKOCYTE INTERFERON: AN EXAMPLE OF THE PROCEDURES

Because recombinant DNA technology offered an opportunity to produce large amounts of human interferons (Hu-IFNs) economically, many scientific teams set out to clone them in bacteria. Several groups achieved the isolation of recombinants for several Hu-IFN-α species. The cloning and expression of Hu-IFN-α as an illustration of these procedures is described in this section.

Isolating Hu-IFN DNA sequences was a formidable task since it meant preparing DNA recombinants from cellular mRNA that was present at a low level. This task had never been accomplished previously from a protein whose structure was unknown. In addition, in order to reconstruct DNA recombinants which would express natural interferon, it is useful to know the partial amino acid sequence of the proteins, particularly at the NH_2- and COOH-terminal ends. Without this information synthesis of natural Hu-IFN in bacterial cells would not have been possible. Thus, our purification of the natural Hu-IFNs and determination of their structure assisted us in these efforts.

To isolate recombinants containing the human DNA corresponding to IFN-α, we used a number of procedures. First, it was necessary to isolate and measure the IFN mRNA. This was accomplished several years earlier when IFN mRNA was translated in cell-free extracts (Fig. 8) and in frog oocytes. The next step was to prepare sufficient mRNA from cells synthesizing IFN, and this was accomplished with both fibroblasts and leukocytes.

A library of complementary DNA (cDNA) was prepared from a template of partially purified mRNA isolated from human leukocytes synthesizing IFN (Fig. 9). The dC-tailed double-stranded (ds) DNA obtained was annealed to dG-tailed DNA from plasmid pBR323 which had been cleaved at the *Pst*I restriction nuclease site and introduced into *Escherichia coli* by transformation. About 14,000 tetracycline-resistant and ampicillin-sensitive transformants were obtained (see Fig. 2). This provided a large group of cDNA recombinants with DNA copies of all the mRNAs extracted from the leukocytes. The next and the hardest part of the procedures was to find in this vast library of recombinant plasmids those which contained DNA encoding interferon. If we had been able to begin with pure IFN mRNA, this would have been a simple task. However, we did not have pure IFN mRNA. We began with a mixture of mRNA molecules and only relatively few were IFN-specific. We could show their presence by translating the RNA and assaying the products for IFN activity (Fig. 8). In

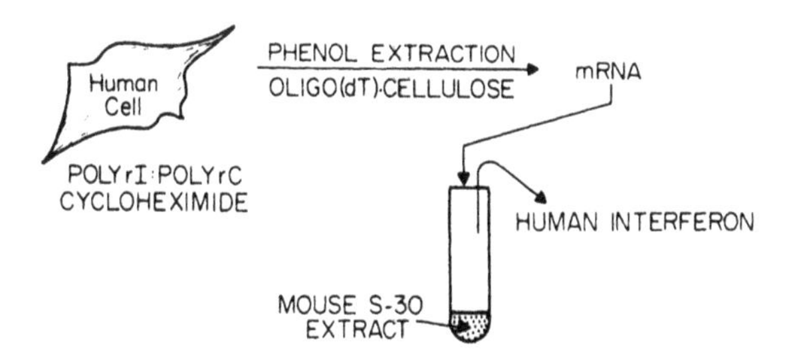

Figure 8 Interferon synthesis in a cell-free system. Human fibroblasts were stimulated to produce interferon (IFN) by treatment with the inducer polyI:polyC. The total RNA present was extracted 4 h later from the cell with phenol, and the fraction enriched in mRNA was selected on an oligo(dT)-cellulose column. An S-30 supernatant translation fraction was prepared from mouse cells. This was able to translate the various mRNA species, including that for IFN, into the corresponding proteins. The IFN formed was detected in a standard antiviral assay. In this way, biologically active human interferon was synthesized in the test tube for the first time. By injecting the mRNA into intact frog oocytes, the mRNA could be measured at 1/100 to 1/1000 the levels that could be detected in the cell-free extracts.

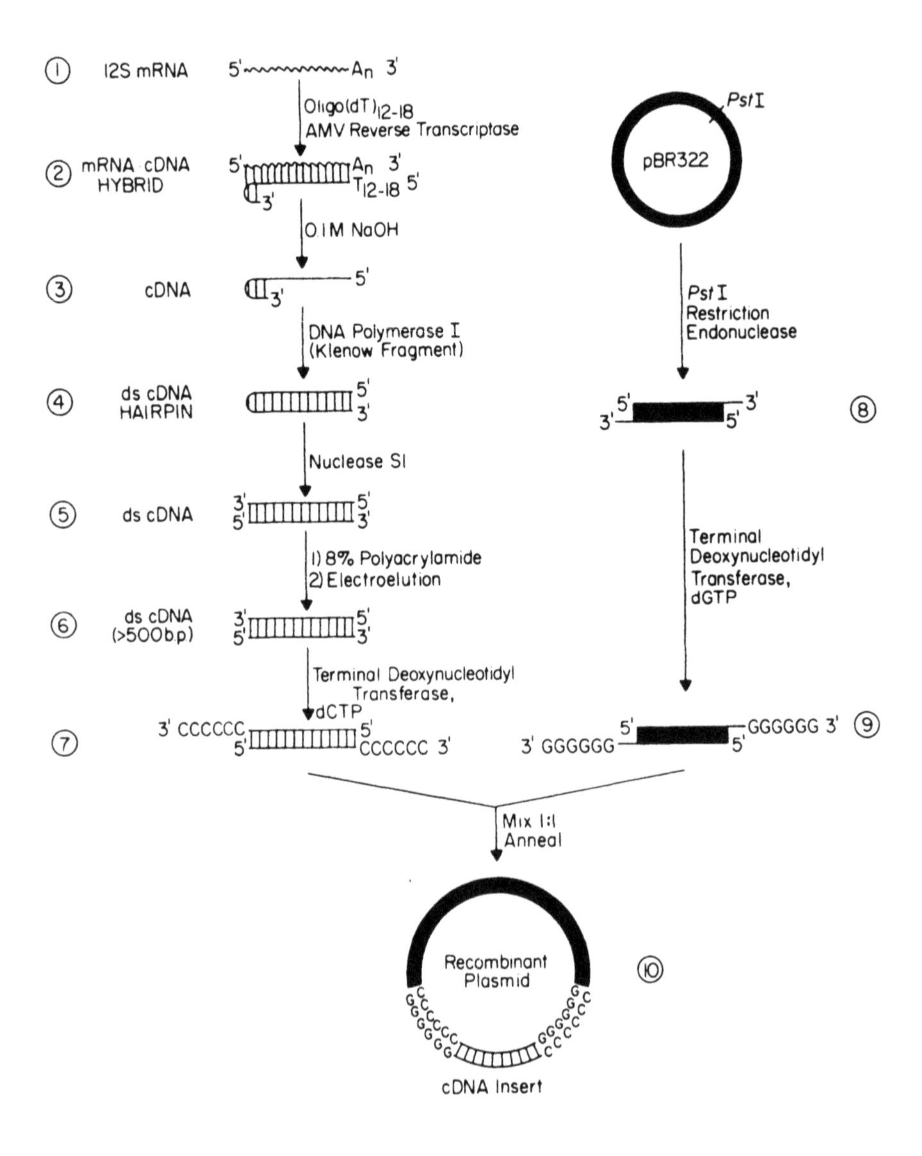
① 12S mRNA
5' An 3'
Oligo(dT)12-18
AMV Reverse Transcriptase
② mRNA cDNA HYBRID
5' An 3'
T12-18 5'
3'
0.1 M NaOH
③ cDNA
3' 5'
DNA Polymerase I
(Klenow Fragment)
④ ds cDNA HAIRPIN
5'
3'
Nuclease SI
⑤ ds cDNA
3' 5'
5' 3'
1) 8% Polyacrylamide
2) Electroelution
⑥ ds cDNA (>500b.p)
3' 5'
5' 3'
Terminal Deoxynucleotidyl
Transferase,
dCTP
⑦ 3' CCCCCC
5'
5'
CCCCCC 3'
PstI
pBR322
PstI
Restriction
Endonuclease
⑧ 5' 3'
3' 5'
Terminal
Deoxynucleotidyl
Transferase,
dGTP
⑨ 5' GGGGGG 3'
3' GGGGGG 5'
Mix 1:1
Anneal
Recombinant
Plasmid
⑩
cDNA Insert

this situation, we had to devise an indirect two-stage procedure. In the first stage, we screened all the bacterial colonies to find those with cDNA made from the RNA of induced cells (see Fig. 6); among these there might have been some carrying IFN cDNA. We, therefore, screened all the recombinants for their ability to bind to mRNA from cells synthesizing IFN (induced cells), but not to mRNA from uninduced cells (those not producing IFN) (Fig. 6). To do this, individual transformant colonies were screened by colony hybridization for the presence of induced-specific sequences with ^{32}P-labeled IFN mRNA (mRNA from induced cells) as probe. In the presence of excess mRNA from uninduced cells (Fig. 6), recombinants that were representative of mRNA sequences existing only in induced cells should be evident on hybridization. This screening procedure allowed us to discard about 90% of the colonies: since their plasmids carried no induced cDNA, these could not encode IFN.

In the second stage, we had to identify those recombinants containing the IFN DNA sequences among the remaining 10%. To do this, we pooled the recombinant plasmids in groups of ten and examined these for the presence of IFN-specific sequences by an assay which depends upon hybridization of IFN mRNA to plasmid DNA (see hybrid selection procedure, Fig. 5). Plasmid

Figure 9 Preparation of Interferon DNA recombinants from mRNA. Step 1: a human leukocyte suspension was induced to form interferon by stimulation with Sendai virus. An extract was made 6 h later from which mRNA was prepared that was enriched in 12S mRNA by differential centrifugation. Step 2: the mRNA fraction was used as a template with AMV reverse transcriptase in the presence of all four deoxynucleotide triphosphates and oligo$(dT)_{12-18}$ as a primer to generate a DNA stand complementary to the mRNA (cDNA) forming mRNA:cDNA hybrids. Step 3: treatment with 0.1 M NaOH digested away the RNA, leaving a cDNA fragment with a self-annealed 3′ end. Step 4: extension of the chain with DNA polymerase (Klenow fragment) produced a ds-cDNA containing a hairpin loop. Step 5: treatment with nuclease S1 opened the hairpin loop. Step 6: the ds-cDNA was sized on an 8% polyacrylamide gel followed by electroelution to give a fraction containing at least 500 base pairs. Step 7: treatment with terminal deoxynucleotidyl transferase in the presence of dCTP added cytosine homoplymer tails at the 3′ ends of each strand. Step 8: the plasmid pBR322 was cleaved (linearized) at the *Pst*I site with *Pst*I restriction endonuclease. Step 9: homopolymer tails of oligodeoxyguanylate were added to the 3′ ends of the linearized plasmid DNA with terminal deoxynucleotidyl transferase in the presence of dGTP. Step 10: the two fragments with complementary sticky ends (G and C are complementary) were mixed in equal proportions and annealed to yield a recombinant plasmid. A larger number of similar plasmids not containing interferon genetic information were also formed during this procedure. This cDNA library contained copies of all the mRNA in these cells including interferon-specific mRNA.

DNA from ten recombinants was isolated and covalently bound to diazobenzloxymethyl (DBM) paper (Fig. 5). The mRNA from induced cells was hybridized to each filter. Unhybridized mRNA was removed by washing. After the specifically hybridized mRNA was eluted, both fractions were translated in *Xenopus laevis* oocytes. Once a positive group had been found (one in which the specifically hybridized mRNA yielded IFN after microinjection into frog oocytes), it was necessary to identify the specific clone or clones containing IFN cDNA. The ten individual colonies were grown, the plasmid DNAs were prepared, and each individual DNA was examined by mRNA hybridization as above (Fig. 5). By these procedures a recombinant, plasmid 104 (p104), containing most of the coding sequence for a Hu-IFN-α was identified. The DNA sequence was determined and found to correspond to what was then known of the amino acid sequence of our purified natural Hu-IFN-α. The cDNA insert in p104 contains the sequence corresponding to more than 80% of the amino acids in IFN-α, but not for those at its amino-terminal end. It was, therefore, used as a probe for finding a full-length copy of the IFN cDNA sequence which could be used for expression of Hu-IFN-α in *E. coli*. In addition, p104 DNA was used to isolate DNA sequences corresponding to other IFN-α species directly from a human gene bank in phage lambda. The phage plaques on agar plates were screened with the use of labeled p104 DNA as a probe after transferring the phage DNA to filters (see Fig. 6).

Examination of the coding regions of the leukocyte IFN genes that have been isolated in our laboratory and others have shown that these correspond to a family of homologous proteins, the IFN-α species which are closely related to each other and yet each unique in amino acid sequence. Thus, the previously discovered heterogeneity in Hu-IFN-α was shown to be at least in part the result of distinct genes representing each expressed Hu-IFN-α sequence. The cloned Hu-IFN-αA, which was the first one we isolated, corresponds to one of the natural IFNs which we purified from the mixture present in IFN-α by high-performance liquid chromatography. By similar procedures to those described for p104, plasmid p101 was shown to contain the sequence for Hu-IFN-β. Thus, the nucleotide sequences coding for Hu-IFN-α and -β were identified.

Construction of an Interferon Expression Vector

As noted above, p104 contained most of the sequence for a Hu-IFN-αA, but did not contain the sequence coding for the amino terminus of the protein. Accordingly, plasmid p104 was used to screen additional cDNA recombinants, and a number of cDNA recombinants which hybridized to its unique IFN-coding sequence were identified. Most contained IFN DNA sequences of sufficient size to code for the entire IFN-αA protein.

Our first full-length recombinant isolated corresponded to IFN-αA and its entire *Pst*I insert was sequenced. Existing knowledge about IFN protein sequences permitted us to determine the correct translational reading frame, and, hence, to predict the entire amino acid sequence for the IFN species encoded by this recombinant. The mature IFN-αA was expressed directly by reconstruction of the recombinant (Fig. 7). The leader sequence of the protein was removed (the sequence for the signal peptide) and an ATG translation-initiation codon was placed immediately preceding the codon for the first amino acid of mature IFN-αA. Next, a 300-base pair *Eco*RI fragment of *E. coli* DNA was constructed, containing the tryptophan (*trp*) promoter-operator and the *trp* leader ribosome-binding site, but stopping short of the ATG sequence needed to initiate translation of the leader peptide of the *trp* regulatory region. This DNA fragment was attached to the reconstructed Hu-IFN-αA in front of the ATG codon or start codon (Fig. 7). Inserted into *E. coli*, the recombinant yielded high levels of activity, about 2×10^8 units (about 1 mg) of IFN per liter of culture. The IFN protein produced in *E. coli* behaves similarly to IFN formed directly by human leukocytes: it is stable to acid treatment, is neutralized by antiserum to Hu-IFN-α, and binds to monoclonal antibodies specific for Hu-IFN-α. It can be purified to homogeneity with the use of monoclonal antibodies (see below). With improvements in fermentation and in the bacterial strains, high levels of Hu-IFN-αA (1×10^{10} units/L, 50 mg/L) can be obtained in large-scale cultures. Similar high levels have been produced with other IFNs and other eukaryotic proteins.

Purification of Recombinant Human Leukocyte Interferon

Monoclonal antibodies to the human leukocyte interferons were used to purify recombinant human leukocyte A interferon (Hu-IFN-αA) produced in bacteria. *E. coli* containing the human leukocyte interferon, IFN-αA, were broken. Unbroken cells and cellular debris were removed by centrifugation. The IFN-αA and soluble bacterial proteins remain in the cell lysate. Nucleic acids (DNA and RNA) that make the lysate viscous and thus difficult to handle easily are precipitated by combination with polymin P. The soluble proteins in the lysate can be passed through a column containing a monoclonal antibody to human leukocyte interferon (Fig. 10). The antibodies bind only the interferon; all the other components and proteins passed through the column. After washing the column, the IFN-αA bound to the column is removed by elution with an acidic solution. A virtually pure interferon solution is eluted from the monoclonal antibody column (Figs. 10 and 11). The column is then washed and neutralized so that it can be used repeatedly. The interferon solution is concentrated by passage over a column of carboxymethyl-cellulose. The activity of the purified interferon made in bacteria is very similar to that of the same human leukocyte interferon species synthesized by human cells.

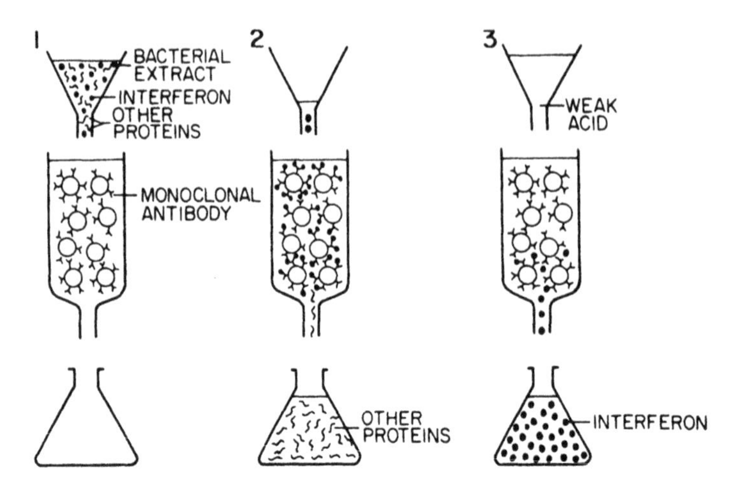

Figure 10 Schematic illustration of purification of interferon with monoclonal antibodies. Monoclonal antibody is attached covalently to a solid support. For the purification of Hu-IFN-αA, monoclonal antibody LI-8 was coupled to Affi-gel, a polyacrylamide solid support. The bacterial extract containing Hu-IFN-αA is passed through the monoclonal antibody column (1). Hu-IFN-αA, but not other proteins, bind to the column which is exhaustively washed to remove other contaminating proteins (2). Purified Hu-IFN-αA is then eluted by adjusting the elution buffer to pH 2.5 with acetic acid (3).

Since repeated use of the monoclonal antibody columns is possible, these affinity columns provide a convenient method for preparing homogeneous human leukocyte interferon from bacterial fermentations. Several biological activities of this purified recombinant interferon have been determined. The recombinant IFN-αA exhibits antiviral activity and antiproliferative activity comparable to those of crude and purified natural leukocyte interferons. IFN-αA also stimulates natural killer cell activity as do the natural species of human leukocyte interferon.

Rapid advances in recombinant DNA technology have made it possible to produce virtually an unlimited supply of interferon. Expression of interferon in bacterial cells and subsequent purification with the use of immobilized monoclonal antibodies as shown above is essentially a one step purification scheme. When combined with high expression in bacteria, this procedure provides high recoveries and low cost of the interferon.

Purification of human leukocyte interferon synthesized in bacteria was accomplished by employing these immobilized monoclonal antibodies. Utilization of this essentially, one-step purification procedure provided the first pure human interferon for clinical evaluation and, on January 15, 1981, our mate-

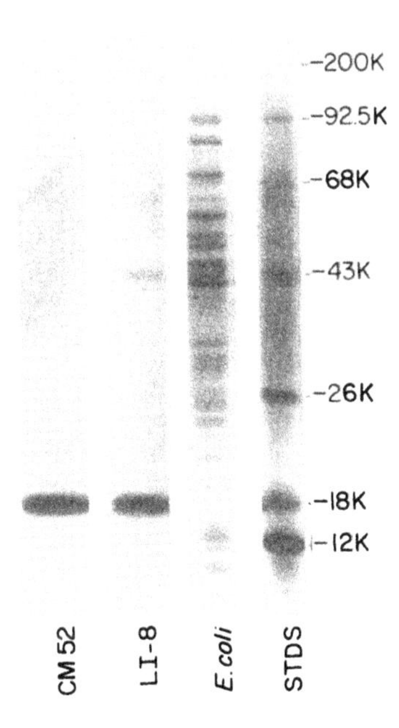

Figure 11 Sodium dodecyl sulfate polyacrylamide gel electrophoresis (SDS-PAGE) of purified interferon. The gel was stained with 2.5% Coomassie Brilliant Blue. Approximately 20 μg of interferon purified by monoclonal antibody column LI-8 was subjected to electrophoresis. Stds, represents the standard molecular weight markers; *E. coli*, represents total soluble proteins from *Escherichia coli*; LI-8 represents the fraction after passage through the monoclonal antibody column; and CM-52, represents the purified interferon after passage through the carboxymethyl-cellulose column.

rial produced at Hoffmann-La Roche was used in a patient for the first time. This purified material is being investigated in clinical studies for efficacy as an antiviral and/or antitumor agent. The development of technology from the laboratory level to a prototype operation for commercial production was essential to provide sufficient material for continued large-scale clinical trials. Since sufficient antibody supplies were a limiting factor, it was imperative to develop a system whereby a column of immobilized monoclonal antibody could be used repeatedly. The use of dilute acetic acid as an eluting agent does not appear to be particularly denaturing to the immobilized antibody. Contamination with *E. coli* proteins does not appear to be a serious problem. Success with the immunoaffinity purification of IFN-αA has prompted the extension of similar purification procedures to other proteins and other affinity chromatography procedures. Nevertheless, nonimmunoaffinity procedures have also been successful.

Immunosorbents made by covalently coupling monoclonal antibodies to agarose gels have high specificities and therefore are powerful tools for the purification of proteins. Specificity was demonstrated in the case of IFN-αA: essentially crude extracts of *E. coli* are passed over a column of immunoabsorbent gel and IFN-αA is purified in essentially one step. Despite initial skepticism, monoclonal antibody immunosorbent chromatography has proven to be a viable procedure for the purification of proteins for pharmaceutical applications. The ability to reuse the same column hundreds of times makes these columns practical because the quantity of monoclonal antibody required is reasonable and readily obtainable without heroic efforts. Accordingly, the commercial manufacture of protein therapeutics by immobilized monoclonal antibodies for purification should be considered when planning the research and development of these pharmaceuticals.

Concluding Comments

Although purification of the interferons to homogeneity remained elusive for nearly two decades after their discovery, they are now available in pure form in many laboratories. The largest amounts available are the species produced in *E. coli*. For purification, immunoaffinity chromatography has proved to be generally useful both on a laboratory as well as a commercial scale. Nevertheless, techniques other than immunoaffinity chromatography have proved to be successful as well. The availability of these proteins for laboratory and clinical studies has already catalyzed extensive new developments with these agents. In the years ahead, it is likely our understanding and use of these agents will be even more expansive and provide new insights into their actions.

THE FUTURE

Recombinant DNA technology has revolutionized our approaches to basic scientific questions and brought new methods to industry. Genes and proteins formerly inaccessible are now readily available for study and use. Our understanding of how organisms develop and cells differentiate into various tissues, how genes are regulated, and how enzymes function to name but three areas is now being approached by mobilizing these techniques. Applications of genetic engineering are being realized in the pharmaceutical and other industries. All this has occurred in under 15 years. The pace of research in these areas continues to be rapid. Enthusiasm and confidence in the future have permitted scientists and the public to debate the advisability and timing, not the feasibility, of sequencing the human genome (3×10^9 base pairs). It is a time to enjoy the intellectual and practical fruits of all these changes. It will take intelligent management of our global resources to do all this well, but I am confident it will be done for the benefit of all mankind.

BIBLIOGRAPHY

Interferon

Pestka, S. The purification and manufacture of human interferons. *Sci Am 249*, 36–43 (1982).

Pestka, S. The human inteferons—from protein purification and sequence to cloning and expression in bacteria: Before, between, and beyond. *Arch. Biochem. Biophys. 221*, 1–37 (1983).

Methods in Enzymology Vols. 78, 79, and 119. Edited by S. Pestka. Academic Press, New York.

General Techniques

From Genes to Clones: Introduction to Gene Technology, Edited by E. L. Winnacker. VCH Verlagsgeskllachaft, Weinlein, FRG, 1987.

Methods in Enzymology: Guide to Molecular Cloning Techniques, Vol 152. Edited by S. L. Berger and A. R. Kimmel. Academic Press, New York, 1987.

Molecular Cloning: A Laboratory Manual, Vols. 1–3 Edited by T. Maniatis, E. F. Fritsch, and J. Sambrook. Cold Spring Harbor Laboratory, Cold Spring Harbor, NY, 1989.

12

Monoclonal Antibodies

James N. Bausch

Schering Corporation
Kenilworth, New Jersey

BACKGROUND

It is more than fifty years since gamma globulins were first designated a distinct group of serum proteins by Tiselius. The term gamma globulin derives from the fact that they migrate more slowly in an electric field than the two other groups referred to as alpha and beta globulins. Several years later, Kabat and Tiselius demonstrated that the "antibodies" of blood serum were restricted to the γ-globulins class. An antibody can be defined as a group of proteins produced by a subset of white blood cells (B cells) that recognize and bind foreign substances (or antigens). This interaction leads to a series of events which serve as a defense mechanism to pathogens. Thus, antibody production constitutes a major portion of the immune response of animals.

STRUCTURE OF ANTIBODIES

Antibodies (Ab) or immunoglobulins (Ig), as they are often called, are globular proteins containing carbohydrate. They occur as monomers and polymers and are divided into classes and subclasses based upon the unique differences in their polypeptide-building blocks (1). Each monomer, in turn, consists of two heavy polypeptide chains and two light polypeptide chains. They are referred to as heavy and light based upon their molecular weights. Each of these chains can be grouped into families of related proteins. The families are designated by the "determinants." Determinants are areas of the protein which can cause an immune reaction when injected into an animal of a different species. In other words, the structural features of antibodies were analyzed by using them as antigens.

The heavy chain family comprises $\gamma 1$, $\gamma 2$, $\gamma 3$, $\gamma 4$, α, μ, δ, and ϵ. The light chains are either κ or λ. Each monomer consists of two identical heavy and two identical light chains and various combinations are possible. These combinations and how they relate to immunoglobulin class are outlined in Table 1. Antibodies from different classes possess different immunological activities as well as unique biological characteristics (see Tables 2 and 3).

Table 1 Classification of Human Immunoglobulin

Ab class	H chain antigenic determinant	L chain antigenic determinant	Molecular combination
IgG_1	γ^1	κ or λ	${\gamma^1}_2 \kappa_2$ or ${\gamma^1}_2 \lambda_2$
IgG_2	γ^2	κ or λ	${\gamma^2}_2 \kappa_2$ or ${\gamma^2}_2 \lambda_2$
IgG_3	γ^3	κ or λ	${\gamma^3}_2 \kappa_2$ or ${\gamma^3}_2 \lambda_2$
IgG_4	γ^4	κ or λ	${\gamma^4}_2 \kappa_2$ or ${\gamma^4}_2 \lambda_2$
IgA (secretory version)	α	κ or λ	$(\alpha_2\kappa_2)_2 + p^a$ and $(\alpha_2\lambda_2)_2 + p^a$
IgA (serum version)	α	κ or λ	$(\alpha_2\kappa_2)_{1-3}$ and $(\alpha_2\lambda_2)_{1-3}$
IgM	μ	κ or λ	$(\mu_2\lambda_2)_5$ and $(\mu_2K_2)_5$
IgD	δ	κ or λ	$\delta_2\kappa_2$ and $\delta_2\lambda_2$
IgE	ϵ	κ or λ	$\epsilon_2\kappa_2$ and $\epsilon_2\lambda_2$

[a] Protein produced by secreting cell.

Table 2 Activities of Human Immunoglobulins

Antibody class	Detected activities
IgG	Late response to antigen; antibacterial; antiviral; antitoxin; antiblood group
IgA (serum)	Blocks bacterial adherence; viral defense
IgA (secretory)	Found in mucus secretions; blocks bacterial adherence; viral defense
IgM	Early response to antigen; antibacterial; antiviral; antiblood group
IgD	Present on lymphocyte surface
IgE	Allergic reactions; respiratory tract defense; mast cell fixation; antiparasite; cytophilic for basophils and mast cells

Table 3 Human Immunoglobulin Properties

Immunoglobulin class	Half life (days)	Serum concentration (mg/ml)	Synthesis rate (mg/kg/day)	Percent of total immunoglobulins	Approx. molecular weight (daltons)
IgG	23	10	33	80	160,000
IgA (serum)	6	2	24	16	170,000
IgM	5	1.2	6	4	900,000
IgD	3	0.03	0.4	0.001	185,000
IgE	3	0.0005	0.02	0.00003	196,000

The class of immunoglobulins termed ''IgG'' has been studied intensively and serves nicely as a basic model for the structure of all the Igs. It is generally the most useful class and its production as a monoclonal is usually preferred.

Figure 1 shows the IgG as a Y-shaped monomeric, 4 polypeptide chain complex, containing 2 identical light (L) chains and 2 identical heavy (H) chains. The four chains are held together by both noncovalent forces and covalent disulfide bonds. The complex nature of the chain folding (secondary and tertiary structures) is not shown. The central region called the hinge region allows the two arms, which contain the antigen binding sites, to bend and

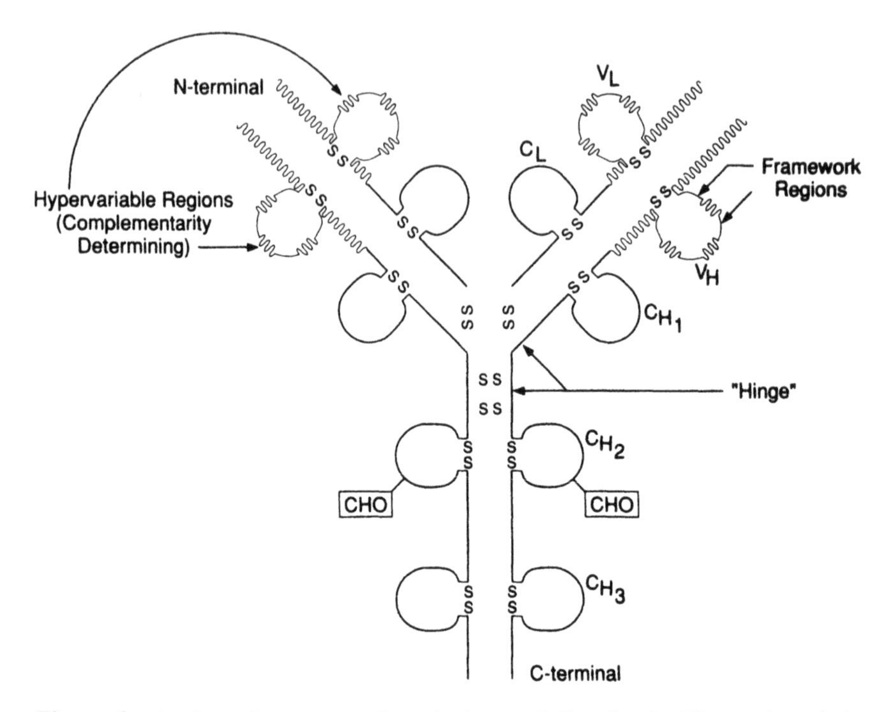

Figure 1 A schematic representation of a human IgG molecule. The portions designated C_L and V_L represent the constant and variable regions of the light chain, respectively. The designations of V_H, C_{H1}, C_{H2}, and C_{H3} represent variable and constant regions of the heavy chains. The box containing "CHO" shows the usual area for attachment of carbohydrates.

flex sufficiently to effect binding of the antibody to the antigen. The protein is subdivided into domains. The domains of the heavy chain are designated V_H, C_H1, hinge, C_H2 and C_H3. The light chain domains are V_L and C_L. The "V" stands for variable region and it is this portion of the protein that determines the interaction with a particular antigen. The "C" represents the constant region which remains the same within a subtype chain. That is, the C regions define the subtype (e.g., $\gamma1$, from $\gamma2$, etc.). A carbohydrate moiety is found attached to the C_H2 domain. The upper half of the molecule is referred to as the Fab fragment and contains the antigen binding site. The lower half of the molecule is designated Fc fragment and it carries out the biological effector functions, for example, complement fixation or Fc receptor binding.

The antibodies in the serum of a normal individual are heterogeneous, even when stimulated by a single antigen. They consist of unique families of different clonal origins. They vary in specificity and binding affinity for antigenic determinants, as well as in the Ig class. Thus, when an antigen is injected into an animal, the production of various classes of antibodies is elicited. Within each class (i.e. having identical C_H and C_L domains) one finds a variety of antibodies differing in their V_H and V_L regions.

The natural production of a wide spectrum of antigen elicited antibodies by an individual results in what is commonly called "polyclonal antiserum" ("poly" meaning "many"). Although this heterogeneity of antibody production is both fascinating and extremely useful in combating infection, it has represented an obstacle to researchers interested in studying antibodies and also in using antibodies for research and diagnostic purposes. At best, antisera produced against the purest natural Ags contain a confusing mixture of Abs of differing affinities and various Ig classes; these may be directed against different regions (antigenic determinants) on the same molecule.

The engineering of a "monoclonal antibody" ("mono" meaning "one") source is an attempt to simplify this chaotic situation. The theory of monoclonal antibody production is based upon the clonal selection hypothesis of Burnet (2). The theory states that (i) antibody diversity is a reflection of the presence of many different lymphocyte clones, that (ii) each individual lymphocyte produces only a single type of antibody, and that (iii) even homogeneous antigens induce a heterogeneous antibody response. This was referred to as the "one lymphocyte-one antibody" concept.

This theory was verified by Kohler and Milstein in 1975. In their extraordinary breakthrough, they succeeded in immortalizing a usually short-lived plasma cell (Ab producing cell derived from the B-cell lineage). They accomplished this by hybridizing (or fusing) it with a neoplastic syngeneic plasma cell line (myeloma cell or "cancer type" cell). Using cell culturing techniques, individual cells can be selected. Each single hybrid cell is grown into a colony which can be expanded indefinitely in cultures, producing a single type of antibody in abundance (i.e., monoclonal antibody). The antibody that is produced is uniform and exquisitely specific for a single antigenic determinant. Justifiably, the contributions of monoclonal antibodies to science and medicine were recognized by the award of the Nobel Prize in Physiology and Medicine in 1984 to Milstein, Kohler, and Jerne.

SECONDARY ACTIVITIES OF ANTIBODIES

Antibodies are known to carry out diverse tasks under a variety of microenvironmental conditions. Therefore, nature has provided much diversity in their subtle characteristics. These qualities can be grouped together under the term "effector functions." One effector function is the ability to fix complement (an enzymatic cascade which leads to lysis of a target cell). Other activities include agglutination of particles, precipitation of soluble antigens, enhancement of phagocytoxic activities (i.e., attack by cytotoxic T cells), as well as promotion of cell cytotoxicity via intercellular crosslinking. In order to effect these different activities, nature has provided for different classes (and subclasses) of antibodies, each having subtle and occasionally large differences. Mouse IgG_2 and IgM fix complement, while IgG_1 does not and, therefore, cannot mediate complement-mediated toxicity. IgG_3 MAbs appear difficult to use and relatively insoluble. IgM and IgA antibodies have short clearance time from the circulation as compared to IgG. In addition, properties such as bivalency can result in large precipitin complexes causing tissue damage. It is also sometimes advantageous to abrogate some of the effector functions of the Fc region, which are involved in cell-mediated immunity. An example of this is the undesirable Fc-mediated binding of monoclonals to receptors on macrophages and certain normal cells, especially if the monoclonals are radioactively tagged or conjugated to cytotoxic drugs.

PRODUCTION OF MONOCLONAL ANTIBODIES (MAbs)

The Production of Mouse MAbs

The most commonly employed MAbs are derived from mice. The immune system of the mouse is still the primary source of the genetic design and resulting protein expression for the MAb. The lab researcher simply separates (i.e., clones) out one (i.e., mono) antibody-producing white blood cell from the multitude which the mouse produces for a given antigen. Figure 2 is a schematic protocol for the production of the MAbs from a mouse.

The mouse is first injected with the antigen of interest. If the antigen is of a low molecular size, it is often coupled to a carrier protein. This conjugated antigen is known as a hapten. In response to this injection, the mouse's immune system produces a variety of antibodies to the antigen. The antibodies are produced by the white blood cells, termed B cells (originally for their *b*one marrow origins), which can be found in the spleen of the mouse. At this stage the mouse antiserum production is still categorized as "polyclonal" by virtue of its multicellular diversity.

The researcher now has three goals. First, he must isolate the antibody-producing B cells. Second, he needs to keep these cells from dying. That is, they must be immortalized. Third, he wants these new immortal cells to proliferate (divide) uncontaminated by other cells and to obtain a high yield of individual monoclonal antibody-producing cell lines.

The first goal is accomplished by surgically removing the spleen from the immunized mouse and physically isolating the white blood cells by differential centrifugation. The second goal of immortalization is satisfied by fusing the spleen cells with an immortal cell line (i.e., a myeloma mutant). In this process the membranes of two cells actually merge in the presence of a chemical

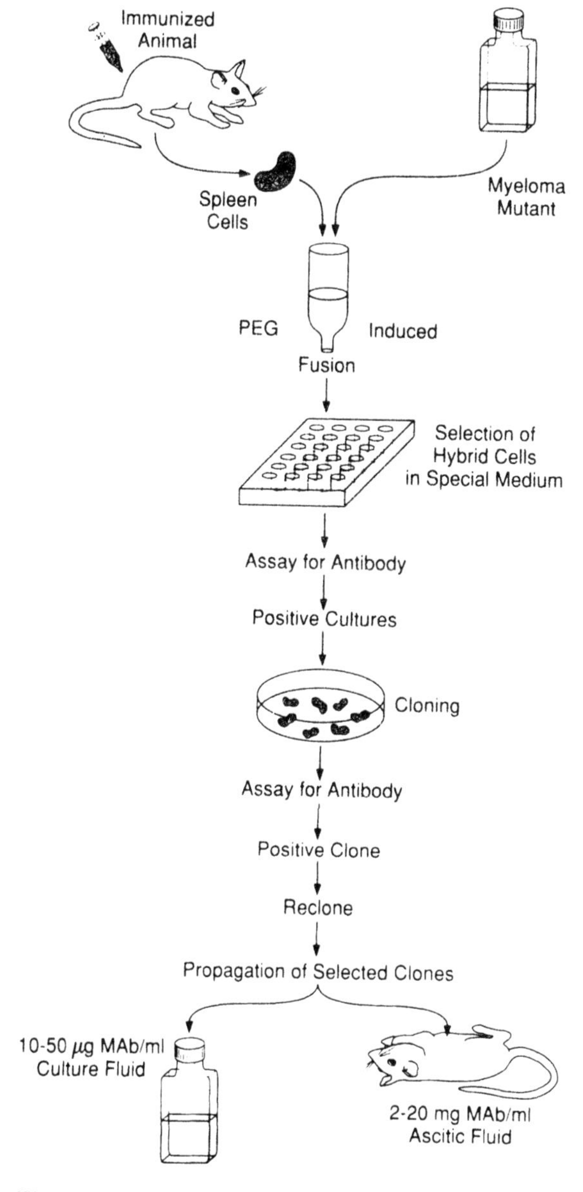

Figure 2 The most common path for production of mouse monoclonal antibodies.

inducer, such as polyethylene glycol (PEG). The resultant hybrid cells (hybridomas) are selected by their ability to grow in special culture fluid called HAT medium, which will only support "fused" cells. Therefore, the leftover spleen and myeloma cells expire. The third goal is achieved by first testing for the production of antibodies by the fused cells using an immunoassay, then by performing a series of "cloning" steps. These cloning steps are essentially diluting steps in which each well of the culture plate is expected to contain a single cell. After growing and testing these individual clones, one obtains groups of cells, each of which traces its origins to one and only one progenitor cell. Once this goal has been achieved, the researcher will reclone the appropriate clone either in a tissue culture system or in the peritoneal cavity of a mouse (i.e., ascites fluid). Either of these systems results in the production of a large amount of monoclonal antibody for subsequent use.

The main advantage of the in vitro tissue culture approach is the purity of the product (if serum-free medium is employed). The advantage of the in vivo production in mice is high quantity and an inexpensive operation. For commercial production, most companies favor the tissue culture route, which can more readily produce a clean product.

The Production of Human MAbs

The potential of monoclonal antibodies as therapeutic agents in humans may hinge upon the future ability to produce human-type monoclonals. Human MAbs offer a number of therapeutic advantages over murine MAbs with regard to both toxicity and effectiveness. The benefits of murine MAb therapy in human cancer patients is quickly abrogated by the patient's own immune response to this "foreign" intrusion. Human MAb therapy may result in slight immune response, but this effect should be relatively minor. In addition, the availability of human MAbs, for example, in Rh immune globulin treatment would preclude the always present possibility of contamination of HIV, hepatitis virus, or cytomegalovirus in the product obtained from human blood.

The infection of B lymphocytes in the resting state by Epstein-Barr virus leads to a permanent stimulation of cell growth, a phenomenon termed "transformation" or "immortalization." The cells that grow out in vitro are termed lymphoblastoid cell lines (LCL). The transformation can be performed before or after selection for hapten binding. The cloning procedure will then result in a clonal cell line capable of producing the desired MAb. LCLs secreting MAbs have been produced against tetanus toxoid, streptococcal carbohydrate A, diphtheria toxoid, melanoma antigens, and many others. The two disadvantages of EBV immortalization are (1) most of the cultures of transformed lines frequently lose their ability to produce specific antibody after long-term culture and (2) very low quantities of antibody (<1 μg/ml) are generally produced.

Investigations are still underway to create human hybridomas by fusing human lymphocytes with (a) murine myeloma cells, (b) human myelomas or (c) heteromyelomas (mouse × human hybrid myelomas). These attempts have met with only spotty success. Scale-up is also a serious limitation as ascites production has not proved successful.

Second-Generation MAbs

One interesting technique investigated was the production of bifunctional antibodies by extension of hybridoma technology (7). In this line of research the investigators are seeking to change the normal bivalency of the MAb. Normally, the MAb will contain two Fab regions with identical specificity. If the two Fabs were different, they would be capable of binding two different antigen epitopes. When properly designed and produced routinely, these bifunctional antibodies will have a tremendous impact on the development of immunodiagnostic assays and immunotherapeutic reagents.

For example, a hybrid MAb containing an antigen-binding site and an enzyme affinity site would dispense with the need for covalent coupling of the enzyme to the antibody. Additional potential applications are for enzyme channeling and fluorescence energy transfer in immunoassays. There is also great hope for utilizing bifunctional antibodies in an in vivo setting as diagnostic and therapeutic agents. Their "dual binding" potential allows the investigator to first dose the patient with unlabeled hybrid MAb. These antibodies will arrive at the desired site in the body as well as bind to that specific site. After this step, the drug or radioisotope could be injected and hopefully localize in the desired organ by virtue of Fab binding to the hybrid antibody.

This concept is not new. Researchers have been pursuing this possibility for 25 years. But the advent of hybridoma technology has provided a new avenue to achieve the desirable bifunctionality of the antibody. The conventional approach was to chemically synthesize hybrid antibodies by reduction of IgG to half molecules followed by renaturation to whole antibody (8). With mouse monoclonal antibodies the yields of this method are low and results are not very reproducible. Burnet et al. (8) describes two new and different approaches in the design of hybrid molecules. In the first approach, a hybridoma producing an antibody of one antigenic specificity is fused to a mouse spleen cell immunized with the second antigen of interest. The resulting MAb is capable of binding to both antigens simultaneously. The second technique, following similar experimental rationale is called the "tetradoma technique." In this protocol two MAb-producing cell lines of known specificity are fused with one another. This again produces a "bifunctional" antibody.

A second group of improvements involves specific biochemical alterations in the currently produced MAbs. These can be put into three categories: (a) changes in antigen binding, (b) alterations via deletions or point mutations in the constant region leading to changes in effector functions, and (c) class and subclass variants. Because monoclonal antibodies have moderate or low affinities, it is often desirable to select cell clones with higher affinities. Studies show that somatic mutations of single amino acid substitutions that cause changes in antigen binding can arise frequently and spontaneously in nature. The phenotypic changes which can result due to somatic mutations are listed in Table 4. The frequency of the spontaneous mutations are as high as 10^{-3}–10^{-4}. For perspective, other genes, such as those involved in drug resistance, mutate at frequencies of about 10^{-6}–10^{-7}. This suggests that the immunoglobulin genes are uniquely unstable in cultured hybridoma cells. This mutation rate although still low indicates that researchers should carefully monitor MAbs for changes in primary structure and activity. On the other hand, selection of biochemically altered clones may be useful for obtaining variants with desired properties.

A third research approach to the design and creation of better monoclonals involves using hybridoma production in combination with recombinant DNA techniques and gene transfer technologies. The genetic information for the antibody is conventionally divided into exons that reflect the structural protein domains of both the heavy and light chain polypeptides. The heavy chain complex, including the Vh gene exons, and the light chain gene complexes are located on three distinct chromosomes in all animals examined. Unique gene rearrangements take place during the maturation of antibody producing B lymphocytes. This event juxtaposes distinct gene segments to create the coding sequence for each particular antibody chain (see Chap. 5). The various possibilities for completing this recombination event are responsible for the existence of antigen-binding variable region exons. When these exons are translated into protein, they differentiate antibody families and comprise the basis for antibody gene diversity and expression (9). In a new branch of MAb research the antigen-binding capabilities provided by mouse, rat and human antibody-producing hybridomas or virus-transformed cell lines are borrowed by cloning the desired immunoglobulin genes, and manipulated using recombinant DNA

Table 4 Phenotypes of Somatic Mutants of Cultured Antibody Producing Cells

Changes in Ig expression	and/or	Changes in Ig structure
Chain alterations (H^-L^+, H^+L^-, H^-L^-)		Class and subclass switches
		Deletions
Quantity produced		Point mutations

techniques. These spliced gene sequences are then transferred into mouse myeloma cell lines where they are transcribed, translated and assembled, finally secreting chimeric antibody molecules.

The diagram of this type of novel transformation is sketched in Figure 3. Briefly, the variable region genes of the mouse light and heavy chain are isolated and cloned up in quantity. They are then placed into transfection vectors juxtaposed proximal to the constant regions of the human light and the constant region of human heavy chain, respectively. These hybrid transfection vectors are then used to cotransfect the nonproducing myeloma cell line. The final result is a transfectoma cell line producing human–mouse chimeric antibody molecules. In this way, the usual mouse system can be utilized for generating the antigen-binding specificity, but the major portion of the antibody protein is of human origin. This new technique of exon shuffling or isotype switching was pioneered by Oi and Morrison (10). It is hoped that the chimeric human/mouse MAb will be more effective in vivo than the usual mouse MAb. An interesting variation of this approach, namely immunoadhesins, is described later in this chapter.

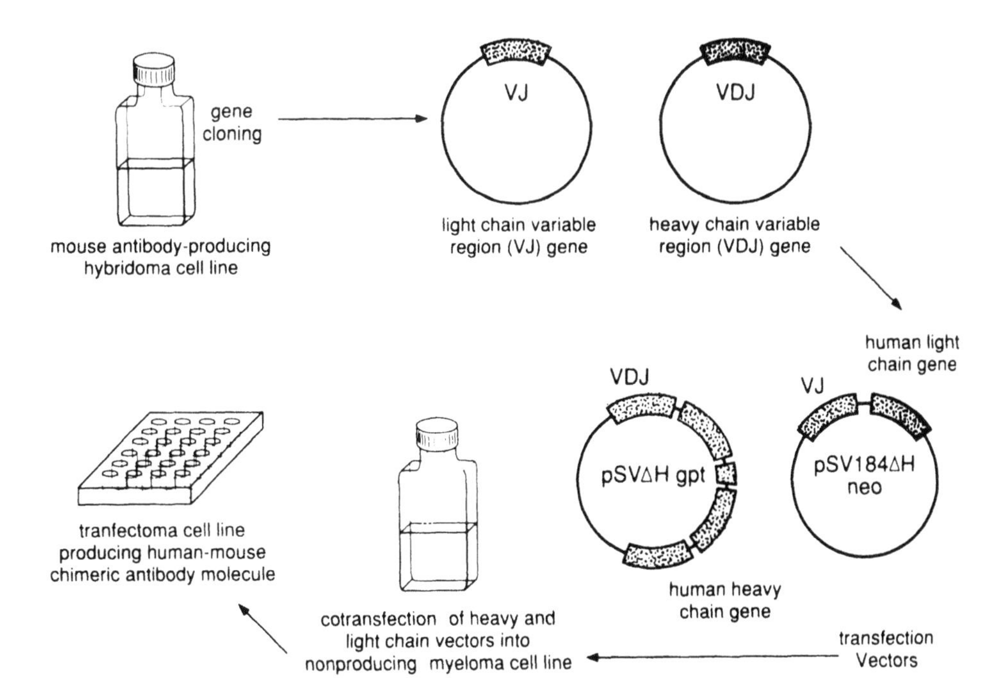

Figure 3 The production of chimeric human/mouse monoclonal antibodies.

PURIFICATION OF MONOCLONALS

The two most common means of scaling up hybridoma/MAb production are (a) in vivo using the ascites method in rodents (i.e. peritoneal cavity tumor), usually mice or (b) an in vitro tissue culture system (11). It is also possible to expand MAb production in vivo by formation of subcutaneous tumors (12), but this method is used less frequently. The steps in purification for ascites derived and tissue culture derived material are shown in Figure 4.

Concentration

This can be performed by a variety of techniques including ammonium sulfate precipitation (used for ascites only), dialysis against polyethylene glycol, vacuum dialysis, or membrane filtration. It is generally suggested that the researcher perform a trial experiment on a small amount of MAb-containing fluid, since MAb derived by these new methods cannot be guaranteed to behave like serum-derived immunoglobulins.

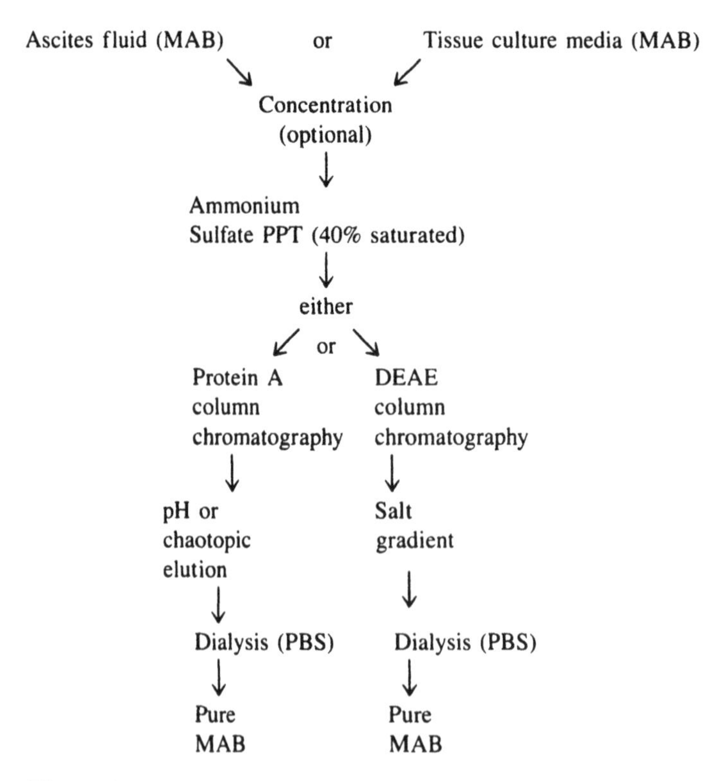

Figure 4 A generalized purification scheme for monoclonal antibodies.

Likely protein contaminants from tissue culture production are those from fetal calf serum. These will be albumins for the most part, but there will also be some bovine immunoglobulins. There may also be some cellular debris, including nucleases and proteases. For ascitic fluid the major contaminants will be irrelevant immunoglobulin with varying amounts of proteases and nucleases. The amount of irrelevant immunoglobulin depends upon the particular class of Ig involved.

Regardless of which source the antibody comes from, it is prudent to determine the class of the antibody which one is about to purify (IgG, IgM, IgA, etc.). This will influence the general outline of the purification scheme. The class can be determined by a variety of methods (11) including enzyme-linked immunosorbent assay, Ouchterlony double diffusion, or labeling in vivo followed by SDS-PAGE gel electrophoresis.

Column Separations

The remaining steps are the same, regardless of whether the source of the Ab is ascites, tissue culture fluid, or even animal serum. Immunoglobulin purification is covered in many texts, usually with emphasis on serum immunoglobulins (13, 14). The principals of separation always remain constant: (1) change, (2) size, (3) hydrophobicity, and (4) anti-immunoglobulin binding (affinity chromatography). The unique concentration of basic amino acids in the Fc region of most immunoglobulins is exploited in the purification step using DEAE (diethylaminoethyl) resins. At low pH and low ionic strength, the IgG MAbs bind poorly, while most other classes (IgM etc.) and impurities such as proteases and nucleases bind strongly to the resin. Loading at low pH and low ionic strength followed by a shallow salt gradient is sufficient to obtain a highly pure MAb (>98% pure).

One interesting approach involves purification by protein A. This protein is synthesized by *Staphylococcus aureus* and binds to the Fc region of many IgG molecules. It can be purchased as soluble protein or "immunobilized" on various matrices. This affinity chromatography medium differs widely in its ability to bind MAb from different species (mouse, human, etc.) as well as in its ability to bind different classes (IgG, IgM, IgA, etc.) and subclasses of MAbs (IgG, IgG_2 etc.). Its use should therefore be evaluated on a pilot scale before devoting large efforts. Sometimes this technique utilizes a pH 3.0 eluate. This is considered one of the drawbacks of the technique as compared to DEAE chromatography.

A variety of other chromatographic procedures have been utilized. For example, DEAE-Affigel Blue columns have been utilized very successfully in eliminating tenacious impurities such as proteases, nucleases and albumin from ascites fluid (15). Antigen affinity chromatography is also used occasionally in cases in which DEAE and protein A methods fall short. The rationale is based on specificity of antigen/antibody reactions. It requires that the researcher have a reasonably large quantity (i.e., milligrams) of fairly pure antigen and that the antigen be immobilized on a matrix such as agarose. Binding is carried out at neutral pH, followed by extensive washing. The MAb is then eluted using either high or low pH buffers or chaotropes such as KSCN. The purity obtained using such a protocol is usually excellent. The only reservation is that the MAb should be stable under the harsh conditions of elution.

If the MAb of interest belongs to the IgM class it is generally purified by a combination of low ionic strength precipitation in 2 mM phosphate buffer pH 6. This is then followed by size exclusion chromatography on Sepharase 6B to rid the IgM of IgG contaminants.

Reverse-phase high-pressure liquid chromatography (RP-HPLC) is currently employed as a purity indicator during purification processes. The use of organic eluting agents still precludes the use of this chromatography medium for antibody purification.

TECHNIQUES FOR DETECTION AND USE OF MONOCLONALS

It is appropriate at this point to list and explain the assays involved in monoclonal antibody research. The immunoassays generally serve two functions. First, they are employed in detecting, selecting, and quantitating the MAbs of interest. Second, with slight changes, these are the assays in which the monoclonals will be employed to carry out their ultimate purpose. They will be examined in order of their usual frequency of application.

Solid Support Assay

It has been approximately 15 years since the development of solid support assays. Their simplicity and wide spread application has led to their enthusiastic acceptance (16,17). The assays can be grouped into two categories: *E*nzyme-*L*inked-*I*mmuno-*S*orbent *A*ssay (ELISA), and *I*mmuno*R*adio-*M*etric *A*ssay (IRMA). The basic understanding of these assays is illustrated in Figures 5 and 6. As these assays have proliferated the acronyms have not yet been standardized. Therefore, when reading a scientific paper that uses one of the immunoassays, the reader should always examine the details of the experimental immunoassay, thereby determining what type of IRMA or ELISA is employed.

Figure 5, on the left shows the binding of the antigen to a "sticky" solid surface. This is accomplished by overlaying the plastic surface (usually polystyrene) with the antigen diluted to approximately 10 μg/ml in a high pH buffer (>pH8). The next layer upward shows the MAb of interest binding to the Ag, usually introduced after thorough surface washing, for at least 4 h between 25 and 37°C. The third layer, introduced after washing, shows the reporting anti-immunoglobulin antibody (i.e., rabbit antibody to mouse immunoglobulin) binding to the MAb. The reporting antibody is radiolabeled. This naturally allows us to visualize or quantitate the reaction. This type of assay is called a radioimmunoassay (RIA). The figure on the right is analogous to the one on the left with the exception that the second antibody (i.e., in the third layer) is enzyme labeled. Therefore, quantitation or visualization involves the addition of the appropriate enzyme substrate. Depending on the substrate, the enzyme can produce either a soluble or insoluble product. This assay is called

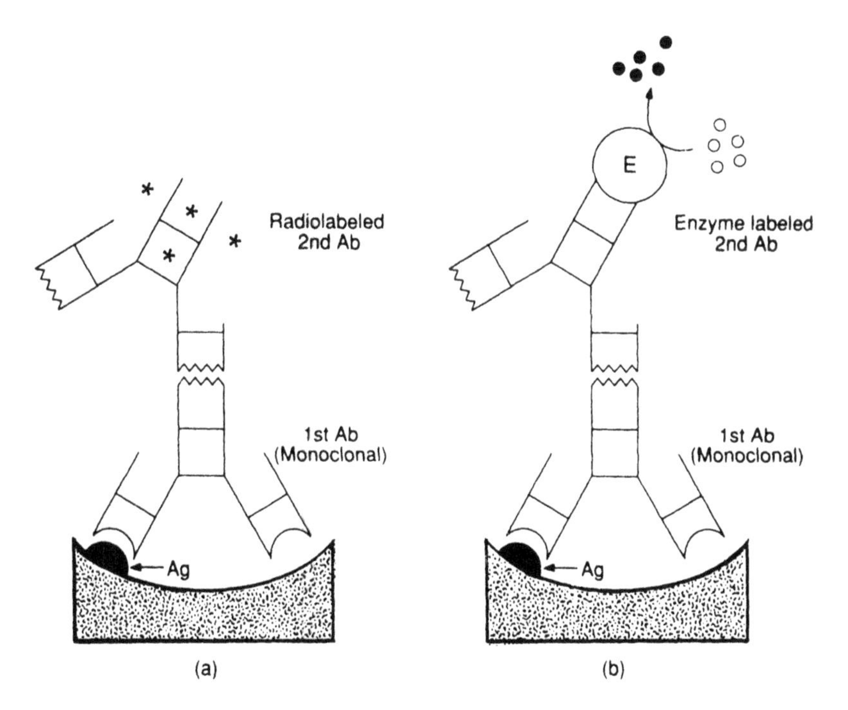

Figure 5 Antigen immobilized/solid phase binding assays employing monoclonal antibodies. (a) the mode which relies on radioisotope; (b) the mode which depends on enzymatic detection. Since antibodies can themselves be antigens, these assays can be used to detect monoclonal antibodies.

an enzyme-linked immunosorbent assay (ELISA). It is generally employed in screening for and quantitating MAbs and is also useful when titering polyclonals (i.e., determining the concentration of specific Abs in serum).

Figure 6 illustrates the second type of assay in which monoclonals may be employed. It is generally used in a routine quantitative assay for an antigen. The first antibody, which may or may not be monoclonal, is bound to a solid support. This is followed by the unknown (or standard) solution containing the antigen. After binding to the first antibody is completed, the second antibody is introduced. This second antibody may or may not be monoclonal. In Figure 6A the second antibody is radiolabeled. The methods of quantitation or visualization are the same as in Figure 5. This type of assay is referred to as an immunoradiometric assay (IRMA). However, in the case of Figure 6B, it is probably a misnomer (since radioactivity is not used). Readers will still find this terminology employed.

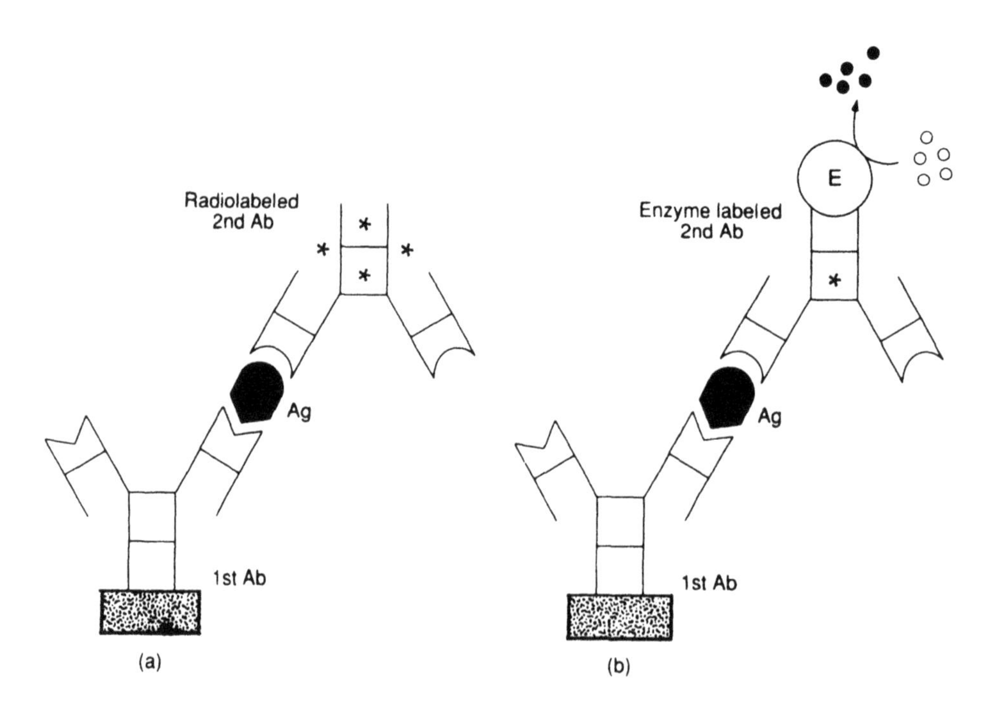

Figure 6 Sandwich mode/solid-phase binding assays employing monoclonal antibodies. (a) Shows the radioisotopic mode and (b) shows the use of enzymatic detection. Since antibodies can themselves by antigens, these assays can be used to detect monoclonal antibodies.

The solid support to which the antibody is attached can be nylon, glass, Sepharose, DBM-activated paper, CNBr-activated paper, cellulose or nitrocellulose paper. More researchers are using polyvinyl or polystyrene wells arranged 96 to a plate. The binding of the antigen or antibody to the solid support is carried out overnight at a neutral to basic pH. Antigen/antibody incubation periods are of the order of 2 h at room temperature to 4°C overnight. The second antibody is usually labeled with ^{125}I in the case of the RIA. For the ELISA the second antibody is usually conjugated to one of the following enzymes: horseradish peroxidase, alkaline phosphatase, or β-galactosidase.

Another type of solid-phase immunoassay is the dot immunobinding assay, euphemistically called the "dot blot." In this assay flat nitrocellulose sheets are usually used rather than the polystyrene surface. The same protocols of antibody/antigen binding, incubation, and detection are utilized as in the previously described ELISA and IRMA. An interesting application of this technique is the screening of colonies of MAb producing hybridoma cells grown in individual wells of a microtitre plate. The microtiter plates are inverted over the sheets, which will become saturated with the volume of MAb-containing media. The wells which contain the MAb can be pinpointed by then incubating the sheet with labeled second antibody, followed by substrate addition.

Another solid-phase assay which is rapidly becoming an invaluable tool in biotechnology laboratories is "Western blotting" or "immunoblotting" (18–20). This is probably the most powerful tool available to the biochemist who is interested in isolation, characterization, and purity assessment of new proteins. It combines the premier technique for protein separation, sodium dodecyl sulfate polyacrylamide gel electrophoresis (SDS-PAGE), with the exquisite selectivity of antibody/antigen reactions. In this procedure, the test sample containing the antigen is subjected to SDS-PAGE (21). This separates the proteins into discrete bands in the gels according to their molecular weight. The proteins are then electrophoretically transferred out of the gel (in a direction perpendicular to the plane of the polyacrylamide slab) onto nitrocellulose paper or nylon sheets. The proteins become tenaciously adsorbed to the sheets and the transfer buffer is now replaced with a more physiological buffer. The sheet is then incubated with MAb raised against the antigen of interest. If the MAb itself is enzyme labeled, then detection is realized by the addition of substrate (ELISA), or if radiolabeled by exposure of x-ray film (autoradiography). If the MAb is not labeled, then the detection employs a labeled second antibody. The band of interest is thus detected by the MAb in the presence of many other proteins. The researcher can use this technique to follow his purification schemes or assess the purity of a protein by comparing the Western blot to a conventionally stained sample (see Chap. 4, Fig. 15). This immunodetection method is capable of detecting as little as 1 ng of antigen (22).

Soluble-Phase Assays

The next general category of assays in which monoclonals become involved are generally referred to as soluble-phase systems. These are assays in which the antigen of interest is labeled either with a radiolabel or an enzyme. The drawbacks in this type of immunoassay are twofold. First, they require a reasonable amount (usually 10 μg or more) of a relatively pure antigen which must be radiolabeled. This radiolabeled or enzyme labeled antigen must be prepared by the researcher and is not generally commercially available. Second, in these assays, a step is needed to separate free labeled antigen from MAb-bound antigen. This requires the use of a second antibody, either in the formation of bulk precipitate or attached to beads. "Immunobilized" protein A may also be used.

The assay does have advantages. First, it does not require that the antigen be bivalent as the IRMA in Figure 5 does. Second, the background is usually good for these assays, since a false positive interfering substance must bind to the antigen and also be precipitable from solution by a second antibody. These requirements lend themselves to good assays with low background.

Cellular Assays

The third general category of immunoassays for MAbs are cellular assays. They are employed, sometimes, in the actual screening process for the MAb, as well as to obtain answers in the usual course of biological research investigations. These assays are often difficult to perform, but can be the most realistic if the final use of the MAb will be an identical type of assay (i.e., blood typing, complement fixation, etc.). They are also useful if the antigen to which the MAb would be directed is a membrane antigen and is, therefore, difficult to isolate and work with in solution.

Bioassays

The fourth category in which MAbs are involved are termed biological assays. They are, generally, time consuming and costly, but often necessary. Examples of these assays might include a MAb in a lymphokine neutralization assay. The neutralizing antibody might bind to either the lymphokine itself, the lymphokine receptor, or a lymphokine cofactor. Another example would be enzyme-neutralizing MAb. Simply screening for the binding to the enzyme may not be sufficient. The actual diminution of enzymic activity must be assessed using a biological assay.

Such assays have been useful in research projects involving the isolation of a previously uncharacterized protein. A partially purified preparation, containing the biological activity, is used for immunization. MAbs (or polyclonals) are selected based upon their ability to neutralize the biological activity. These Abs then become powerful reagents for use in the purification of the new protein.

Histochemical Assays

The last general category is that of immunohistochemical assays. In these assays the researcher ascertains not only the presence of the antigen, but also its tissue or intracellular distribution. These assays are used for screening hybridomas if the purpose of the MAb would be immunohistochemical staining. The MAbs themselves are generally not labeled. A second antibody, usually tagged with biotin, horseradish peroxidase, or fluorescein, is employed to detect the MAb binding (23, 24). The immunocytochemical staining can be carried out on cell monolayers as well as frozen tissue sections. The visualization is carried out by either locating fluorescent patches under a fluorescent microscopic or in the case of enzyme-labeled antibodies, a substrate is added which polymerizes in situ resulting in darkly-hued areas easily seen in light microscopy.

MONOCLONAL ANTIBODY APPLICATIONS

Diagnosis

In diagnoses of infections, laboratories use combinations of microscopic examination of bodily fluids and tissues, culture methods, and immunological antigen or antibody identification. The microscopic examination is sometimes ambiguous or incomplete. The second method of culturing is difficult, expensive, and time consuming. Immunologic techniques utilizing conventional antisera were plagued by inconsistency and cross reactivities (i.e., detection of a structurally similar, but unrelated substance). The recent development of monoclonal antibody systems has provided a substantial boost to this area of microbiological testing.

MAbs are endowed with two attributes, making them ideal for diagnostic procedures. The first is high specificity. The antibodies are uniform and directed against one epitope on the antigen. The second is their reproducibility. They can be produced in quantity to yield a standard, uniform titer over an indefinite period of time. Therefore, MAbs are continually replacing polyclonal Abs in testing for common serum components and bacterial antigens. Monoclonal testing is currently in use to detect bacterial infections (25). More specifically, the bacteria *Haemophilus influenzae* (26), *Streptococcus pneumo-*

niae (27), *Escherichia coli* (28), and *Neisseria meningitidis* (29) and their resultant infections can be traced by MAb assay designed to detect their capsular antigens. MAbs to membrane antigens of *Neisseria meningitidis* (30), *Pseudomonas aeruginosa* (25), *Neisseria gonorrhoea* (31), *Leptospira* (32), *Vibris chlorae* (33), and *Legionella pneumophila* (34) have also been utilized in detection of and research on the diseases caused by these bacteria. Even MAbs to pili have been generated for *E. coli* (28) in the research of virulence and colonization patterns of *E. coli*. MAbs to bacterial exotoxins and endotoxins, the Lac carrier protein of *E. coli* as well as the Rec A protein of *E. coli*, and numerous bacterial enzymes have furthered our understanding of bacterial infections. Researchers at Genetics System Corporation (Seattle, WA) have developed three panels of monoclonal antibodies for the detection of human sexually transmitted infections. These include *Chlamydia trachomatis* bacteria, *Neisseria gonorrhoeae* (bacteria), and herpesvirus types I and II (virus) (35).

MAbs have also found application in the food industry, especially a simple, rapid, and accurate assay for *Salmonella* (36). The potential of MAb have also been realized in the diagnoses and treatment of parasitic diseases such as malaria (37), schistosomiasis (38), and leishmania (39) and viral diseases such as influenza (40), hepatitis (41), polio (42), and rabies (43).

MAbs are now also used in tissue typing and blood typing (44–46). Diagnosis of cancer utilizing MAb is making great strides. There are MAbs for detection of tumor associated antigens (47, 48) as well as antibodies to tissue or cell-type-specific antigens (49, 50). The uses here cross the lines of immunoradiometric assay, immunocytochemistry and in vivo scintigraphy. In general terms, scintigraphy is the use of a radiolabeled compound (in this case, a MAb) to seek out and bind to a region of interest in a patient. It might be used, for instance, to locate small metastatic tumors in patient, so they could be surgically removed (51). MAbs are also contributing to our understanding of cardiology (via myocardial infarction and thrombi imaging), disease, and errors in metabolic pathways.

Digressing from the monoclonal theme, the detection of certain antibodies in a patient's blood can be used as a diagnostic assay. Human immunodeficiency virus (HIV) is present at extremely low levels during the early stages of AIDS infection. Rather than trying to detect these viral antigens, diagnostic procedures are based on the detection of antibodies generated against these viral problems. An example of a commonly used immunoblot procedure is shown in Figure 7. The HIV proteins, p24, p31, gp41, and gp120 have been immobilized on test strips. In this figure, results from testing of blood of four different patients (panels A-D) at different stages of the disease can be compared with the positive (left strip) and negative (right strip) controls in panel E (52).

Therapy

The MAbs are, of course, excellent candidates for use as therapeutic agents. Limitations in their in vivo application center around their immunogenicity as foreign proteins, and also their rather long half-life in the blood. These factors eventually result in an immune response which is mounted by the host followed by immune complexes and abrogation of therapy. This problem is being addressed in various ways. For example, the IgG can be cleaved by the enzyme papain to give Fab fragments. The result is a free floating antigen-binding site, capable of binding one molecule of antigen. The Fab which is significantly smaller than the IgG molecule has more penetrating diffusion properties. It has a comparatively short half-life in the blood. The Fab is less immunogenic than the IgG molecules. Finally, since the Fc region is missing, no complement is fixed. The Ag/Ab complexes that do form are generally small enough so that nephrotoxicity is not a problem. As discussed earlier, human MAbs are preferable to mouse MAbs.

Another avenue to create MAbs is to engineer them using the molecular biological gene splicing techniques and the cloning of immunoglobulin sequences discussed earlier. One goal is to produce very small molecules which mimic the action at the immunoglobulin-binding site only. Recombinant DNA methods can also permit the introduction of human immunoglobulin framework sequences, so that the mechanism for tolerance to intrinsic proteins prevents an immune response.

Another type of treatment which side steps the issue of foreign protein rejection takes place external to the patient. Two examples of this are: use of monoclonals in treating autologous or allogenic bone marrow transplants and purging autografts from leukemic cells. In this procedure, bone marrow is removed from the patient (or a genetically matched donor), cultured in the lab, and treated with MAbs to remove certain unwanted populations of cells (49,

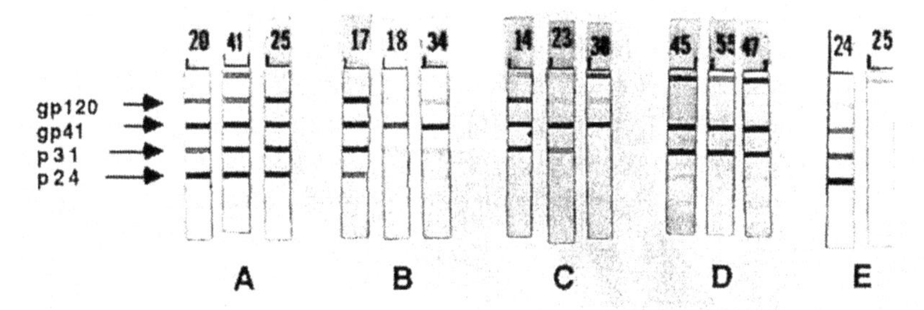

Figure 7 Quantitative evaluation of antibody patterns in blood of HIV-infected hemophiliac patients. Assays were done on the Chiron RIBA-HIV 216 Blot from Ortho Diagnostics.

50, 53, 54). A human monoclonal antibody has been reported by Phillips et al. (55) to successfully treat a glioma. In general, human tumor therapy has been confined to tumors of the lymphoid due to the variety of antibodies available. An impressive success was reported by Miller et al. (56) in which an anti-idiotype MAb of human origin was employed in treating a patient's tumor.

Use of murine antibodies is usually not successful because of host rejection. However, even human monoclonals can be rendered ineffective by a variety of circumstances. The tumor may undergo antigenic changes and lose the target antigen. New populations of tumor cells may not contain the target antigen. The MAb may combine with free circulating tumor antigen and never reach the primary site (57, 58). The MAb may be prevented from destroying the tumor by an access problem (for example, the IgM MAb is five times as large as a IgG MAb). One might also envision the solid mass of tumor inhibiting MAb penetration.

Intense interest has focused on the use of MAbs as directed transport vehicles for toxic drugs or radionuclides to tumor cells. This methodology should leave normal cells unharmed. There have been reports of linking MAbs to ricin protein or diphtheria toxin and the employment of these conjugates to kill a select population of cells in a mixed cell culture (59–61). Cytotoxic drugs may also be conjugated to MAbs (62). The use of any cytotoxin-MAb conjugate must be carefully researched. Any reaction with nontarget tissue, even low nonspecific interactions can preclude this approach. Research may also extend to MAb-directed delivery of cytokines such as interferons and interleukins.

MAbs are being used therapeutically to prevent the rejection phenomenon in heart and kidney transplantation (63). In addition, MAbs have been utilized under controlled circumstances in vitro to deplete specific target cells from bone marrow intended for either autologous or allogeneic transplantation. One might envision this type of anti-immune cell MAb to be a great potential in autoimmune diseases. In any of these cases it would be of tremendous advantage to work with a human MAb.

Another immediate application for MAb therapy is to replace the current method of treatment for Rh^- (rhesus negative) mothers. A human MAb which was produced by Epstein-Barr (EB) virus transformation is in clinical trials (64, 65). Other novel applications for MAbs are in the areas of viral disease (i.e., hepatitis), neutralization of snake or insect venom in affected patients, as well as other forms of systemic poisonings (i.e., digitalis intoxication), and adrenergic receptor blockade (66).

An interesting variation on the theme of genetic engineering of antibody molecules is being applied to AIDS therapy (67). In a new molecular construction, the variable region of a human heavy chain is replaced by the sequence of the CD_4 receptor protein. HIV, the causative virus in AIDS, infects target lymphocytes specifically through interaction with this receptor. Thus, the antigen recognition function of the variable domain now resides in a genetically engineered receptor sequence. It is presumed that this "immunoadhesin" will bind to HIV and that the Fc portion of the bound immunoadhesin will then elicit the biological consequences typical of an immune response.

Agents for Purification

A major area in which MAbs have gained rapid prominence is immunoaffinity purification chromatography. This area exploits the unique specificity, high affinity and the reversibility of the antigen–antibody complex. In short, the technique involves passing the antigen through a column composed of the antibody immobilized on an insoluble matrix. The antibody binds the target antigen and all extraneous proteins can be washed away. The purified antigen can then be recovered by dissociation of the immune complex. Often this antigen is a protein.

For many years only polyclonal antibodies were available to implement such a technique. MAbs have three main advantages over the more conventional polyclonal antisera. First, since specificity is achieved through the cloning and screening procedure, there is no need for an ultrapure antigen when the researcher is attempting to generate the antibodies. Second, a polyclonal antiserum consists of an entire spectrum of antibodies, many of which bind to different epitopes on the antigen. This multiplicity may lead to binding of the antigen at more than one site, which in turn may mean denaturation of the molecule if harsh eluants are needed to effect the breaking of the antigen/antibody complex. MAbs have the advantage that they are directed against a single epitope on the antigen, thus avoiding multipoint attachment and, therefore, harsh dissolution conditions.

Third, the MAb represents an unlimited, well defined source of reagent for the purification chemist. The antibody can be grown up in quantity with no batch to batch variation in affinity or specificity. Applications include interferons (see Chap. 8), hormones, blood clotting factors, enzymes as well as the difficult-to-purify membrane proteins and membrane receptors (Table 5) (68–84). It can be seen that the substances being purified are very diverse. The methods of eluting them can also be very diverse. The complex can be dissociated by acids, bases, salts, protein denaturants, or even organic solvents (85).

Currently, the major drawback to using MAbs in purification is the expense of producing a large amount (sometimes even up to gram quantities) necessary for industrial purification purposes. The second drawback, for the pharmaceutical industry in particular, is the problems involved in using a biological pu-

Table 5 Biological Substances Purified Using Immobilized Monoclonal Antibodies

Substance	Reference
Alphafetoprotein	83
Cyclic nucleotide phosphodiesterase	74
Factor VIIIc	84
Human C3b inactivator	75
Human HLA antigens	77
Interferon-α	68, 69
Interferon-β	70
Interferon-γα	71
Murine MHC antigens	78, 79
Nicotinic acetycholine receptors	80
Plasminogen activator zymogen	73
Prolactin	81
Pregnancy specific β-1-glycoprotein	82
Urokinase	72

rification step. The possibility exists for contamination of the therapeutic product with the MAb leached from the column, as well as any cellular biomolecules or viruses which may have copurified with the MAbs during their own production.

Research Uses

MAbs (and polyclonals) are being employed by many researchers for the creation of antipeptide antibodies (86). The advantages are twofold. First, since the researcher usually synthesizes the immunizing peptide, the changes in the immunogen are controllable and can be very subtle in origin. The second advantage is that there is no need to isolate pure "whole" antigen for immunization. The researchers can simply assemble a peptide based on the gene sequence even if that protein has not been previously isolated.

In this approach, the primary structure of the protein is scanned to select the most suitable epitopes, as shown in Figure 8. The concept is that hydrophobic amino acids tend to coalesce toward the interior of the protein, whereas hydrophilic amino acids tend to be on the outer surface of the protein. Additionally, bends in the protein backbone tend to be exposed in the protein. Figure 8 represents a composite (87) of secondary structure (88) (sawtooth is beta-sheet, sine wave is alpha-helix) overlayed with symbols for hydrophilicity and hydrophobicity (89). The general rule is that synthetic peptides of 16–20 residues in length, comprising relatively hydrophilic regions of the protein and including bends in the backbone would best serve to generate antibodies that recognize the native protein.

Uses include (a) identification of proteins predicted by a nucleic acid sequence, (b) aid in discriminating proteins of similar sequences, (c) demonstrating similarities between related proteins, (d) elucidating proteolyte processing, (e) for tissue or intracellular localization, (f) analysis of protein-protein associations, (g) probing protein function, (h) use as immunogens to develop neutralizing sera or disease protection. MAbs have also proved extremely useful for selection of transfectants expressing genes coding for cell surface antigens

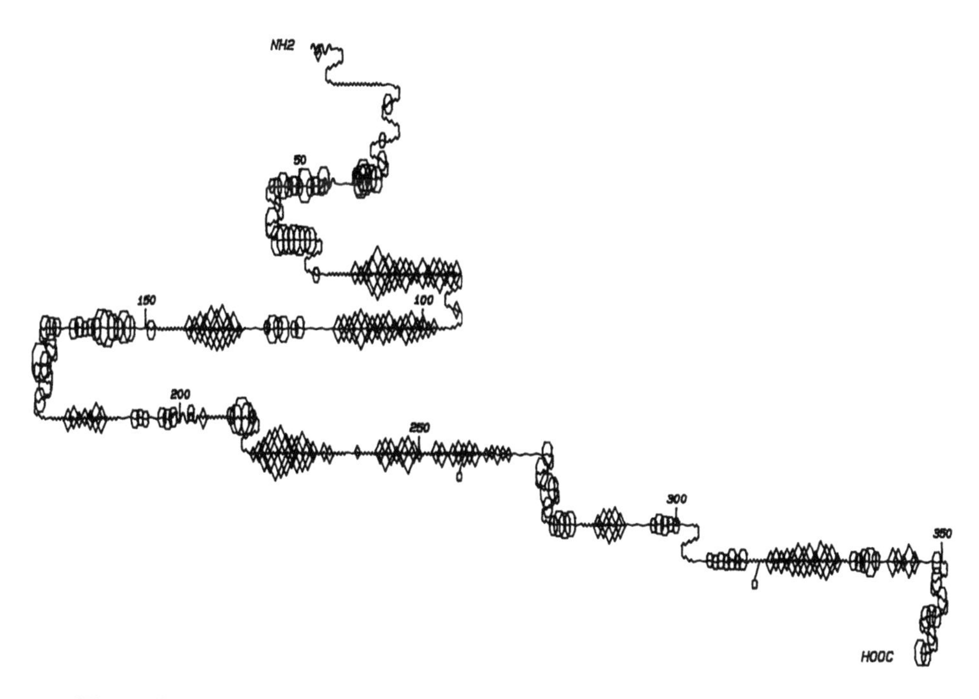

Figure 8 Representation of the structure of a protein. Changes in the direction of the line are indicative of bends in the protein. Also shown are the predicted secondary structure and the hydropathicity in different regions of the protein.

or receptors (90). This technique usually involves a fluorescent-labeled MAb against the antigen of interest, and a fluorescent-activated cell sorter (FACS). Gene libraries can be screened for the gene of interest by transfecting cells and assaying for the expression of the gene product using a MAb for the gene product. Another example of this is the use of MAbs in the more conventional DNA library screening which employs bacteriophage λ and *E. coli* cells. A recombinant DNA library is first constructed in the bacteriophage expression vector and foreign antigens are expressed in the *E. coli* cells from these DNA inserts. The antigens which are produced are transferred onto nitrocellulose filters and then probed with the MAb to determine the desired transformant. This technique is useful in locating genetic sequences which are present in low copy number and therefore are difficult to locate using hybrid selection methods or nucleic acid probe technology.

MAbs have proven invaluable in the isolation of certain scarce mRNAs. In order to accomplish this, polysomes (consisting of ribosomes, mRNA, and nacent polypeptide strands) are immunopurified utilizing a MAb directed against determinants on the nacent polypeptide chain. The MAb-polysome complex is in turn isolated by protein A Sepharose and the mRNA is dissociated from the polysome using EDTA. Oligo-dT cellulose is then used as a final cleansing step on the poly A tailed mRNA. Several examples can be found employing similar sequence of steps (91–93).

In addition to detecting the presence of proteins in partially purified mixtures, or assisting on the purification of proteins' mRNA, MAbs can be useful tools for studying posttranslational events (i.e., glycosylation modifications for secretion etc.). An elegant use of MAbs was demonstrated by Unkeless and co-workers in studying Fc receptor expression on genetically impregnated *Xenopus* oocytes (94).

Catalytic Antibodies

A prominent example of the creative manipulation of monoclonal antibody structure is the design of catalytic antibodies (i.e., antibodies that function as enzymes). This area is being pioneered by the research groups of R. A. Lerner (95) and P. G. Schultz (96). Essentially, MAbs are prepared against a compound that favors formation of the transition state in a particular chemical reaction. The reactant (or reactants) has a reasonable affinity for the binding domain of the MAb. The reaction then proceeds in a favorable manner due to transition state stabilization (95–98), catalysis by approximation (99, 100), or introduction of cofactors (101, 102). Using this approach, MAbs can be designed for catalyzing specific chemical reactions. This factor, combined with the ability to produce MAbs inexpensively and on a large scale, should have an impact on the industrial production of fine chemicals.

COMPARISON OF MONOCLONALS AND CONVENTIONAL ANTISERA

Thus far, we have thoroughly described monoclonal antibodies, presented methods for their production, and more exotic avenues for alterations and improvements. Let us now summarize the advantages and disadvantages of monoclonal production as compared with conventional antiserum production. It must be remembered that MAb production is costly and time consuming. The researchers should first assess whether conventional antisera might not be sufficient. Table 6 gives a direct comparison of the MAb and conventional antisera in several categories which normally concern the immunologist. To begin with, conventional antisera elicited against an antigen will cover many determinants, whereas a monoclonal antibody will be directed at a single antigenic

Table 6 Comparison Between Conventional Antiserum and Monoclonal Antibodies

	Conventional antiserum	Monoclonal antibody
Determinant (s)	Several epitopes	Single epitope
Specificity for Antigen	Variable with animal and bleed Partial cross reactions with common determinants Seldom too specific	Constant Unexpected cross-reactions may occur May be too specific for requirements
Affinity	Variable with bleed	May be selected during cloning; non-variable
Yield of useful antibody	Up to 1 mg/ml	Up to 100 g/ml in tissue culture Up to 20 mg/ml in ascitic fluid
Contaminating immunoglobulin	Up to 100%	None in culture; up to 10% in ascitic fluid
Purity of antigen	Usually pure antigen	Some degree of antigen purification desirable but not essential
Approximate minimum cost	Usually below $250	Capital cost $25,000 Running costs $10,000/yr
Isotype Production	Many; difficult to separate	Many but already separated
Standardization	No	Yes

determinant. The second consideration is that of specificity. Partial cross-reactions with common determinants may take place with conventional antisera. The cross-reactions, however, should represent only a small percentage of the total antibody/antigen reactions. With a monoclonal, cross-reaction is a less likely event, due to its narrow focus, but if it occurs it can be as great as 100%. This double-edged sword is best illustrated by thinking of a case where the MAb would be used diagnostically. Due to its exquisite specificity the MAb might allow the researcher to differentiate between two similar strains of bacterium. A conventional antisera would not be able to differentiate, since it considers all of the many determinants the two bacterium have in common. On the other hand, if the MAb is employed in a general screening assay for the bacterium, the MAb may give more false negatives than would conventional antisera. Its unique specificity may result in a false negative if the bacterium is polymorphic for or has lost that one determinant. Conventional antisera would not be as open to this error.

The comparison of affinities in conventional antisera and MAbs shows mixed advantages. Since they are separate entities, MAbs allow the researchers to choose the one of the appropriate affinity. The use of single monoclonals, however, preclude the cooperative aspects found in the interaction of antibodies binding at different epitopes on the antigen (13). The categories of antibody yield and contamination should be considered together. Equal quantities of useful antibody can be produced using either conventional or MAb techniques. In general, the MAb production is more time consuming and more expensive, but the final product is much cleaner. When discussing the purity of the antigen required to produce the antisera, monoclonals have a definite advantage. For conventional antisera, purity is obtained either by first having a pure immunogen (antigen) or absorption of the antiserum against unwanted components (a very laborious process). For monoclonals, however, the cloning and screening steps used for their generation, preclude the necessity for a pure antigen. An impure antigen can be used as long as one is willing to screen the MAb exhaustively. MAbs generally have an edge when isotyping is considered. Conventional antisera will contain many isotypes, each in small amounts and difficult to separate. MAbs will each be one already distinct isotype, which is hopefully the one required or, at least, of no immediate concern.

One of the strong arguments for MAbs is standardization. MAb producing cell lines can live indefinitely, providing uniform reagents for many different labs. Conventional antisera will vary from bleed to bleed in the animal and, of course, only extends to the animal's life span.

In terms of cost (for one animal or antibody source) the MAb may cost 100 times as much in terms of capital investment and running costs, but the results may not be achievable through conventional production. Therefore, it is obvious that a careful cost effective assessment is necessary before undertaking the production of MAbs. The advantages inherent in MAbs, however, may be incalcuable. The commercial importance of MAb is illustrated by the many companies which have seized upon the opportunity to enter into the production and sale of monoclonal antibodies and peripheral items (103).

RECENT ADVANCES

Recombinant DNA technology may be on the verge of superseding hybridoma technology for MAb production. For example, an Fab fragment library of V_H and V_L was made from spleen cell mRNA with the aid of PCR (polymerase chain reaction) (104). Constructs with these gene fragments were prepared and then recombined into a phage vector system. *E. coli* transfectants were screened using labeled antigen. Numerous high affinity, disulfide linked Fabs of 50,000 Da were generated within 2 weeks.

REFERENCES

1. Myrvik, Q. N. and Weiser, R. S. *Fundamentals of Immunology*, 2nd ed. Lea & Febiger, Philadelphia, 1984.
2. Burnet, F. M. *The Clonal Selection Theory of Acquired Immunity*. Cambridge University Press, Cambridge, 1959.
3. Spira, G., et al. The generation of better monoclonal antibodies through somatic mutation. In *Hybridoma Technology in Biosciences and Medicine*. Edited by T. A. Springer. Plenum Press, New York, 1986.
4. Liesegang, B., Radbruch, A., and Rajewsky, K. *Proc. Natl. Acad. Sci.* (USA) 75, 3901–3905 (1978).
5. Dangl, J. F. and Herzenberg, L. A. *J. Immunol. Meth.* 52, 1–14 (1982).
6. Muller, C. E. and Rajewsky, K., *J. Immunol.* 131, 877–881 (1983).
7. Burnet, K. Production of bifunctional antibodies by hybridoma technology. In *Biotechnology*. Edited by P. N. Cheremisinoff. Techromic Publishing Co., 1986.
8. Palmer, J. L. and Nisonoff, A., *Science* 143, 376 (1963).
9. Honjo, T. Immunoglobulin genes. *Ann. Rev. Immunol.* 2;239–256 (1983).
10. Oi, V. and Morrison, S. *Biotechniques* 4(3), 215–220 (1986).
11. Campbell, A. M. *Monoclonal Antibody Technology*. Edited by R. H. Burdon and P. H. Van Knippenberg. Elsevier, New York, 1984.
12. Galfre, G. and Milstein, C., *Meth. Enzymol.* 738 (1981).
13. Weir, D. M. Ed. *Handbook of Experimental Immunology and Immunochemistry*, Vol. 1. Academic Press, New York, 1978.
14. Johnstone, A. and Thorpe, R. *Immunochemistry in Practice*. Blackwell, London, 1982.
15. Bruck, C., Portelle, D., Gliner, R. C., and Bollen, A. *J. Immunol. Meth.* 53, 313 (1982).
16. Engvall, E. and Perlman, P., *Immunochemistry* 8; 871 (1971).

17. Miles, L. E. M. and Hales, C. N., *Nature (London)* 219, 186 (1968).
18. Towbin, H., Staehelin, T., and Gordon, J. *Proc. Natl. Acad. Sci (USA)* 76, 4350 (1979).
19. Clark, W. A., Frogner, K. S., and Zak, R. *J. Cell. Biol.* 95;369a (1982).
20. Yurchenko, P. D., Speicher, D. W., Morrow, J. S., Knowles, W. J., and Marchesi, V. T. *J. Biol. Chem.* 257;9103 (1982).
21. Laemmli, U. K. and Favre, M. *J. Mol. Biol.* 80;575 (1973).
22. Perides et al. *Anal. Biochem.* 152;94–99 (1986).
23. Farr, A. G. and Nakane, P. K. *J. Immunol. Meth.* 47;129 (1981).
24. Bayer, E. A. and Wilchek, M., *Meth. Biochem. Anal.* 26;1 (1980).
25. Mackie, E. B., Longenecker, B. M., and Bryan, T. E. Monoclonal antibodies to bacterial antigents. In *Biotechnology*. Edited by P. N. Cheremisinoff and R. P. Ouellette, Technomic Publishing Co. Inc., Lancaster, PA, 1985.
26. Hansen, E. J., Gulig, P. A., Robertson, S. M., Frasch, C. F., and Haanes, E. J. *Lancet* 13;366 (1982).
27. Schroer, K. R., Jinkim, K., Prescott, B., and Baker, B. J., *J. Exp. Med.* 150;698 (1979).
28. Suderstrom, T., Stein, K., Brinton, C. C., Hosea, S., Burch, H. A., Hansson, H. A., Karpas, A., Schneirson, R., Sutton, A., Vann, W. I., Hanson, L. A., *Prog. Allergy* 33;259 (1983).
29. Cross, A. S., Zollinger, W., Mandrel, R., Gemski, P., and Sadoff, J. *J. Infect. Dis.* 147;68 (1983).
30. Mackie, E. B., Longenecker, B. M., Rabin, H. R., DiNinno, V. L., and Bryan, L. E. *J. Immunol.* 129;829 (1982).
31. Nachamkin, I., Cannon, J. G., and Mettler, R. S., *Infect. Immun.* 32;641 (1981).
32. Ono, E., Naiki, M., and Yanagawa, R., *Zbl. Bakt. Hyg. I Abt. Orig. A.* 252;414 (1982).
33. Gustafson, B., Rose, A., and Holme, T., *Infect. Immun.* 38;449 (1982).
34. Sethi, K. K., Drueke, V., and Brandis, H., *J. Clin. Microbial.* 17;953 (1983).
35. Tam, M. R., Goldstein, L. C., and Nowinski, R. C. In *Biotechnology*. Edited by P. N. Cheremisinoff, Technomic Publishing Co., Lancaster, PA, 1985, Chap. 29.
36. Mattingly, J. A. and Robison, B. J. In *Biotechnology*. Edited by P. N. Cheremisinoff. Technomic Publishing Co., Lancaster, PA, 1986, Chap. 39.
37. Yoshida, N., Nussenweig, R. S., Potocnjak, P., Nussenweig, V., and Aikawa, M. *Science* 207;71 (1980).
38. Taylor, D. W. and Butterworth, A. E. *Parasitology* 84;83 (1982).
39. McMahon-Pratt, D. and David, J. *Nature* 284;366 (1981).
40. Gerhard, W., Yewdall, J., Frenkel, M. E., and Webster, R. *Nature* 290;713 (1981).
41. Shek, W. J., Cole, P. J., Dapolito, G. M., and Berin, J. F., *J. Virol. Methods* 1;257 (1980).
42. Ferguson, M., Scheld, G. C., Minar, P. D., Yates, P. J., and Spitz, M., *J. Gen. Virol.* 54;437 (1980).
43. Wiktor, T. J. and Koprowski, H., *Proc. Natl. Acad. Sci. (USA)* 75;3938 (1978).

44. Brodsky, F. M., Parham, P., Barnstable, C. J., Crumpton, M. J., and Bodmer, W. T. *Immunol. Rev.* 47;3 (1979).
45. Trucco, M. M., Garotta, G., Stocker, J. W., and Ceppellini, R. *Immunol. Rev.* 47;219 (1979).
46. Howard, J. C., Butcher, G. W., Galfre, G., Milstein, C., and Melstein, C. P., *Immunol. Rev.* 47;137 (1979).
47. Ashall, F., Bramwell, M. E., and Harris, H., *Lancet* (1982).
48. McGee, J. O., Ashall, F., Bramwell, M. W., Woods, J. C., and Harris, H. *Lancet* (1982).
49. Janossy, G., Cosimi, A. B., and Goldstein, G. In *Monoclonal Antibodies in Clinical Medicine*. Edited by A. McMichael and J. W. Fabre. Academic Press, New York, 1982.
50. Taylor-Papadimitriou, J., Peterson, J. A., Arklie, J., Burchell, J., Cariani, R. C., and Bodmer, W. F., *Int. J. Cancer* 28;17 (1981).
51. Epenetos, A. A., Britton, K. E., Mather, S., Sheperd, J., Granowska, M., Taylor-Papdimitrou, J., Nimmon, C. C., Durbin, H., Hawkins, T. R., Malpas, J. S., and Bodmer, W. F., *Lancet* (1982).
52. Raska, Jr., K., Kim, H. C., Martin, E., Raska, III, K., Raskova, J., and Saidi, P. *Clin. Exp. Immunol.* (1989).
53. Janossy, G. *Proc. Roy. Soc. Edin.* 81B;233 (1982).
54. Kemshead, J. T., Goldman, A., Fritschy, J., Malpas, J. S., and Pritchard, J., *Lancet* (1983).
55. Phillips J., Sikora, K., and Watson, J. V. *Lancet* 1215 (1982).
56. Miller, R. A., Maloney, D. G., Warnke, R., and Levy, R. N. Engl. J. Med. 306;517 (1982).
57. Hamblin, T. J., Abdul-Ahad, A. K., Gordon, J., Stephenson, F. K., and Stephenson, G. T., *Br. J. Cancer* 42;495 (1980).
58. Nadler, T. M., Stashenko, P., Hardy, R., Kaplan, W. D., Button, F. N., Tufe, D. W., Antiman, K. H., and Schlossman, S. F. *Cancer Res.* 40, 3147 (1980).
59. Krolick, K. A., Villemoy, C., Isakson, P., Uhr, J. W., and Vitetta, E. S. *Proc. Natl. Acad. Sci. (USA)* 77;5419 (1980).
60. Gilliland, D. G., Stephewski, Z., Collier, R. J., Mitchell, K. F., Cheng, T. H., and Koprowski, H., *Proc. Natl. Acad. Sci. (USA)* 77;4539 (1980).
61. Youle, R. J. and Neville, D. M. *Proc. Natl. Acad. Sci. (USA)* 77;5483 (1980).
62. Hurwitz, E., Levy, R., Maron, R., Wichek, M., Arnon, R., and Sela, M. *Cancer Res.* 35;1175 (1975).
63. Carpenter, C. B. "Manipulation of T-cell populations to abrogate allograft rejection. In *Hybridoma Technology in Biosciences and Medicine*. Edited by T. A. Springer. Plenum, New York, 1985, Chap. 30.
64. Crawford, D., Barlow, N. S., Harrison, J. F., Winger, L., and Huckins, E. R., *Lancet* (8321);386 (1982).
65. Crawford, D. H., Huchns, E. R., and Epstein, M. A., *Lancet* 1040 (1983).
66. Haber, E. Antibodies in vivo. In *Hybridoma Technology in Bioscience and Medicine*. Edited by T. A. Springer. Plenum, New York, 1986, Chap. 28.

67. Capon, D., Chamow, S. M., Mordenti, J., Marsters, S. A., Gregory, T., Mitsuya, H., Byrn, R. A., Lucas, C., Wurm, F. M., Groopman, J. E., Broder, S., and Smith, D. H. *Nature* 337, 525–531 (1989).
68. Secher, D. S. and Burke, D. C, *Nature* 285;446 (1980).
69. Novick, D., Eshhar, Z., and Rubinstein, M. *J. Immunol.*, 129;2244 (1982).
70. Novick, D., Eschar, Z., Gigi, O., Marks, Z., Revel, M., and Rubinstein, M. *J. Gen. Virol.* 64;905 (1983).
71. Novick, D., Eshhar, Z., Fischer, D. G., Friedlander, J., and Rubinstein, M. *EMBO J* 2;1527 (1983).
72. Vetterlein D. and Calton, G. J. *Thromb. Haemostas.* 49;24 (1983).
73. Nielson, L. S., Hansen, J. G., Skriver, L., Wilson, E. L., Kaltoft, K., Zeuthen, J., and Danoe, K. *Biochemistry* 21;6410 (1982).
74. Hansen, R. S., and Beavo, J. A. *Proc. Natl. Acad. Sci. (USA)* 79;2788 (1982).
75. Hsiung, L. M., Barclay, A. N., Brandon, M. R.., Smith, E., and Porter, R. R., *Biochem. J.* 203;293 (1982).
76. Frackelton, A. R. Jr., Ross, A. H., and Eisen, H. N. *Molec. Cell Biol* 3;1343 (1983).
77. Parham, P. In *Methods in Enzymology*. Edited by J. J. Langone and H. V. Vunakis. Academic Press, New York, 1983), Vol. 92, p. 110.
78. Zecher, R. and Reske, K., *Mol. Immunol.* 19;1037 (1982).
79. Mescher, M. F., Stallcup, K. C., Sullivan, C. P., Turkewitz, A. P., and Herrmann, S. H. In *Methods in Enzymology*. Edited by J. J. Lagone and H. V. Vunakis. Academic Press, New York, 1983, Vol. 92, p. 86.
80. Lennon, V. A., Thompson, M., and Chen, J. *J. Biol.Chem.* 255;4395 (1980).
81. Stuart, M. C., Boscato, L. M., and Underwood, P. A. *Clin. Chem.* 29;241 (1983).
82. Heikinheimo, M., Stenman, U-H., Bang, B., Hurme, M., Makela, O., and Bohn, H. *J. Immunol. Methods* 60;25 (1983).
83. Stenman, U-H., Sutinen, M. L., Selander, R. K., et al. *J. Immunol. Methods* 46;337 (1981).
84. Fulcher, C. A. and Zimmerman, T. S. *Proc. Natl. Acad. Sci. (USA)* 79;1648 (1982).
85. Birch, J. R., Hill, C. R., and Kenney, A. C. Affinity chromatography: Its role in industry. In *Biotechnology*. Edited by P. N. Cheremisinoff and R. P. Ouellette. Techromic Publishing, Lancaster, PA, 1986, Chap. 44.
86. Schaffhansen, B. S. Designing and using site-specific antibodies to synthetic peptides. In *Hybridoma Technology in Biosciences and Medicine*. Edited by T. A. Springer. Plenum, New York, 1985, Chap. 21.
87. Starcich, et al. *Cell* 45;637–648 (1986).
88. Chou and Fasman. *Adv. Enzymes* 47;45–147 (1978).
89. Kyte and Doolittle. *J. Mol. Biol.* 157;105–132 (1982).
90. Kavathas, P. Amplification and molecular cloning of transfected genes. In *Hybridoma Technology in Biosciences and Medicine*. Edited by T. A. Springer. Plenum, New York, 1985, Chap. 23.
91. Korman, A. J., Knudsen, P. J., Kaufman, J. F., and Strominger, J. L., *Proc. Nat. Acad. Sci. (USA)* 79;1844–1848 (1982).

92. Oren, M. and Levine, A. J., *Proc. Nat. Acad. Sci. (USA)* 79;1844–1848 (1983).
93. Brown, J. P., Rose, T. M., and Plowman, G. D., Purification of messenger RNA by polysome isolation with monoclonal antibodies. In *Hybridoma Technology in Bioscience and Medicine*. Edited by T. A. Springer. Plenum Press, New York, 1985, Chap. 26.
94. Pure, E., Luster, A. D., and Unkeless, J. C. *J. Exp. Med.* 160;606–611 (1984).
95. Tramantano, A., Janda, K. D., and Lerner, R. A. *Science* 234;1566 (1986).
96. Pollack, S., Jacobs, J., and Schultz, P. G. *Science* 234;1570 (1986).
97. Schultz, P. G. *Science* 240;426 (1988).
98. Janda, K. D., Schloeder, D., Benkovic, S., and Lerner, R. A. *Science* 241;1188 (1988).
99. Benkovic, S. J., Napper, A. D., and Lerner, R. A. *Proc. Natl. Acad. Sci. (USA)* 85;5355 (1988).
100. Jackson, D. Y. et al. *J. Am. Chem. Soc.* 110;4841 (1988).
101. Cochran, A. G., Sugasawara, R., and Schultz, P. G. *Science J. Am. Chem. Soc.* 110;7888 (1988).
102. Pollack, S., Nakayama, G., and Schultz, P. G. *Science* 242;1038 (1988).
103. *Science* 239-Part II, G28 (1988).
104. Huse, W., Sastry, L., Iverson, S. A., Kang, A. S., Alting-Mees, M., Burton, D. R., Benkovic, S. J. and Lerner, R. A. *Science* 246; 1275 (1989).

Index